A Brief History of Physics

Edited by Paul F. Kisak

Contents

Chapter 1

History of physics

"If I have seen further, it is only by standing on the shoulders of giants." – Isaac Newton [*][1]

Physics (from the Ancient Greek φύσις *physis* meaning "nature") is the fundamental branch of science that developed out of the study of nature and philosophy known, until around the end of the 19th century, as "natural philosophy". Today, physics is ultimately defined as the study of matter, energy and the relationships between them. Physics is, in some senses, the oldest and most basic pure science; its discoveries find applications throughout the natural sciences, since matter and energy are the basic constituents of the natural world. The other sciences are generally more limited in their scope and may be considered branches that have split off from physics to become sciences in their own right. Physics today may be divided loosely into classical physics and modern physics.

1.1 Ancient history

Further information: History of astronomy

Elements of what became physics were drawn primarily from the fields of astronomy, optics, and mechanics, which were methodologically united through the study of geometry. These mathematical disciplines began in antiquity with the Babylonians and with Hellenistic writers such as Archimedes and Ptolemy. Ancient philosophy, meanwhile – including what was called "physics" – focused on explaining nature through ideas such as Aristotle's four types of "cause".

1.1.1 Ancient Greece

The move towards a rational understanding of nature began at least since the Archaic period in Greece (650–480 BCE) with the Pre-Socratic philosophers. The philosopher Thales of Miletus (7th and 6th centuries BCE), dubbed "the Father of Science" for refusing to accept various supernatural, religious or mythological explanations for natural phenomena, proclaimed that every event had a natural cause.[*][2] Thales also made advancements in 580 BCE by suggesting that water is the basic element, experimenting with the attraction between magnets and rubbed amber and formulating the first recorded cosmologies. Anaximander, famous for his proto-evolutionary theory, disputed the Thales' ideas and proposed that rather than water, a substance called *apeiron* was the building block of all matter. Around 500 BCE, Heraclitus proposed that the only basic law governing the Universe was the principle of change and that nothing remains in the same state indefinitely. This observation made him one of the first scholars in ancient physics to address the role of time in the universe, a key and sometimes contentious concept in modern and present-day physics. The early physicist Leucippus (fl. first half of the 5th century BCE) adamantly opposed the idea of direct divine intervention in the universe, proposing instead that natural phenomena had a natural cause. Leucippus and his student Democritus were the first to develop the theory of atomism, the idea that everything is composed entirely of various imperishable, indivisible elements called atoms.

During the classical period in Greece (6th, 5th and 4th centuries BCE) and in Hellenistic times, natural philos-

Aristotle
(384–322 BCE)

ophy slowly developed into an exciting and contentious field of study. Aristotle (Greek: Ἀριστοτέλης, *Aristotélēs*) (384 – 322 BCE), a student of Plato, promoted the concept that observation of physical phenomena could ultimately lead to the discovery of the natural laws governing them. Aristotle's writings cover physics, metaphysics, poetry, theater, music, logic, rhetoric, linguistics, politics, government, ethics, biology and zoology. He wrote the first work which refers to that line of study as "Physics" – in the 4th century BCE, Aristotle founded the system known as Aristotelian physics. He attempted to explain ideas such as motion (and gravity) with the theory of four elements. Aristotle believed that all matter was made up of aether, or some combination of four elements: earth, water, air, and fire. According to Aristotle, these four terrestrial elements are capable of inter-transformation and move toward their natural place, so a stone falls downward toward the center of the cosmos, but flames rise upward toward the circumference. Eventually, Aristotelian physics became enormously popular for many centuries in Europe, informing the scientific and scholastic developments of the Middle Ages. It remained the mainstream scientific paradigm in Europe until the time of Galileo Galilei and Isaac Newton.

Early in Classical Greece, knowledge that the Earth is spherical ("round") was common. Around 240 BCE,

as the result a seminal experiment, Eratosthenes (276–194 BCE) accurately estimated its circumference. In contrast to Aristotle's geocentric views, Aristarchus of Samos (Greek: Ἀρίσταρχος; c.310 – c.230 BCE) presented an explicit argument for a heliocentric model of the Solar system, i.e. for placing the Sun, not the Earth, at its centre. Seleucus of Seleucia, a follower of Aristarchus' heliocentric theory, stated that the Earth rotated around its own axis, which, in turn, revolved around the Sun. Though the arguments he used were lost, Plutarch stated that Seleucus was the first to prove the heliocentric system through reasoning.

The ancient Greek mathematician Archimedes, famous for his ideas regarding fluid mechanics and buoyancy.

In the 3rd century BCE, the Greek mathematician Archimedes of Syracuse (Greek: Ἀρχιμήδης (287–212 BCE) – generally considered to be the greatest mathematician of antiquity and one of the greatest of all time – laid the foundations of hydrostatics, statics and calculated the underlying mathematics of the lever. A leading scientist of classical antiquity, Archimedes also developed elaborate systems of pulleys to move large objects with a minimum of effort. The Archimedes' screw underpins modern hydroengineering, and his machines of war helped to hold back the armies of Rome in the First Punic War. Archimedes even tore apart the arguments of Aristotle and his metaphysics, pointing out that it was impossible to separate mathematics and nature and proved it by converting mathematical theories into practical inventions. Furthermore, in his work *On Floating Bodies*, around 250 BCE, Archimedes developed the law of buoyancy, also known as Archimedes' Principle. In mathematics, Archimedes used the method of exhaustion to calculate the area under the arc of a parabola with the summation of an infinite series, and gave a remarkably accurate approximation of pi. He also defined the spiral bearing his name, formulae for the volumes of surfaces of revolution and an ingenious system for expressing very large numbers. He also developed the principles of equilibrium states and centers of gravity, ideas that would influence the well known scholars, Galileo, and Newton.

Hipparchus (190–120 BCE), focusing on astronomy and mathematics, used sophisticated geometrical techniques to map the motion of the stars and planets, even predicting the times that Solar eclipses would happen. In addition, he added calculations of the distance of the Sun and Moon from the Earth, based upon his improvements to the observational instruments used at that time. Another of the most famous of the early physicists was Ptolemy (90–168 CE), one of the leading minds during the time of the Roman Empire. Ptolemy was the author of several scientific treatises, at least three of which were of continuing importance to later Islamic and European science. The first is the astronomical treatise now known as the *Almagest* (in Greek, Ἡ Μεγάλη Σύνταξις, "The Great Treatise", originally Μαθηματικὴ Σύνταξις, "Mathematical Treatise"). The second is the *Geography*, which is a thorough discussion of the geographic knowledge of the Greco-Roman world.

Much of the accumulated knowledge of the ancient world was lost. Even of the works of the better known thinkers, few fragments survived. Although he wrote at least fourteen books, almost nothing of Hipparchus' direct work survived. Of the 150 reputed Aristotelian works, only 30 exist, and some of those are "little more than lecture notes".

Value	0	1	2	3	4	5	6	7	8	9
Western Arabic	٠	١	٢	٣	٤	٥	٦	٧	٨	٩
Eastern Arabic	۰	۱	۲	۳	۴	۵	۶	۷	۸	۹
Devanagari	०	१	२	३	४	५	६	७	८	९
Gujarati	૦	૧	૨	૩	૪	૫	૬	૭	૮	૯
Gurmukhi	੦	੧	੨	੩	੪	੫	੬	੭	੮	੯
Limbu	□	□	□	□	□	□	□	□	□	□
Bengali	০	১	২	৩	৪	৫	৬	৭	৮	৯
Oriya	୦	୧	୨	୩	୪	୫	୬	୭	୮	୯
Telugu	౦	౧	౨	౩	౪	౫	౬	౭	౮	౯
Kannada	೦	೧	೨	೩	೪	೫	೬	೭	೮	೯
Malayalam	൦	൧	൨	൩	൪	൫	൬	൭	൮	൯
Tamil (Grantha)	0	௧	௨	௩	௪	௫	௬	௭	௮	௯
Tibetan	༠	༡	༢	༣	༤	༥	༦	༧	༨	༩
Burmese	□	□	□	□	□	□	□	□	□	□
Thai	๐	๑	๒	๓	๔	๕	๖	๗	๘	๙
Khmer	០	១	២	៣	៤	៥	៦	៧	៨	៩
Lao	໐	໑	໒	໓	໔	໕	໖	໗	໘	໙

The Hindu-Arabic numeral system. The inscriptions on the edicts of Ashoka (3rd century BCE) display this number system being used by the Imperial Mauryas.

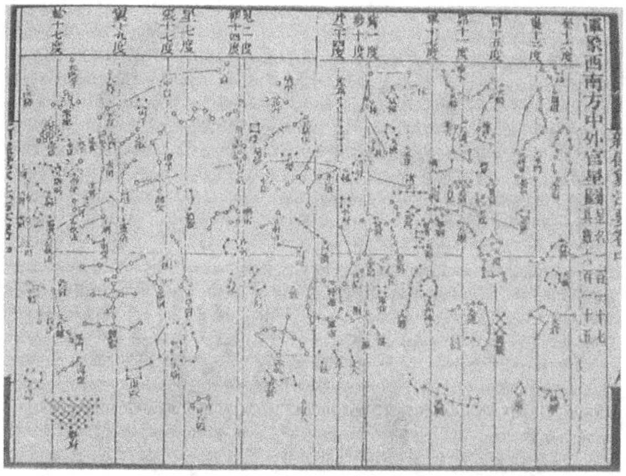

Star maps by the 11th-century Chinese polymath Su Song are the oldest known woodblock-printed star maps to have survived to the present day. This example, dated 1092,[note 1] employs cylindrical projection.*

1.1.2 India and China

Further information: History of science and technology in China and History of Indian science and technology

Important physical and mathematical traditions also existed in ancient Chinese and Indian sciences.

In Indian philosophy, Maharishi Kanada was the first to systematically develop a theory of atomism around 200 BCE*[3] though some authors have allotted him an earlier era in the 6th century BCE.*[4]*[5] It was further elaborated by the Buddhist atomists Dharmakirti and Dignāga during the 1st millennium CE.*[6] Pakudha Kaccayana, a 6th-century BCE Indian philosopher and contemporary of Gautama Buddha, had also propounded ideas about the atomic constitution of the material world. These philosophers believed that other elements (except ether) were physically palpable and hence comprised minuscule particles of matter. The last minuscule particle of matter that could not be subdivided further was termed Parmanu. These philosophers considered the atom to be indestructible and hence eternal. The Buddhists thought atoms to be minute objects unable to be seen to the naked eye that come into being and vanish in an instant. The Vaisheshika school of philosophers believed that an atom was a mere point in space. Indian theories about the atom are greatly abstract and enmeshed in philosophy as they were based on logic and not on personal experience or experimentation. In Indian astronomy, Aryabhata's *Aryabhatiya* (499 CE) proposed the Earth's rotation, while Nilakantha Somayaji (1444–1544) of the Kerala school of astronomy and mathematics proposed a semi-heliocentric model resembling the Tychonic system.

The study of magnetism in Ancient China dates back to the 4th century BCE. (in the *Book of the Devil Valley Master*),[7] A main contributor to this field was Shen Kuo (1031–1095), a polymath and statesman who was the first to describe the magnetic-needle compass used for navigation, as well as establishing the concept of true north. In optics, Shen Kuo independently developed a camera obscura.[8]

1.1.3 Muslim scientists

Main articles: Physics in medieval Islam and Science in the medieval Islamic world
See also: List of Muslim scientists
In the 5th to 15th centuries, scientific progress occurred in

Ibn al-Haytham ("Alhazen")
(965–1039)

the Muslim world. Many classic works in Latin and Greek were translated into Arabic. Ibn Sīnā (980–1037), known as "Avicenna", was a polymath from Bukhara (now in present-day Uzbekistan) responsible for important contributions to physics, optics, philosophy and medicine. He is

most famous for writing *The Canon of Medicine*, a text that was used to teach student doctors in Europe until the 1600s.

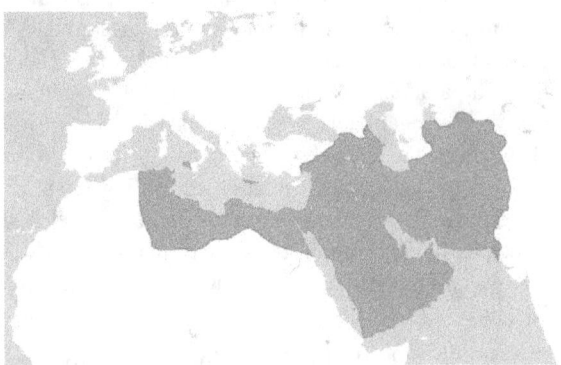

830 CE: the Abbasid Caliphate at its height.

Important contributions were made by Ibn al-Haytham (965–1040), a mathematician from Basra (in present-day Iraq) considered one of the founders of modern optics. Ptolemy and Aristotle theorised that light either shone from the eye to illuminate objects or that light emanated from objects themselves, whereas al-Haytham (known by the Latin name Alhazen) suggested that light travels to the eye in rays from different points on an object. The works of Ibn al-Haytham and Abū Rayhān Bīrūnī eventually passed on to Western Europe where they were studied by scholars such as Roger Bacon and Witelo.[9][note 2] Omar Khayyám (1048–1131), a Persian scientist, calculated the length of a solar year and was only out by a fraction of a second when compared to our modern day calculations. He used this to compose a calendar considered more accurate than the Gregorian calendar that came along 500 years later. He is classified as one of the world's first great science communicators, said, for example to have convinced a Sufi theologian that the world turns on an axis.

Nasir al-Din al-Tusi (1201–1274), an astronomer and mathematician from Baghdad, authored the *Treasury of Astronomy*, a remarkably accurate table of planetary movements that reformed the existing planetary model of Roman astronomer Ptolemy by describing a uniform circular motion of all planets in their orbits. This work led to the later discovery, by one of his students, that planets actually have an elliptical orbit. Copernicus later drew heavily on the work of al-Din al-Tusi and his students, but without acknowledgment.[10] The gradual chipping away of the Ptolemaic system paved the way for the revolutionary idea that the Earth actually orbited the Sun (heliocentrism).

1.1.4 Medieval Europe

Further information: Theory of impetus

A page from al-Khwārizmī's Algebra.

Awareness of ancient works re-entered the West through translations from Arabic to Latin. Their re-introduction, combined with Judeo-Islamic theological commentaries, had a great influence on Medieval philosophers such as Thomas Aquinas. Scholastic European scholars, who sought to reconcile the philosophy of the ancient classical philosophers with Christian theology, proclaimed Aristotle the greatest thinker of the ancient world. In cases where they didn't directly contradict the Bible, Aristotelian physics became the foundation for the physical explanations of the European Churches. Quantification became a core element of medieval physics.*[11]

Based on Aristotelian physics, Scholastic physics described things as moving according to their essential nature. Celestial objects were described as moving in circles, because perfect circular motion was considered an innate property of objects that existed in the uncorrupted realm of the celestial spheres. The theory of impetus, the ancestor to the concepts of inertia and momentum, was developed along similar lines by medieval philosophers such as John Philoponus and Jean Buridan. Motions below the lunar sphere

were seen as imperfect, and thus could not be expected to exhibit consistent motion. More idealized motion in the "sublunary" realm could only be achieved through artifice, and prior to the 17th century, many did not view artificial experiments as a valid means of learning about the natural world. Physical explanations in the sublunary realm revolved around tendencies. Stones contained the element earth, and earthly objects tended to move in a straight line toward the centre of the earth (and the universe in the Aristotelian geocentric view) unless otherwise prevented from doing so.*[12]

1.2 Scientific revolution

During the 16th and 17th centuries, a large advancement of scientific progress known as the Scientific revolution took place in Europe. Dissatisfaction with older philosophical approaches had begun earlier and had produced other changes in society, such as the Protestant Reformation, but the revolution in science began when natural philosophers began to mount a sustained attack on the Scholastic philosophical program and supposed that mathematical descriptive schemes adopted from such fields as mechanics and astronomy could actually yield universally valid characterizations of motion and other concepts.

1.2.1 Nicolaus Copernicus

Main articles: Nicolaus Copernicus, Tycho Brahe and Johannes Kepler

A breakthrough in astronomy was made by Polish astronomer Nicolaus Copernicus (1473–1543) when, in 1543, he proposed a heliocentric model of the Solar system, ostensibly as a means to render tables charting planetary motion more accurate and to simplify their production. In heliocentric models of the Solar system, the Earth orbits the Sun along with other bodies in Earth's galaxy, a contradiction according to the Greek-Egyptian astronomer Ptolemy (2nd century CE; see above), whose system placed the Earth at the center of the Universe and had been accepted for over 1,400 years. The Greek astronomer Aristarchus of Samos (c.310 – c.230 BCE) had suggested that the Earth revolves around the Sun, but Copernicus' theory was the first to be accepted as a valid scientific possibility. Copernicus' book presenting the theory (*De revolutionibus orbium coelestium*, "On the Revolutions of the Celestial Spheres") was published just before his death in 1543 and, as it is now generally considered to mark the beginning of modern astronomy, is also considered to mark the beginning of the Scientific revolution. Copernicus' new perspective, along with the accurate observations made by Tycho Brahe, enabled German astronomer Johannes Kepler (1571–1630) to

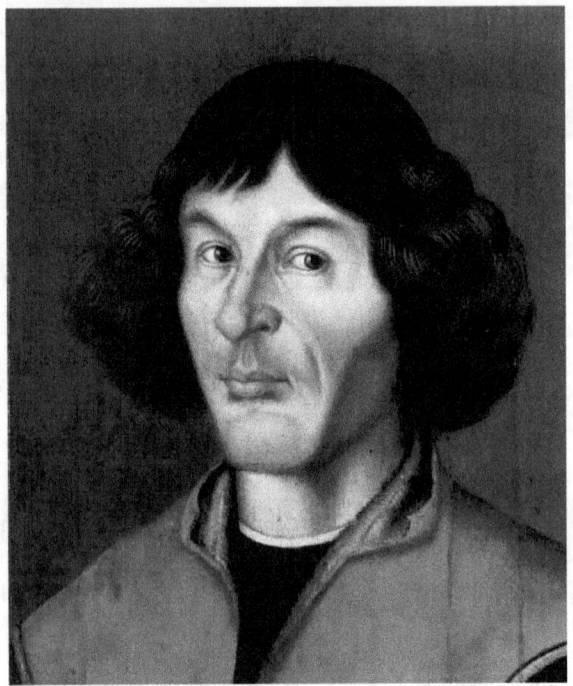

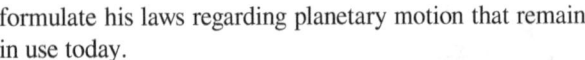

The Polish astronomer Nicolaus Copernicus (1473–1543) is re-membered for his development of a heliocentric model of the Solar system.

Galileo Galilei
(1564–1642)

formulate his laws regarding planetary motion that remain in use today.

1.2.2 Galileo Galilei

Main article: Galileo Galilei

The Italian mathematician, astronomer, and physicist Galileo Galilei (1564–1642) was the central figure in the Scientific revolution and famous for his support for Copernicanism, his astronomical discoveries, empirical experiments and his improvement of the telescope. As a mathematician, Galileo's role in the university culture of his era was subordinated to the three major topics of study: law, medicine, and theology (which was closely allied to philosophy). Galileo, however, felt that the descriptive content of the technical disciplines warranted philosophical interest, particularly because mathematical analysis of astronomical observations – notably, Copernicus' radical analysis of the relative motions of the Sun, Earth, Moon, and planets – indicated that philosophers' statements about the nature of the universe could be shown to be in error. Galileo also performed mechanical experiments, insisting that motion itself – regardless of whether it was produced "naturally" or "artificially" (i.e. deliberately) – had universally consistent characteristics that could be described mathematically.

Galileo's early studies at the University of Pisa were in medicine, but he was soon drawn to mathematics and physics. At 19, he discovered (and, subsequently, verified) the isochronal nature of the pendulum when, using his pulse, he timed the oscillations of a swinging lamp in Pisa's cathedral and found that it remained the same for each swing regardless of the swing's amplitude. He soon became known through his invention of a hydrostatic balance and for his treatise on the center of gravity of solid bodies. While teaching at the University of Pisa (1589–92), he initiated his experiments concerning the laws of bodies in motion that brought results so contradictory to the accepted teachings of Aristotle that strong antagonism was aroused. He found that bodies do not fall with velocities proportional to their weights. The famous story in which Galileo is said to have dropped weights from the Leaning Tower of Pisa is apocryphal, but he did find that the path of a projectile is a parabola and is credited with conclusions that anticipated Newton's laws of motion (e.g. the notion of inertia). Among these is what is now called Galilean relativity, the first precisely formulated statement about properties of space and time outside three-dimensional geometry.

Galileo has been called the "father of modern observational astronomy",[13] the "father of modern physics",[14] the "father of science",[14] and "the father of modern science".[15] According to Stephen Hawking, "Galileo, perhaps more than any other single person, was responsible for the birth of modern science."[16] As religious ortho-

A composite montage comparing Jupiter (lefthand side) and its four Galilean moons (top to bottom: Io, Europa, Ganymede, Callisto).

doxy decreed a geocentric or Tychonic understanding of the Solar system, Galileo's support for heliocentrism provoked controversy and he was tried by the Inquisition. Found "vehemently suspect of heresy", he was forced to recant and spent the rest of his life under house arrest.

The contributions that Galileo made to observational astronomy include the telescopic confirmation of the phases of Venus; his discovery, in 1609, of Jupiter's four largest moons (subsequently given the collective name of the "Galilean moons"); and the observation and analysis of sunspots. Galileo also pursued applied science and technology, inventing, among other instruments, a military compass. His discovery of the Jovian moons was published in 1610 and enabled him to obtain the position of mathematician and philosopher to the Medici court. As such, he was expected to engage in debates with philosophers in the Aristotelian tradition and received a large audience for his own publications such as the *Discourses and Mathematical Demonstrations Concerning Two New Sciences* (published abroad following his arrest for the publication of *Dialogue Concerning the Two Chief World Systems*) and *The Assayer.*[17][18] Galileo's interest in experimenting with and formulating mathematical descrip-

tions of motion established experimentation as an integral part of natural philosophy. This tradition, combining with the non-mathematical emphasis on the collection of "experimental histories" by philosophical reformists such as William Gilbert and Francis Bacon, drew a significant following in the years leading up to and following Galileo's death, including Evangelista Torricelli and the participants in the Accademia del Cimento in Italy; Marin Mersenne and Blaise Pascal in France; Christiaan Huygens in the Netherlands; and Robert Hooke and Robert Boyle in England.

1.2.3 René Descartes

Main article: René Descartes
The French philosopher René Descartes (1596–1650) was

René Descartes
(1596–1650)

well-connected to, and influential within, the experimental philosophy networks of the day. Descartes had a more ambitious agenda, however, which was geared toward replacing the Scholastic philosophical tradition altogether. Questioning the reality interpreted through the senses, Descartes sought to re-establish philosophical explanatory schemes by reducing all perceived phenomena to being attributable to the motion of an invisible sea of "corpuscles". (Notably, he reserved human thought and God from his scheme, holding these to be separate from the physical universe). In proposing this philosophical framework, Descartes supposed that different kinds of motion, such as that of planets versus that of terrestrial objects, were not fundamentally different, but were merely different manifestations of an endless chain of corpuscular motions obeying universal principles. Particularly influential were his explanations for circular astronomical motions in terms of the vortex motion of corpuscles in space (Descartes argued, in accord with the beliefs, if not the methods, of the Scholastics, that a vacuum could not exist), and his explanation of gravity in terms of corpuscles pushing objects downward.[*][19][*][20][*][21]

Descartes, like Galileo, was convinced of the importance of mathematical explanation, and he and his followers were key figures in the development of mathematics and geometry in the 17th century. Cartesian mathematical descriptions of motion held that all mathematical formulations had to be justifiable in terms of direct physical action, a position held by Huygens and the German philosopher Gottfried Leibniz, who, while following in the Cartesian tradition, developed his own philosophical alternative to Scholasticism, which he outlined in his 1714 work, *The Monadology*. Descartes has been dubbed the 'Father of Modern Philosophy', and much subsequent Western philosophy is a response to his writings, which are studied closely to this day. In particular, his *Meditations on First Philosophy* continues to be a standard text at most university philosophy departments. Descartes' influence in mathematics is equally apparent; the Cartesian coordinate system —allowing algebraic equations to be expressed as geometric shapes in a two-dimensional coordinate system —was named after him. He is credited as the father of analytical geometry, the bridge between algebra and geometry, important to the discovery of calculus and analysis.

1.2.4   Isaac Newton

Main articles: Isaac Newton and History of classical mechanics

The late 17th and early 18th centuries saw the achievements of the greatest figure of the Scientific revolution: Cambridge University physicist and mathematician Sir Isaac Newton (1642-1727), considered by many to be the greatest and most influential scientist who ever lived. Newton, a fellow of the Royal Society of England, combined his own discoveries in mechanics and astronomy to earlier ones to create a single system for describing the workings of the universe. Newton formulated three laws of motion and the law of universal gravitation, the latter of which could be used to explain the behavior not only of falling bodies on the earth but also planets and other celestial bodies in the heavens. To arrive at his results, Newton invented one form of an entirely new branch of mathematics: calculus (also invented independently by Gottfried Leibniz), which was to become an essential tool in much of the later de-

Sir Isaac Newton
(1642–1727)

velopment in most branches of physics. Newton's findings were set forth in his *Philosophiæ Naturalis Principia Mathematica* ("Mathematical Principles of Natural Philosophy"), the publication of which in 1687 marked the beginning of the modern period of mechanics and astronomy.

Newton was able to refute the Cartesian mechanical tradition that all motions should be explained with respect to the immediate force exerted by corpuscles. Using his three laws of motion and law of universal gravitation, Newton removed the idea that objects followed paths determined by natural shapes and instead demonstrated that not only regularly observed paths, but all the future motions of any body could be deduced mathematically based on knowledge of their existing motion, their mass, and the forces acting upon them. However, observed celestial motions did not precisely conform to a Newtonian treatment, and Newton, who was also deeply interested in theology, imagined that God intervened to ensure the continued stability of the solar system.

Newton's principles (but not his mathematical treatments) proved controversial with Continental philosophers, who found his lack of metaphysical explanation for movement and gravitation philosophically unacceptable. Beginning around 1700, a bitter rift opened between the Continen-

*Gottfried Leibniz
(1646–1716)*

tal and British philosophical traditions, which were stoked by heated, ongoing, and viciously personal disputes between the followers of Newton and Leibniz concerning priority over the analytical techniques of calculus, which each had developed independently. Initially, the Cartesian and Leibnizian traditions prevailed on the Continent (leading to the dominance of the Leibnizian calculus notation everywhere except Britain). Newton himself remained privately disturbed at the lack of a philosophical understanding of gravitation, while insisting in his writings that none was necessary to infer its reality. As the 18th century progressed, Continental natural philosophers increasingly accepted the Newtonians' willingness to forgo ontological metaphysical explanations for mathematically described motions.[*][22][*][23][*][24]

Newton built the first functioning reflecting telescope[*][25] and developed a theory of color, published in *Opticks*, based on the observation that a prism decomposes white light into the many colours forming the visible spectrum. While Newton explained light as being composed of tiny particles, a rival theory of light which explained its behavior in terms of waves was presented in 1690 by Christiaan Huygens. However, the belief in the mechanistic philosophy coupled with Newton's reputation meant that the wave theory saw relatively little support until the 19th century. Newton also formulated an empirical law of cooling, studied the

speed of sound, investigated power series, demonstrated the generalised binomial theorem and developed a method for approximating the roots of a function. His work on infinite series was inspired by Simon Stevin's decimals.[*][26] Most importantly, Newton showed that the motions of objects on Earth and of celestial bodies are governed by the same set of natural laws, which were neither capricious nor malevolent. By demonstrating the consistency between Kepler's laws of planetary motion and his own theory of gravitation, Newton also removed the last doubts about heliocentrism. By bringing together all the ideas set forth during the Scientific revolution, Newton effectively established the foundation for modern society in mathematics and science.

1.2.5 Other achievements

Other branches of physics also received attention during the period of the Scientific revolution. Wilbert Gilbert, court physician to Queen Elizabeth I, published an important work on magnetism in 1600, describing how the earth itself behaves like a giant magnet. Robert Boyle (1627–91) studied the behavior of gases enclosed in a chamber and formulated the gas law named for him; he also contributed to physiology and to the founding of modern chemistry. Another important factor in the scientific revolution was the rise of learned societies and academies in various countries. The earliest of these were in Italy and Germany and were short-lived. More influential were the Royal Society of England (1660) and the Academy of Sciences in France (1666). The former was a private institution in London and included such scientists as John Wallis, William Brouncker, Thomas Sydenham, John Mayow, and Christopher Wren (who contributed not only to architecture but also to astronomy and anatomy); the latter, in Paris, was a government institution and included as a foreign member the Dutchman Huygens. In the 18th century, important royal academies were established at Berlin (1700) and at St. Petersburg (1724). The societies and academies provided the principal opportunities for the publication and discussion of scientific results during and after the scientific revolution. In 1690, James Bernoulli showed that the cycloid is the solution to the tautochrone problem; and the following year, in 1691, Johann Bernoulli showed that a chain freely suspended from two points will form a catenary, the curve with the lowest possible center of gravity available to any chain hung between two fixed points. He then showed, in 1696, that the cycloid is the solution to the brachistochrone problem.

Early thermodynamics

A precursor of the engine was designed by the German scientist Otto von Guericke who, in 1650, designed and built the world's first vacuum pump and created the world's first ever vacuum known as the Magdeburg hemispheres experiment. He was driven to make a vacuum to disprove Aristotle's long-held supposition that 'Nature abhors a vacuum'. Shortly thereafter, Irish physicist and chemist Boyle had learned of Guericke's designs and in 1656, in coordination with English scientist Robert Hooke, built an air pump. Using this pump, Boyle and Hooke noticed the pressure-volume correlation for a gas: $PV = k$, where P is pressure, V is volume and k is a constant: this relationship is known as Boyle's Law. In that time, air was assumed to be a system of motionless particles, and not interpreted as a system of moving molecules. The concept of thermal motion came two centuries later. Therefore, Boyle's publication in 1660 speaks about a mechanical concept: the air spring.[27] Later, after the invention of the thermometer, the property temperature could be quantified. This tool gave Gay-Lussac the opportunity to derive his law, which led shortly later to the ideal gas law. But, already before the establishment of the ideal gas law, an associate of Boyle's named Denis Papin built in 1679 a bone digester, which is a closed vessel with a tightly fitting lid that confines steam until a high pressure is generated.

Later designs implemented a steam release valve to keep the machine from exploding. By watching the valve rhythmically move up and down, Papin conceived of the idea of a piston and cylinder engine. He did not however follow through with his design. Nevertheless, in 1697, based on Papin's designs, engineer Thomas Savery built the first engine. Although these early engines were crude and inefficient, they attracted the attention of the leading scientists of the time. Hence, prior to 1698 and the invention of the Savery Engine, horses were used to power pulleys, attached to buckets, which lifted water out of flooded salt mines in England. In the years to follow, more variations of steam engines were built, such as the Newcomen Engine, and later the Watt Engine. In time, these early engines would eventually be utilized in place of horses. Thus, each engine began to be associated with a certain amount of "horse power" depending upon how many horses it had replaced. The main problem with these first engines was that they were slow and clumsy, converting less than 2% of the input fuel into useful work. In other words, large quantities of coal (or wood) had to be burned to yield only a small fraction of work output. Hence the need for a new science of engine dynamics was born.

1.3   18th-century developments

During the 18th century, the mechanics founded by Newton was developed by several scientists as more mathematicians learned calculus and elaborated upon its initial formulation. The application of mathematical analysis to problems of motion was known as rational mechanics, or mixed mathematics (and was later termed classical mechanics).

1.3.1   Mechanics

Daniel Bernoulli
(1700–1782)

In 1714, Brook Taylor derived the fundamental frequency of a stretched vibrating string in terms of its tension and mass per unit length by solving a differential equation. The Swiss mathematician Daniel Bernoulli (1700–1782) made important mathematical studies of the behavior of gases, anticipating the kinetic theory of gases developed more than a century later, and has been referred to as the first mathematical physicist.[28] In 1733, Daniel Bernoulli derived the fundamental frequency and harmonics of a hanging chain by solving a differential equation. In 1734, Bernoulli solved the differential equation for the vibrations of an elastic bar clamped at one end. Bernoulli's treatment of fluid dynamics and his examination of fluid flow was introduced in his 1738 work *Hydrodynamica*.

Rational mechanics dealt primarily with the development of elaborate mathematical treatments of observed motions, using Newtonian principles as a basis, and emphasized improving the tractability of complex calculations and developing of legitimate means of analytical approximation. A representative contemporary textbook was published by Johann Baptiste Horvath. By the end of the century analytical treatments were rigorous enough to verify the stability of the solar system solely on the basis of Newton's laws without reference to divine intervention—even as deterministic treatments of systems as simple as the three body problem in gravitation remained intractable.*[29] In 1705, Edmond Halley predicted the periodicity of Halley's Comet, William Herschel discovered Uranus in 1781, and Henry Cavendish measured the gravitational constant and determined the mass of the Earth in 1798. In 1783, John Michell suggested that some objects might be so massive that not even light could escape from them.

In 1739, Leonhard Euler solved the ordinary differential equation for a forced harmonic oscillator and noticed the resonance phenomenon. In 1742, Colin Maclaurin discovered his uniformly rotating self-gravitating spheroids. British work, carried on by mathematicians such as Taylor and Maclaurin, fell behind Continental developments as the century progressed. Meanwhile, work flourished at scientific academies on the Continent, led by such mathematicians as Bernoulli, Euler, Lagrange, Laplace, and Legendre. In 1743, Jean le Rond d'Alembert published his "Traite de Dynamique", in which he introduces the concept of generalized forces for accelerating systems and systems with constraints. In 1747, Pierre Louis Maupertuis applied minimum principles to mechanics. In 1759, Euler solved the partial differential equation for the vibration of a rectangular drum. In 1764, Euler examined the partial differential equation for the vibration of a circular drum and found one of the Bessel function solutions. In 1776, John Smeaton published a paper on experiments relating power, work, momentum and kinetic energy, and supporting the conservation of energy. In 1788, Joseph Louis Lagrange presented Lagrange's equations of motion in *Mécanique Analytique*. In 1789, Antoine Lavoisier states the law of conservation of mass. Newton's mechanics received brilliant exposition in both Lagrange's 1788 work and the *Celestial Mechanics* (1799–1825) of Pierre-Simon Laplace.

1.3.2 Thermodynamics

During the 18th century, thermodynamics was developed through the theories of weightless "imponderable fluids", such as heat ("caloric"), electricity, and phlogiston (which was rapidly overthrown as a concept following Lavoisier's identification of oxygen gas late in the century). Assuming that these concepts were real fluids, their flow could be traced through a mechanical apparatus or chemical reactions. This tradition of experimentation led to the development of new kinds of experimental apparatus, such as the Leyden Jar; and new kinds of measuring instruments, such as the calorimeter, and improved versions of old ones, such as the thermometer. Experiments also produced new concepts, such as the University of Glasgow experimenter

Joseph Black's notion of latent heat and Philadelphia intellectual Benjamin Franklin's characterization of electrical fluid as flowing between places of excess and deficit (a concept later reinterpreted in terms of positive and negative charges). Franklin also showed that lightning is electricity in 1752.

The accepted theory of heat in the 18th century viewed it as a kind of fluid, called caloric; although this theory was later shown to be erroneous, a number of scientists adhering to it nevertheless made important discoveries useful in developing the modern theory, including Joseph Black (1728–99) and Henry Cavendish (1731–1810). Opposed to this caloric theory, which had been developed mainly by the chemists, was the less accepted theory dating from Newton's time that heat is due to the motions of the particles of a substance. This mechanical theory gained support in 1798 from the cannon-boring experiments of Count Rumford (Benjamin Thompson), who found a direct relationship between heat and mechanical energy.

While it was recognized early in the 18th century that finding absolute theories of electrostatic and magnetic force akin to Newton's principles of motion would be an important achievement, none were forthcoming. This impossibility only slowly disappeared as experimental practice became more widespread and more refined in the early years of the 19th century in places such as the newly established Royal Institution in London. Meanwhile, the analytical methods of rational mechanics began to be applied to experimental phenomena, most influentially with the French mathematician Joseph Fourier's analytical treatment of the flow of heat, as published in 1822.*[30]*[31]*[32] Joseph Priestley proposed an electrical inverse-square law in 1767, and Charles-Augustin de Coulomb introduced the inverse-square law of electrostatics in 1798.

At the end of the century, the members of the French Academy of Sciences had attained clear dominance in the field.*[24]*[33]*[34]*[35] At the same time, the experimental tradition established by Galileo and his followers persisted. The Royal Society and the French Academy of Sciences were major centers for the performance and reporting of experimental work. Experiments in mechanics, optics, magnetism, static electricity, chemistry, and physiology were not clearly distinguished from each other during the 18th century, but significant differences in explanatory schemes and, thus, experiment design were emerging. Chemical experimenters, for instance, defied attempts to enforce a scheme of abstract Newtonian forces onto chemical affiliations, and instead focused on the isolation and classification of chemical substances and reactions.*[36]

1.4 19th century

Michael Faraday
(1791–1867)

In 1800, Alessandro Volta invented the electric battery (known of the voltaic pile) and thus improved the way electric currents could also be studied. A year later, Thomas Young demonstrated the wave nature of light — which received strong experimental support from the work of Augustin-Jean Fresnel — and the principle of interference. In 1813, Peter Ewart supported the idea of the conservation of energy in his paper *On the measure of moving force*. In 1820, Hans Christian Ørsted found that a current-carrying conductor gives rise to a magnetic force surrounding it, and within a week after Ørsted's discovery reached France, André-Marie Ampère discovered that two parallel electric currents will exert forces on each other. In 1821, William Hamilton began his analysis of Hamilton's characteristic function. In 1821, Michael Faraday built an electricity-powered motor, while Georg Ohm stated his law of electrical resistance in 1826, expressing the relationship between voltage, current, and resistance in an electric circuit. A year later, botanist Robert Brown discovered Brownian motion: pollen grains in water undergoing movement resulting from their bombardment by the fast-moving atoms or molecules in the liquid. In 1829, Gaspard Coriolis introduced the terms of work (force times distance) and

kinetic energy with the meanings they have today.

In 1831, Faraday (and independently Joseph Henry) discovered the reverse effect, the production of an electric potential or current through magnetism – known as electromagnetic induction; these two discoveries are the basis of the electric motor and the electric generator, respectively. In 1834, Carl Jacobi discovered his uniformly rotating self-gravitating ellipsoids. In 1834, John Russell observed a nondecaying solitary water wave (soliton) in the Union Canal near Edinburgh and used a water tank to study the dependence of solitary water wave velocities on wave amplitude and water depth. In 1835, William Hamilton stated Hamilton's canonical equations of motion. In the same year, Gaspard Coriolis examined theoretically the mechanical efficiency of waterwheels, and deduced the Coriolis effect. In 1841, Julius Robert von Mayer, an amateur scientist, wrote a paper on the conservation of energy but his lack of academic training led to its rejection. In 1842, Christian Doppler proposed the Doppler effect. In 1847, Hermann von Helmholtz formally stated the law of conservation of energy. In 1851, Léon Foucault showed the Earth's rotation with a huge pendulum (Foucault pendulum).

There were important advances in continuum mechanics in the first half of the century, namely formulation of laws of elasticity for solids and discovery of Navier–Stokes equations for fluids.

1.4.1 Laws of thermodynamics

Further information: History of thermodynamics

In the 19th century, the connection between heat and mechanical energy was established quantitatively by Julius Robert von Mayer and James Prescott Joule, who measured the mechanical equivalent of heat in the 1840s. In 1849, Joule published results from his series of experiments (including the paddlewheel experiment) which show that heat is a form of energy, a fact that was accepted in the 1850s. The relation between heat and energy was important for the development of steam engines, and in 1824 the experimental and theoretical work of Sadi Carnot was published. Carnot captured some of the ideas of thermodynamics in his discussion of the efficiency of an idealized engine. Sadi Carnot's work provided a basis for the formulation of the first law of thermodynamics — a restatement of the law of conservation of energy — which was stated around 1850 by William Thomson, later known as Lord Kelvin, and Rudolf Clausius. Lord Kelvin, who had extended the concept of absolute zero from gases to all substances in 1848, drew upon the engineering theory of Lazare Carnot, Sadi Carnot, and Émile Clapeyron–as well as the experimentation of James Prescott Joule on the interchangeability of mechan-

William Thomson (Lord Kelvin)
(1824–1907)

James Clerk Maxwell
(1831–1879)

1.4.2 James Clerk Maxwell

ical, chemical, thermal, and electrical forms of work—to formulate the first law.

Kelvin and Clausius also stated the second law of thermodynamics, which was originally formulated in terms of the fact that heat does not spontaneously flow from a colder body to a hotter. Other formulations followed quickly (for example, the second law was expounded in Thomson and Peter Guthrie Tait's influential work *Treatise on Natural Philosophy*) and Kelvin in particular understood some of the law's general implications. The second Law was the idea that gases consist of molecules in motion had been discussed in some detail by Daniel Bernoulli in 1738, but had fallen out of favor, and was revived by Clausius in 1857. In 1850, Hippolyte Fizeau and Léon Foucault measured the speed of light in water and find that it is slower than in air, in support of the wave model of light. In 1852, Joule and Thomson demonstrated that a rapidly expanding gas cools, later named the Joule–Thomson effect or Joule–Kelvin effect. Hermann von Helmholtz puts forward the idea of the heat death of the universe in 1854, the same year that Clausius established the importance of dQ/T (Clausius's theorem) (though he did not yet name the quantity).

In 1859, James Clerk Maxwell discovered the distribution law of molecular velocities. Maxwell showed that electric and magnetic fields are propagated outward from their source at a speed equal to that of light and that light is one of several kinds of electromagnetic radiation, differing only in frequency and wavelength from the others. In 1859, Maxwell worked out the mathematics of the distribution of velocities of the molecules of a gas. The wave theory of light was widely accepted by the time of Maxwell's work on the electromagnetic field, and afterward the study of light and that of electricity and magnetism were closely related. In 1864 James Maxwell published his papers on a dynamical theory of the electromagnetic field, and stated that light is an electromagnetic phenomenon in the 1873 publication of Maxwell's *Treatise on Electricity and Magnetism*. This work drew upon theoretical work by German theoreticians such as Carl Friedrich Gauss and Wilhelm Weber. The encapsulation of heat in particulate motion, and the addition of electromagnetic forces to Newtonian dynamics established an enormously robust theoretical underpinning to physical observations.

The prediction that light represented a transmission of energy in wave form through a "luminiferous ether", and the seeming confirmation of that prediction with Helmholtz student Heinrich Hertz's 1888 detection of electromagnetic radiation, was a major triumph for physical theory and raised the possibility that even more fundamental theories based on the field could soon be developed.[37][38][39][40] Experimental confirmation of Maxwell's theory was provided by Hertz, who generated and detected electric waves in 1886 and verified their properties, at the same time foreshadowing their application in radio, television, and other devices. In 1887, Heinrich Hertz discovered the photoelectric effect. Research on the electromagnetic waves began soon after, with many scientists and inventors conducting experiments on their properties. In the mid to late 1890s Guglielmo Marconi developed a radio wave based wireless telegraphy system[41] (see invention of radio).

The atomic theory of matter had been proposed again in the early 19th century by the chemist John Dalton and became one of the hypotheses of the kinetic-molecular theory of gases developed by Clausius and James Clerk Maxwell to explain the laws of thermodynamics. The kinetic theory in turn led to the statistical mechanics of Ludwig Boltzmann (1844–1906) and Josiah Willard Gibbs (1839–1903), which held that energy (including heat) was a measure of the speed of particles. Interrelating the statistical likelihood of certain states of organization of these particles with the energy of those states, Clausius reinterpreted the dissipation of energy to be the statistical tendency of molecular configurations to pass toward increasingly likely, increasingly disorganized states (coining the term "entropy" to describe the disorganization of a state). The statistical versus absolute interpretations of the second law of thermodynamics set up a dispute that would last for several decades (producing arguments such as "Maxwell's demon"), and that would not be held to be definitively resolved until the behavior of atoms was firmly established in the early 20th century.[42][43] In 1902, James Jeans found the length scale required for gravitational perturbations to grow in a static nearly homogeneous medium.

1.5 20th century: birth of modern physics

At the end of the 19th century, physics had evolved to the point at which classical mechanics could cope with highly complex problems involving macroscopic situations; thermodynamics and kinetic theory were well established; geometrical and physical optics could be understood in terms of electromagnetic waves; and the conservation laws for energy and momentum (and mass) were widely accepted. So profound were these and other developments that it was generally accepted that all the important laws of physics had been discovered and that, henceforth, research would be concerned with clearing up minor problems and particularly

Marie Skłodowska-Curie
(1867–1934)

with improvements of method and measurement. However, around 1900 serious doubts arose about the completeness of the classical theories—the triumph of Maxwell's theories, for example, was undermined by inadequacies that had already begun to appear—and their inability to explain certain physical phenomena, such as the energy distribution in blackbody radiation and the photoelectric effect, while some of the theoretical formulations led to paradoxes when pushed to the limit. Prominent physicists such as Hendrik Lorentz, Emil Cohn, Ernst Wiechert and Wilhelm Wien believed that some modification of Maxwell's equations might provide the basis for all physical laws. These shortcomings of classical physics were never to be resolved and new ideas were required. At the beginning of the 20th century a major revolution shook the world of physics, which led to a new era, generally referred to as modern physics.[44]

1.5.1 Radiation experiments

In the 19th century, experimenters began to detect unexpected forms of radiation: Wilhelm Röntgen caused a sensation with his discovery of X-rays in 1895; in 1896 Henri Becquerel discovered that certain kinds of matter emit radiation on their own accord. In 1897, J. J. Thomson discovered the electron, and new radioactive elements found by

J. J. Thomson (1856–1940) discovered the electron and isotopy and also invented the mass spectrometer. He was awarded the Nobel Prize in Physics in 1906.

Marie and Pierre Curie raised questions about the supposedly indestructible atom and the nature of matter. Marie and Pierre coined the term "radioactivity" to describe this property of matter, and isolated the radioactive elements radium and polonium. Ernest Rutherford and Frederick Soddy identified two of Becquerel's forms of radiation with electrons and the element helium. Rutherford identified and named two types of radioactivity and in 1911 interpreted experimental evidence as showing that the atom consists of a dense, positively charged nucleus surrounded by negatively charged electrons. Classical theory, however, predicted that this structure should be unstable. Classical theory had also failed to explain successfully two other experimental results that appeared in the late 19th century. One of these was the demonstration by Albert A. Michelson and Edward W. Morley — known as the Michelson–Morley experiment — which showed there did not seem to be a preferred frame of reference, at rest with respect to the hypothetical luminiferous ether, for describing electromagnetic phenomena. Studies of radiation and radioactive decay con-

tinued to be a preeminent focus for physical and chemical research through the 1930s, when the discovery of nuclear fission opened the way to the practical exploitation of what came to be called "atomic" energy.

1.5.2 Albert Einstein's theory of relativity

In 1905 a young, 26-year-old German physicist (then a Bern patent clerk) named Albert Einstein (1879–1955), showed how measurements of time and space are affected by motion between an observer and what is being observed. To say that Einstein's radical theory of relativity revolutionized science is no exaggeration. Although Einstein made many other important contributions to science, the theory of relativity alone represents one of the greatest intellectual achievements of all time. Although the concept of relativity was not introduced by Einstein, his major contribution was the recognition that the speed of light in a vacuum is constant, i.e. the same for all observers, and an absolute physical boundary for motion. This does not impact a person's day-to-day life since most objects travel at speeds much slower than light speed. For objects traveling near light speed, however, the theory of relativity shows that clocks associated with those objects will run more slowly and that the objects shorten in length according to measurements of an observer on Earth. Einstein also derived the famous equation, $E = mc^2$, which expresses the equivalence of mass and energy.

Special relativity

Further information: History of special relativity

Einstein argued that the speed of light was a constant in all inertial reference frames and that electromagnetic laws should remain valid independent of reference frame — assertions which rendered the ether "superfluous" to physical theory, and that held that observations of time and length varied relative to how the observer was moving with respect to the object being measured (what came to be called the "special theory of relativity"). It also followed that mass and energy were interchangeable quantities according to the equation $E=mc^2$. In another paper published the same year, Einstein asserted that electromagnetic radiation was transmitted in discrete quantities ("quanta"), according to a constant that the theoretical physicist Max Planck had posited in 1900 to arrive at an accurate theory for the distribution of blackbody radiation — an assumption that explained the strange properties of the photoelectric effect.

The special theory of relativity is a formulation of the relationship between physical observations and the concepts of space and time. The theory arose out of contradictions between electromagnetism and Newtonian mechanics and

Albert Einstein (1879–1955) proposed that gravitation was a result of masses (or their equivalent energies) curving ("bending") the spacetime in which they exist, altering the paths they follow within it.

make different but equally valid (and reconcilable) measurements. What remains absolute is stated in Einstein's relativity postulate: "The basic laws of physics are identical for two observers who have a constant relative velocity with respect to each other."

Special Relativity had a profound effect on physics: started as a rethinking of the theory of electromagnetism, it found a new symmetry law of nature, now called *Poincaré symmetry*, that replaced the old Galilean (see above) symmetry.

Special Relativity exerted another long-lasting effect on dynamics. Although initially it was credited with the "unification of mass and energy", it became evident that relativistic dynamics established a firm *distinction* between rest mass, which is an invariant (observer independent) property of a particle or system of particles, and the energy and momentum of a system. The latter two are separately conserved in all situations but not invariant with respect to different observers. The term *mass* in particle physics underwent a semantic change, and since the late 20th century it almost exclusively denotes the rest (or *invariant*) mass. See mass in special relativity for additional discussion.

General relativity

Further information: History of general relativity

By 1916, Einstein was able to generalize this further, to deal with all states of motion including non-uniform acceleration, which became the general theory of relativity. In this theory Einstein also specified a new concept, the curvature of space-time, which described the gravitational effect at every point in space. In fact, the curvature of space-time completely replaced Newton's universal law of gravitation. According to Einstein, gravitational force in the normal sense is a kind of illusion caused by the geometry of space. The presence of a mass causes a curvature of space-time in the vicinity of the mass, and this curvature dictates the space-time path that all freely-moving objects must follow. It was also predicted from this theory that light should be subject to gravity - all of which was verified experimentally. This aspect of relativity explained the phenomena of light bending around the sun, predicted black holes as well as properties of the Cosmic microwave background radiation —a discovery rendering fundamental anomalies in the classic Steady-State hypothesis. For his work on relativity, the photoelectric effect and blackbody radiation, Einstein received the Nobel Prize in 1921.

had great impact on both those areas. The original historical issue was whether it was meaningful to discuss the electromagnetic wave-carrying "ether" and motion relative to it and also whether one could detect such motion, as was unsuccessfully attempted in the Michelson–Morley experiment. Einstein demolished these questions and the ether concept in his special theory of relativity. However, his basic formulation does not involve detailed electromagnetic theory. It arises out of the question: "What is time?" Newton, in the *Principia* (1686), had given an unambiguous answer: "Absolute, true, and mathematical time, of itself, and from its own nature, flows equably without relation to anything external, and by another name is called duration." This definition is basic to all classical physics.

Einstein had the genius to question it, and found that it was incomplete. Instead, each "observer" necessarily makes use of his or her own scale of time, and for two observers in relative motion, their time-scales will differ. This induces a related effect on position measurements. Space and time become intertwined concepts, fundamentally dependent on the observer. Each observer presides over his or her own space-time framework or coordinate system. There being no absolute frame of reference, all observers of given events

The gradual acceptance of Einstein's theories of relativity and the quantized nature of light transmission, and of Niels Bohr's model of the atom created as many problems as they solved, leading to a full-scale effort to reestablish physics on new fundamental principles. Expanding relativity to cases of accelerating reference frames (the "general theory of relativity") in the 1910s, Einstein posited an equivalence between the inertial force of acceleration and the force of gravity, leading to the conclusion that space is curved and finite in size, and the prediction of such phenomena as gravitational lensing and the distortion of time in gravitational fields. m

1.5.3 Quantum mechanics

Further information: History of quantum mechanics
Although relativity resolved the electromagnetic phenom-

*Max Planck
(1858–1947)*

ena conflict demonstrated by Michelson and Morley, a second theoretical problem was the explanation of the distribution of electromagnetic radiation emitted by a black body; experiment showed that at shorter wavelengths, toward the ultraviolet end of the spectrum, the energy approached zero, but classical theory predicted it should become infinite. This glaring discrepancy, known as the ultraviolet catastrophe, was solved by the new theory of quantum mechanics. Quantum mechanics is the theory of atoms and subatomic systems. Approximately the first 30 years of the 20th century represent the time of the conception and evolution of the theory. The basic ideas of quantum theory

were introduced in 1900 by Max Planck (1858–1947), who was awarded the Nobel Prize for Physics in 1918 for his discovery of the quantified nature of energy. The quantum theory (which previously relied in the ˝correspondence˝ at large scales between the quantized world of the atom and the continuities of the "classical" world) was accepted when the Compton Effect established that light carries momentum and can scatter off particles, and when Louis de Broglie asserted that matter can be seen as behaving as a wave in much the same way as electromagnetic waves behave like particles (wave–particle duality).

*Werner Heisenberg
(1901–1976)*

In 1905, Einstein used the quantum theory to explain the photoelectric effect, and in 1913 the Danish physicist Niels Bohr used the same constant to explain the stability of Rutherford's atom as well as the frequencies of light emitted by hydrogen gas. The quantized theory of the atom gave way to a full-scale quantum mechanics in the 1920s. New principles of a ˝quantum˝ rather than a ˝classical˝ mechanics, formulated in matrix-form by Werner Heisenberg, Max Born, and Pascual Jordan in 1925, were based on the probabilistic relationship between discrete ˝states˝ and denied the possibility of causality. Quantum mechanics was extensively developed by Heisenberg, Wolfgang

Pauli, Paul Dirac, and Erwin Schrödinger, who established an equivalent theory based on waves in 1926; but Heisenberg's 1927 "uncertainty principle" (indicating the impossibility of precisely and simultaneously measuring position and momentum) and the "Copenhagen interpretation" of quantum mechanics (named after Bohr's home city) continued to deny the possibility of fundamental causality, though opponents such as Einstein would metaphorically assert that "God does not play dice with the universe" .[*][45] The new quantum mechanics became an indispensable tool in the investigation and explanation of phenomena at the atomic level. Also in the 1920s, the Indian scientist Satyendra Nath Bose's work on photons and quantum mechanics provided the foundation for Bose–Einstein statistics, the theory of the Bose–Einstein condensate.

The spin–statistics theorem established that any particle in quantum mechanics may be either a boson (statistically Bose–Einstein) or a fermion (statistically Fermi–Dirac). It was later found that all fundamental bosons transmit forces, such as the photon that transmits light.

Fermions are particles "like electrons and nucleons" and are the usual constituents of matter. Fermi–Dirac statistics later found numerous other uses, from astrophysics (see Degenerate matter) to semiconductor design.

1.6 Contemporary and particle physics

Further information: History of subatomic physics

1.6.1 Quantum field theory

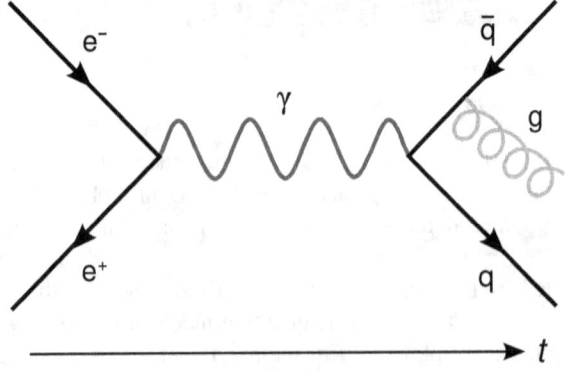

A Feynman diagram representing (left to right) the production of a photon (blue sine wave) from the annihilation of an electron and its complementary antiparticle, the positron. The photon becomes a quark–antiquark pair and a gluon (green spiral) is released.

As the philosophically inclined continued to debate the fundamental nature of the universe, quantum theories continued to be produced, beginning with Paul Dirac's formulation of a relativistic quantum theory in 1928. However, attempts to quantize electromagnetic theory entirely were stymied throughout the 1930s by theoretical formulations yielding infinite energies. This situation was not considered adequately resolved until after World War II ended, when Julian Schwinger, Richard Feynman and Sin-Itiro Tomonaga independently posited the technique of renormalization, which allowed for an establishment of a robust quantum electrodynamics (QED).[*][46]

Meanwhile, new theories of fundamental particles proliferated with the rise of the idea of the quantization of fields through "exchange forces" regulated by an exchange of short-lived "virtual" particles, which were allowed to exist according to the laws governing the uncertainties inherent in the quantum world. Notably, Hideki Yukawa proposed that the positive charges of the nucleus were kept together courtesy of a powerful but short-range force mediated by a particle with a mass between that of the electron and proton. This particle, the "pion", was identified in 1947 as part of what became a slew of particles discovered after World War II. Initially, such particles were found as ionizing radiation left by cosmic rays, but increasingly came to be produced in newer and more powerful particle accelerators.[*][47]

Outside particle physics, significant advances of the time were:

- the invention of the laser (1964 Nobel Prize in Physics);
- the theoretical and experimental research of superconductivity, especially the invention of a quantum theory of superconductivity by Vitaly Ginzburg and Lev Landau (2002 Nobel Prize in Physics) and, later, its explanation via Cooper pairs (1972 Nobel Prize in Physics). The Cooper pair was an early example of quasiparticles.

1.6.2 Unified field theories

Main article: Unified field theory

Einstein deemed that all fundamental interactions in nature can be explained in a single theory. Unified field theories were numerous attempts to "merge" several interactions. One of formulations of such theories (as well as field theories in general) is a *gauge theory*, a generalization of the idea of symmetry. Eventually the Standard Model (see below) succeeded in unification of strong, weak, and electromagnetic interactions. All attempts to unify gravitation with something else failed.

1.6.3   Standard Model

Main article: Standard Model
The interaction of these particles by scattering and decay

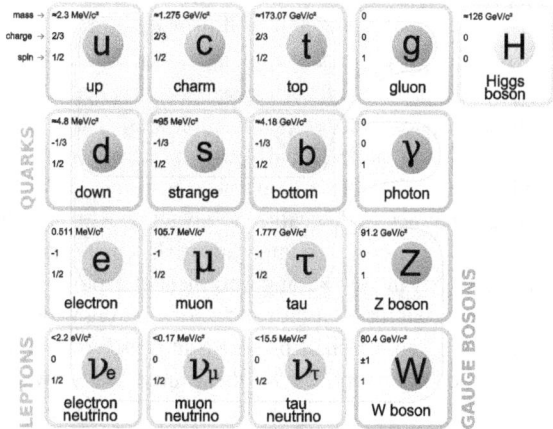

The Standard Model.

provided a key to new fundamental quantum theories. Murray Gell-Mann and Yuval Ne'eman brought some order to these new particles by classifying them according to certain qualities, beginning with what Gell-Mann referred to as the "Eightfold Way". While its further development, the quark model, at first seemed inadequate to describe strong nuclear forces, allowing the temporary rise of competing theories such as the S-Matrix, the establishment of quantum chromodynamics in the 1970s finalized a set of fundamental and exchange particles, which allowed for the establishment of a "standard model" based on the mathematics of gauge invariance, which successfully described all forces except for gravitation, and which remains generally accepted within its domain of application.*[45]

The Standard Model groups the electroweak interaction theory and quantum chromodynamics into a structure denoted by the gauge group SU(3)×SU(2)×U(1). The formulation of the unification of the electromagnetic and weak interactions in the standard model is due to Abdus Salam, Steven Weinberg and, subsequently, Sheldon Glashow. Electroweak theory was later confirmed experimentally (by observation of neutral weak currents),*[48]*[49]*[50]*[51] and distinguished by the 1979 Nobel Prize in Physics.*[52]

Since the 1970s, fundamental particle physics has provided insights into early universe cosmology, particularly the Big Bang theory proposed as a consequence of Einstein's general theory of relativity. However, starting in the 1990s, astronomical observations have also provided new challenges, such as the need for new explanations of galactic stability ("dark matter") and the apparent acceleration in the expansion of the universe ("dark energy").

While accelerators have confirmed most aspects of the

Standard Model by detecting expected particle interactions at various collision energies, no theory reconciling general relativity with the Standard Model has yet been found, although supersymmetry and string theory were believed by many theorists to be a promising avenue forward. The Large Hadron Collider, however, which began operating in 2008, has failed to find any evidence whatsoever that is supportive of supersymmetry and string theory.*[53]

1.6.4   Cosmology

Main article: Physical cosmology

Cosmology may be said to have become a serious research question with the publication of Einstein's General Theory of Relativity in 1916 [1915?] although it did not enter the scientific mainstream until the period known as the "Golden age of general relativity".

About a decade later, in the midst of what was dubbed the "Great Debate", Hubble and Slipher discovered the expansion of universe in the 1920s measuring the redshifts of Doppler spectra from galactic nebulae. Using Einstein's general relativity, Lemaître and Gamow formulated what would become known as the big bang theory. A rival, called the steady state theory was devised by Hoyle, Gold, Narlikar and Bondi.

Cosmic background radiation was verified in the 1960s by Penzias and Wilson, and this discovery favoured the big bang at the expense of the steady state scenario. Later work was by Smoot et al. (1989), among other contributors, using data from the Cosmic Background explorer (CoBE) and the Wilkinson Microwave Anistropy Probe (WMAP) satellites that refined these observations. The 1980s (the same decade of the COBE measurements) also saw the proposal of inflation theory by Guth.

Recently the problems of dark matter and dark energy have risen to the top of the cosmology agenda.

1.6.5   Higgs boson

On July 4, 2012, physicists working at CERN's Large Hadron Collider announced that they had discovered a new subatomic particle greatly resembling the Higgs boson, a potential key to an understanding of why elementary particles have mass and indeed to the existence of diversity and life in the universe.*[54] For now, some physicists are calling it a "Higgslike" particle.*[54] Joe Incandela, of the University of California, Santa Barbara, said, "It's something that may, in the end, be one of the biggest observations of any new phenomena in our field in the last 30 or 40 years, going way back to the discovery of quarks, for exam-

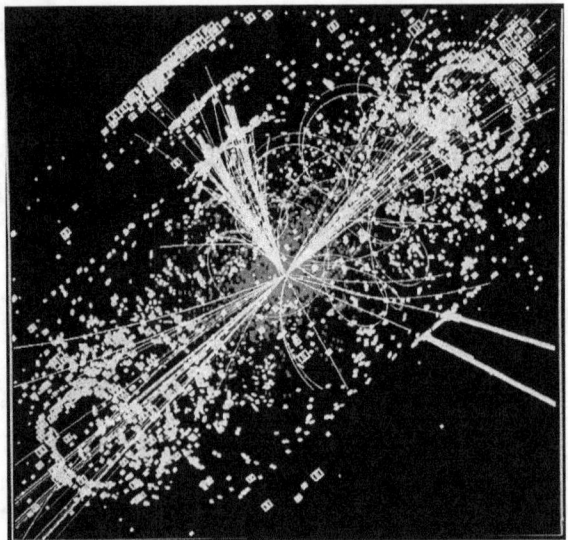

One possible signature of a Higgs boson from a simulated proton–proton collision. It decays almost immediately into two jets of hadrons and two electrons, visible as lines.

1.7 Physical sciences

With increased accessibility to and elaboration upon advanced analytical techniques in the 19th century, physics was defined as much, if not more, by those techniques than by the search for universal principles of motion and energy, and the fundamental nature of matter. Fields such as acoustics, geophysics, astrophysics, aerodynamics, plasma physics, low-temperature physics, and solid-state physics joined optics, fluid dynamics, electromagnetism, and mechanics as areas of physical research. In the 20th century, physics also became closely allied with such fields as electrical, aerospace and materials engineering, and physicists began to work in government and industrial laboratories as much as in academic settings. Following World War II, the population of physicists increased dramatically, and came to be centered on the United States, while, in more recent decades, physics has become a more international pursuit than at any time in its previous history.

1.8 Seminal physics publications

1.9 Influential physicists

The following is a gallery of highly influential and important figures in the history of physics. For a list that includes even more people, see list of physicists.

ple." *[54] Michael Turner, a cosmologist at the University of Chicago and the chairman of the physics center board, said:

> "*This is a big moment for particle physics and a crossroads —will this be the high water mark or will it be the first of many discoveries that point us toward solving the really big questions that we have posed?*"
> —Michael Turner, University of Chicago*[54]

Peter Higgs was one of six physicists, working in three independent groups, who, in 1964, invented the notion of the Higgs field ("cosmic molasses"). The others were Tom Kibble of Imperial College, London; Carl Hagen of the University of Rochester; Gerald Guralnik of Brown University; and François Englert and Robert Brout, both of Université libre de Bruxelles.*[54]

Although they have never been seen, Higgslike fields play an important role in theories of the universe and in string theory. Under certain conditions, according to the strange accounting of Einsteinian physics, they can become suffused with energy that exerts an antigravitational force. Such fields have been proposed as the source of an enormous burst of expansion, known as inflation, early in the universe and, possibly, as the secret of the dark energy that now seems to be speeding up the expansion of the universe.*[54]

- Alhazen (965–1040): made significant improvements in optics, physical science, and the scientific method. In his book, Book of Optics, he showed through experiment that light travels in straight lines, and carried out various experiments with lenses, mirrors, refraction, and reflection, which earned him the title of the "Father of Modern Optics" .

- Nicolaus Copernicus (1473–1543): published *De revolutionibus orbium coelestium* (On the Revolutions of the Celestial Spheres) in 1543—often considered the starting point of modern astronomy—in which he argued that the Earth and the other planets revolved around the Sun (heliocentrism)

- Galileo Galilei (1564–1642): discovered the uniform acceleration rate of falling bodies, improved on the refracting telescope, discovered the four largest moons of Jupiter, described projectile motion and the concept of weight, described the motion of pendulums; known for championing of the Copernican theory of heliocentrism against Church opposition.

- Johannes Kepler (1571-1630): used the accurate observations of Tycho Brahe to formulate three fundamental laws of planetary motion, described elliptical motion of planets around the sun, developed early telescopes, invented the convex eyepiece, discovered a means of determining the magnifying power of lenses.

- Evangelista Torricelli (1608–47): invented the barometer (a glass tube of mercury inverted into a dish), found that the change of height of the mercury each day was from atmospheric pressure, worked in geometry and developed integral calculus, published findings on fluid and projectile motion in his 1644 *Opera Geometrica* (Geometric Works)

- Blaise Pascal (1623–62): experimented with fluids, formulated Pascal's law in the 1650s stating that the pressure applied to a fluid taken in a closed container is transmitted with equal force throughout the container, proved that air has weight and that air pressure can produce a vacuum, namesake of the unit of pressure: the pascal (Pa)

- Christiaan Huygens (1629–95): studied the rings of Saturn and discovered its moon Titan, invented the pendulum clock, studied optics and centrifugal force, theorized that light consists of waves (Huygens–Fresnel principle) which became instrumental in the understanding of wave-particle duality.

- Robert Hooke (1635–1703): formulated the law of elasticity, invented the balance spring, the spiral spring wheel in watches, the Gregorian telescope, and the first screw-divided quadrant, constructed first arithmetical machine, improved cell theory with the microscope

- Sir Isaac Newton (1642–1727): established three laws of motion and a law of universal gravitation in his *Philosophiæ Naturalis Principia Mathematica* (1687), laid foundations for classical mechanics, built the first practical reflecting telescope (the Newtonian telescope), observed that a prism splits white light into the colors of the visible spectrum, formulated a law of cooling, co-invented calculus

- Henry Cavendish (1731–1810): greatest English chemist and physicist of his age, researched composition of the atmosphere, the properties of different gases, the synthesis of water, the law of electrical attraction and repulsion, a mechanical theory of heat, calculated the weight of the Earth in the Cavendish experiment, determined the universal gravitational constant

- Charles-Augustin de Coulomb (1736–1806): formulated a law in 1785 which described the electrostatic interaction between electrically charged particles (attraction and repulsion) and was essential to the development of the theory of electromagnetism, namesake of the unit of electric charge: the coulomb (C)

- Alessandro Volta (1745–1827): built the first electric battery (the voltaic pile) in the 19th century, did substantial work with electric currents, namesake of the unit of electric potential: the volt (V)

- Thomas Young (1773–1829): established the principle of interference of light, resurrected the century-old theory that light is a wave, helped decipher the Rosetta Stone

- Hans Christian Ørsted (1777–1851): discovered that electric currents create magnetic fields (an important aspect of electromagnetism), shaped advances in science in the late 19th century, namesake of the oersted (Oe) (the cgs unit of magnetic H-field strength)

- André-Marie Ampère (1777–1836): main founder of electrodynamics, showed how an electric current produces a magnetic field, stated that the mutual action of two lengths of current-carrying wire is proportional to their lengths and to the intensities of their currents (Ampère's law), namesake of the unit of electric current (the ampere)

- Joseph von Fraunhofer, (1787–1826): first to studied the dark lines of the Sun's spectrum, now known as Fraunhofer lines, first to use extensively the diffraction grating (a device that disperses light more effectively than a prism does), set the stage for the development of spectroscopy, making optical glass and achromatic telescope objectives.

- Georg Ohm (1789–1854): found that there is a direct proportionality between the electric current I and the potential difference (voltage) V applied across a conductor, and that this current is inversely proportional to the resistance R in the circuit, or $I = V/R$, known as Ohm's law, namesake of the unit of electrical resistance (the ohm)

- Michael Faraday (1791–1867): showed how a changing magnetic field can be used to generate an electric current (Faraday's law of induction), applied this knowledge to the development of several electrical machines, described principles of electrolysis, early pioneer in the field of low temperature study

- Christian Doppler (1803–53): first described how the observed frequency of light and sound waves is affected by the relative motion of the source and the detector, a phenomenon which became known as the Doppler effect.

- James Prescott Joule (1818–89): discovered that heat is a form of energy, ideas led to the theory of conservation of energy, worked with Lord Kelvin to develop the absolute scale of temperature, made observations on magnetostriction, found the relationship between current through resistance and the heat dissipated, now called Joule's law.

- William Thomson, 1st Baron Kelvin (1824–1907): major figure in the history of thermodynamics, helped develop law of conservation of energy, studied wave motion and vortex motion in hydrodynamics and produced a dynamical theory of heat, formulated of the first and second laws of thermodynamics

- James Clerk Maxwell (1831–79): united electricity, magnetism, and optics into a consistent electromagnetic theory, formulated Maxwell's equations to show that electricity, magnetism and light are manifestations of the electromagnetic field, developed the Maxwell–Boltzmann distribution (statistical means of describing aspects of the kinetic theory of gases)

- Ernst Mach (1838–1916): contributed the Mach number, studied shock waves and how airflow is disturbed at the speed of sound, influenced logical positivism, forerunner of Einstein's relativity through his criticism of Newton

- Ludwig Boltzmann (1844–1906): developed statistical mechanics (how the properties of atoms – mass, charge, and structure – determine the visible properties of matter, such as viscosity, thermal conductivity, and diffusion), developed the kinetic theory of gases.

- Wilhelm Röntgen (1845–1923): produced and detected electromagnetic radiation in a wavelength range of X-rays or Röntgen rays in 1895, for which he earned the first Nobel Prize in Physics in 1901, namesake of element 111, Roentgenium

- Henri Becquerel (1852–1908): discovered radioactivity along with Marie Skłodowska-Curie and Pierre Curie, for which all three won the 1903 Nobel Prize in Physics.

- Hendrik Lorentz (1853–1928): clarified electromagnetic theory of light, shared the 1902 Nobel Prize in Physics with Pieter Zeeman for the discovery and theoretical explanation of the Zeeman effect, developed concept of local time, derived the transformation equations subsequently used by Albert Einstein to describe space and time.

- J. J. Thomson (1856–1940): showed in 1897 that cathode rays were composed of a previously unknown negatively charged particle (later named the electron), discovered isotopes, invented the mass spectrometer, awarded the 1906 Nobel Prize in Physics for the discovery of the electron and for his work on the conduction of electricity in gases.

- Nikola Tesla (1856–1943): contributed to alternating current (AC) engineering, developed an AC induction motor. Invented the Tesla coil.

- Heinrich Hertz (1857–1894): clarified and expanded Maxwell's electromagnetic theory of light, first to prove the existence of electromagnetic waves by engineering instruments to transmit and receive radio pulses

- Max Planck (1858–1947): founded quantum mechanics in 1900, showed how the energy of a photon is directly proportional to its frequency, won him the 1918 Nobel Prize in Physics. He then used his quantum hypothesis to formulate Planck's Law, thereby resolving the ultraviolet catastrophe.

- Pieter Zeeman (1865–1943): shared the 1902 Nobel Prize in Physics with Hendrik Lorentz for discovering the Zeeman effect (splitting a spectral line into several components in the presence of a static magnetic field)

- Marie Curie (1867–1934): discovered the existence of radioactivity with Henri Becquerel and her husband Pierre Curie, awarded the Nobel Prize in Physics (1903) and the Nobel Prize in Chemistry (1911), found techniques for isolating radioactive isotopes, isolated plutonium and radium

- Robert Andrews Millikan (1868–1953): measured the charge on the electron, worked on the photoelectric effect, performed vital research pertaining to cosmic rays.

- Ernest Rutherford (1871–1937): considered "Father of Nuclear Physics", showed how the atomic nucleus has a positive charge, first to change one element into another by an artificial nuclear reaction, differentiated and named alpha and beta radiation, awarded Nobel Prize for Chemistry in 1908

- Lise Meitner (1878–1968): worked on radioactivity and nuclear physics, gave the first theoretical explanation for nuclear fission, for which her colleague, chemist Otto Hahn, was awarded the Nobel Prize. She is often mentioned, with Ida Noddack, as one of the most glaring examples of women's scientific achievement overlooked by the Nobel committee.

- Albert Einstein (1879–1955): revolutionized physics due to his theories of special and general relativity, described Brownian motion, awarded the Nobel Prize in

Physics in 1921 for his work on the photoelectric effect, formulated mass–energy equivalence formula $E = mc^2$, published more than 300 scientific papers and over 150 non-scientific works, considered the "Father of Modern Physics"

- Niels Bohr (1885–1962): used quantum mechanical model (known as the Bohr model) of the atom which theorized that electrons travel in discrete orbits around the nucleus, showed how electron energy levels are related to spectral lines, received the Nobel Prize in Physics in 1922.

- Erwin Schrödinger (1887–1961): formulated the Schrödinger equation in 1926 describing how the quantum state of a physical system changes with time, awarded the Nobel Prize in Physics in 1933, two years later proposed the thought experiment known as Schrödinger's cat

- James Chadwick (1891–1974): James Chadwick's major work is the discovery of the neutron for which received the Nobel Prize in Physics in 1935. He was one of the primary British scientists who worked in the Manhattan Project in the United States during World War II. He was knighted in 1945 for achievements in physics.

- Louis de Broglie (1892–1987): researched quantum theory, discovered the wave nature of electrons, awarded the 1929 Nobel Prize in Physics, ideas on the wave-like behavior of particles used by Erwin Schrödinger in his formulation of wave mechanics.

- Georges Lemaître (1894–1966): first person to propose the theory of the expansion of the Universe, first to derive what is now known as Hubble's law, made the first estimation of what is now called the Hubble constant which he published in 1927 (two years before Hubble's article), proposed the Big Bang theory of the origin of the Universe

- Wolfgang Pauli (1900–1958): pioneers of quantum physics, received the Nobel Prize in Physics in 1945 (nominated by Albert Einstein), formulated the Pauli exclusion principle involving spin theory (underpinning the structure of matter and the whole of chemistry), published the Pauli–Villars regularization, formulated the Pauli equation, coined the phrase 'not even wrong'

- Werner Heisenberg (1901–1976): developed method to express ideas of quantum mechanics in terms of matrices in 1925, published his famous uncertainty principle in 1927, awarded Nobel Prize in Physics in 1932

- Enrico Fermi (1901–1954): developed first nuclear reactor (Chicago Pile-1), contributed to quantum theory, nuclear and particle physics, and statistical mechanics, awarded the 1938 Nobel Prize in Physics for his work on induced radioactivity.

- Paul Dirac (1902–1984): made fundamental contributions to the early development of quantum mechanics and quantum electrodynamics, formulated the Dirac equation describing the behavior of fermions, predicted the existence of antimatter, shared the 1933 Nobel Prize in Physics with Erwin Schrödinger

- John Bardeen (1908–1991): awarded Nobel Prize in Physics in 1956 with William Shockley and Walter Brattain for the invention of the transistor and again in 1972 with Leon Cooper and John Robert Schrieffer for a fundamental theory of conventional superconductivity known as the BCS theory.

- John Wheeler (1911–2008): revived interest in general relativity in the United States after World War II, worked with Niels Bohr to explain principles of nuclear fission, tried to achieve Einstein's vision of a unified field theory, coined the terms black hole, quantum foam, wormhole, and the phrase "it from bit".

- Richard Feynman (1918–1988): developed the path integral formulation of quantum mechanics, the theory of quantum electrodynamics, and the physics of the superfluidity of supercooled liquid helium, awarded the Nobel Prize in Physics in 1965 with Julian Schwinger and Sin-Itiro Tomonaga, developed the Feynman diagram representing subatomic particle behavior.

- Gerardus 't Hooft (1946–present): a Dutch theoretical physicist and professor at Utrecht University, he shared the 1999 Nobel Prize in Physics with his thesis advisor Martinus J. G. Veltman "for elucidating the quantum structure of electroweak interactions". His work on electroweak theory was crucial to Peter Higgs in the development of higgs boson theory.

- Peter Higgs (1929–present): Along with François Englert, Robert Brout, Gerald Guralnik, C. R. Hagen, and Tom Kibble, he developed the theory of Higgs field and Higgs boson, which together form the higgs mechanism that explains how subatomic particles gain their mass. However CERN have been cautious with the results, stating that new tests are needed to confirm the discovery. He received the Nobel Prize in Physics in 2013 for his work on the mentioned mechanism.

- Stephen Hawking (1942–present): provided, with Roger Penrose, theorems of general relativity regarding the occurrence of gravitational singularities (black holes) and theoretically predicted that black holes should emit radiation (Hawking radiation).

1.10   See also

- History of optics
- History of electrical engineering
- History of electromagnetism
- List of physicists
- Nobel Prize in physics
- Timeline of fundamental physics discoveries

1.11   Notes

[1] Click the image to see further details.

[2] Mariam Rozhanskaya and I. S. Levinova (1996), "Statics", p. 642, in Rashed & Morelon (1996, pp. 614–642):

> "Using a whole body of mathematical methods (not only those inherited from the antique theory of ratios and infinitesimal techniques, but also the methods of the contemporary algebra and fine calculation techniques), Islamic scientists raised statics to a new, higher level. The classical results of Archimedes in the theory of the centre of gravity were generalized and applied to three-dimensional bodies, the theory of ponderable lever was founded and the 'science of gravity' was created and later further developed in medieval Europe. The phenomena of statics were studied by using the dynamic approach so that two trends – statics and dynamics – turned out to be inter-related within a single science, mechanics."
> "The combination of the dynamic approach with Archimedean hydrostatics gave birth to a direction in science which may be called medieval hydrodynamics."
> "Archimedean statics formed the basis for creating the fundamentals of the science on specific weight. Numerous fine experimental methods were developed for determining the specific weight, which were based, in particular, on the theory of balances and weighing. The classical works of al-Biruni and al-Khazini can by right be considered as the beginning of the application of experimental methods in medieval science."
> "Arabic statics was an essential link in the

progress of world science. It played an important part in the prehistory of classical mechanics in medieval Europe. Without it classical mechanics proper could probably not have been created."

1.12   References

[1] Letter to Robert Hooke (15 February 1676 by Gregorian reckonings with January 1 as New Year's Day). equivalent to 5 February 1675 using the Julian calendar with March 25 as New Year's Day

[2] "This shift from ecclesiastical reasoning to scientific reasoning marked the beginning of scientific methodology." Singer, C., *A Short History of Science to the 19th Century*, Streeter Press, 2008, p. 35.

[3] Oliver Leaman, *Key Concepts in Eastern Philosophy.* Routledge, 1999, page 269.

[4] Chattopadhyaya 1986, pp. 169–70

[5] Radhakrishnan 2006, p. 202

[6] (Stcherbatsky 1962 (1930). Vol. 1. P. 19)

[7] Li Shu-hua, "Origine de la Boussole 11. Aimant et Boussole", *Isis*, Vol. 45, No. 2. (Jul., 1954), p.175

[8] Joseph Needham, Volume 4, Part 1, 98.

[9] Glick, Livesey & Wallis (2005, pp. 89–90)

[10] "Top 10 ancient Arabic scientists". COSMOS magazine. 2011-01-06. Retrieved 2013-04-20.

[11] Alistair C. Crombie, "Quantification in medieval physics." *Isis* (1961): 143-160. in JSTOR

[12] David C. Lindberg, and Elspeth Whitney, eds., *The beginnings of Western science: The European scientific tradition in philosophical, religious, and institutional context, 600 BC to AD 1450* (University of Chicago Press, 1992)

[13] Singer, Charles (1941), *A Short History of Science to the Nineteenth Century*, Clarendon Press, page 217.

[14] Weidhorn, Manfred (2005), *The Person of the Millennium: The Unique Impact of Galileo on World History*, iUniverse, p. 155, ISBN 0-595-36877-8

[15] Finocchiaro (2007).

[16] Stephen Hawking, "Galileo and the Birth of Modern Science", *American Heritage's Invention & Technology*, Vol. 24, No. 1 (Spring 2009), p. 36.

[17] Drake (1978)

[18] Biagioli (1993)

[19] Shea (1991)

[20] Garber (1992)

[21] Gaukroger (2002)

[22] Hall (1980)

[23] Bertolini Meli (1993)

[24] Guicciardini (1999)

[25] James R. Graham (webpage). The Early Period (1608–1672) [Retrieved 3 February 2009].

[26] Błaszczyk, Piotr; Katz, Mikhail; Sherry, David (2012), "Ten misconceptions from the history of analysis and their debunking", *Foundations of Science*, arXiv:1202.4153, doi:10.1007/s10699-012-9285-8

[27] New Experiments physico-mechanicall, Touching the Spring of the Air and its Effects (1660).

[28] Darrigol (2005)

[29] Bos (1980)

[30] Heilbron (1979)

[31] Buchwald (1989)

[32] Golinski (1999)

[33] Greenberg (1986)

[34] Guicciardini (1989)

[35] Garber (1999)

[36] Ben-Chaim (2004)

[37] Buchwald (1985)

[38] Jungnickel and McCormmanch (1986)

[39] Hunt (1991)

[40] Buchwald (1994)

[41] Michael Windelspecht, Groundbreaking Scientific Experiments, Inventions, and Discoveries of the 19th Century, Greenwood Publishing Group, 2003 page 195

[42] Smith & Wise (1989)

[43] Smith (1998)

[44] Agar (2012)

[45] Kragh (1999)

[46] Schweber (1994)

[47] Galison (1997)

[48] F. J. Hasert *et al. Phys. Lett.* **46B** 121 (1973).

[49] F. J. Hasert *et al. Phys. Lett.* **46B** 138 (1973).

[50] F. J. Hasert *et al. Nucl. Phys.* **B73** 1(1974).

[51] *The discovery of the weak neutral currents*, CERN courier, 2004-10-04, retrieved 2008-05-08

[52] *The Nobel Prize in Physics 1979*, Nobel Foundation, retrieved 2008-09-10

[53] Woit, Peter (20 October 2013). "Last Links For a While". *Not Even Wrong*. Retrieved 2 November 2013.

[54] Overbye, Dennis (4 July 2012). "Physicists Find Particle That Could Be the Higgs Boson". *The New York Times*.

1.13 Sources

- Agar, Jon (2012), *Science in the Twentieth Century and Beyond*, Cambridge: Polity Press, ISBN 978-0-7456-3469-2.

- Aristotle *Physics* translated by Hardie & Gaye

- Ben-Chaim, Michael (2004), *Experimental Philosophy and the Birth of Empirical Science: Boyle, Locke and Newton*, Aldershot: Ashgate, ISBN 0-7546-4091-4, OCLC 53887772 57202497.

- Bertoloni Meli, Domenico (1993), *Equivalence and Priority: Newton versus Leibniz*, New York: Oxford University Press.

- Biagioli, Mario (1993), *Galileo, Courtier: The Practice of Science in the Culture of Absolutism*, Chicago: University of Chicago Press, ISBN 0-226-04559-5, OCLC 185632037 26767743.

- Bos, Henk (1980), "Mathematics and Rational Mechanics", in Rousseau, G. S.; Porter, Roy, *The Ferment of Knowledge: Studies in the Historiography of Eighteenth Century Science*, New York: Cambridge University Press.

- Buchwald, Jed (1985), *From Maxwell to Microphysics: Aspects of Electromagnetic Theory in the Last Quarter of the Nineteenth Century*, Chicago: University of Chicago Press, ISBN 0-226-07882-5, OCLC 11916470.

- Buchwald, Jed (1989), *The Rise of the Wave Theory of Light: Optical Theory and Experiment in the Early Nineteenth Century*, Chicago: University of Chicago Press, ISBN 0-226-07886-8, OCLC 18069573 59210058.

- Buchwald, Jed (1994), *The Creation of Scientific Effects: Heinrich Hertz and Electric Waves*, Chicago: University of Chicago Press, ISBN 0-226-07888-4, OCLC 29256963 59866377.

- Darrigol, Olivier (2005), *Worlds of Flow: A History of Hydrodynamics from the Bernoullis to Prandtl*, New York: Oxford University Press, ISBN 0-19-856843-6, OCLC 237027708 60839424.

- Dear, Peter (1995), *Discipline and Experience: The Mathematical Way in the Scientific Revolution*, Chicago: University of Chicago Press, ISBN 0-226-13943-3, OCLC 32236425.

- Dijksterhuis, Fokko Jan (2004), *Lenses and Waves: Christiaan Huygens and the Mathematical Science of Optics in the Seventeenth Century*, Springer, ISBN 1-4020-2697-8, OCLC 228400027 56533625

- Drake, Stillman (1978), *Galileo at Work: His Scientific Biography*, Chicago: University of Chicago Press, ISBN 0-226-16226-5, OCLC 185633608 3770650 8235076.

- Galison, Peter (1997), *Image and Logic: A Material Culture of Microphysics*, Chicago: University of Chicago Press, ISBN 0-226-27917-0, OCLC 174870621 231708164 36103882.

- Garber, Daniel (1992), *Descartes' Metaphysical Physics*, Chicago: University of Chicago Press.

- Garber, Elizabeth (1999), *The Language of Physics: The Calculus and the Development of Theoretical Physics in Europe, 1750–1914*, Boston: Birkhäuser Verlag.

- Gaukroger, Stephen (2002), *Descartes' System of Natural Philosophy*, New York: Cambridge University Press.

- Glick, Thomas F.; Livesey, Steven John; Wallis, Faith (2005), *Medieval Science, Technology, and Medicine: An Encyclopedia*, Routledge, ISBN 0-415-96930-1, OCLC 218847614 58829023 61228669

- Greenberg, John (1986), "Mathematical Physics in Eighteenth-Century France", *Isis* **77**: 59–78, doi:10.1086/354039.

- Golinski, Jan (1999), *Science as Public Culture: Chemistry and Enlightenment in Britain, 1760–1820*, New York: Cambridge University Press.

- Guicciardini, Niccolò (1989), *The Development of Newtonian Calculus in Britain, 1700–1800*, New York: Cambridge University Press.

- Guicciardini, Niccolò (1999), *Reading the Principia: The Debate on Newton's Methods for Natural Philosophy from 1687 to 1736*, New York: Cambridge University Press.

- Hall, A. Rupert (1980), *Philosophers at War: The Quarrel between Newton and Leibniz*, New York: Cambridge University Press.

- Heilbron, J. L. (1979), *Electricity in the 17th and 18th Centuries*, Berkeley: University of California Press.

- Hunt, Bruce (1991), *The Maxwellians*, Ithaca: Cornell University Press.

- Jungnickel, Christa; McCormmach, Russell (1986), *Intellectual Mastery of Nature: Theoretical Physics from Ohm to Einstein*, Chicago: University of Chicago Press.

- Kragh, Helge (1999), *Quantum Generations: A History of Physics in the Twentieth Century*, Princeton: Princeton University Press.

- Rashed, R.; Armstrong, Angela (1994), *The Development of Arabic Mathematics*, Springer, ISBN 0-7923-2565-6, OCLC 29181926.

- Rashed, R.; Morelon, Régis (1996), *Encyclopedia of the History of Arabic Science* **2**, Routledge, ISBN 0-415-12410-7, OCLC 34731151 38122983 61834045 61987871.

- Rashed, R. (2007), "The Celestial Kinematics of Ibn al-Haytham", *Arabic Sciences and Philosophy* (Cambridge University Press) **17**: 7–55, doi:10.1017/S0957423907000355.

- Sabra, A. I. (1989), *Ibn al-Haytham, The Optics of Ibn al-Haytham* **I**, London: The Warburg Institute, pp. 90–1.

- Sabra, A. I. (1998), "Configuring the Universe: Aporetic, Problem Solving, and Kinematic Modeling as Themes of Arabic Astronomy", *Perspectives on Science* **6** (3): 288–330.

- Sabra, A. I.; Hogendijk, J. P. (2003), *The Enterprise of Science in Islam: New Perspectives*, MIT Press, pp. 85–118, ISBN 0-262-19482-1, OCLC 237875424 50252039.

- Schweber, Silvan (1994), *QED and the Men Who Made It: Dyson, Feynman, Schwinger, and Tomonaga*, Princeton: Princeton University Press.

- Shea, William (1991), *The Magic of Numbers and Motion: The Scientific Career of René Descartes*, Canton, Massachusetts: Science History Publications.

- Smith, A. Mark (1996), *Ptolemy's Theory of Visual Perception: An English Translation of the Optics with Introduction and Commentary*, Diane Publishing, ISBN 0-87169-862-5, OCLC 185537531 34724889.

- Smith, Crosbie (1998), *The Science of Energy: A Cultural History of Energy Physics in Victorian Britain*, Chicago: University of Chicago Press.

- Smith, Crosbie; Wise, M. Norton (1989), *Energy and Empire: A Biographical Study of Lord Kelvin*, New York: Cambridge University Press.

- Thiele, Rüdiger (August 2005a), "In Memoriam: Matthias Schramm, 1928–2005", *Historia Mathematica* **32** (3): 271–4, doi:10.1016/j.hm.2005.05.002.

- Thiele, Rüdiger (2005b), "In Memoriam: Matthias Schramm", *Arabic Sciences and Philosophy* (Cambridge University Press) **15**: 329–331, doi:10.1017/S0957423905000214.

- Toomer, G. J. (December 1964), "Review: *Ibn al-Haythams Weg zur Physik* by Matthias Schramm", *Isis* **55** (4): 463–465, doi:10.1086/349914.

- Tybjerg, Karin (2002), "Book Review: Andrew Barker, *Scientiic Method in Ptolemy's Harmonics*", *The British Journal for the History of Science* (Cambridge University Press) **35** (3): 347–379, doi:10.1017/S0007087402224784.

1.14 Further reading

- Buchwald, Jed Z. and Robert Fox, eds. *The Oxford Handbook of the History of Physics* (2014) 976pp; excerpt

- Byers, Nina and Williams, Gary (2006), *Out of the Shadows: Contributions of Twentieth-Century Women to Physics*, Cambridge University Press, ISBN 0-521-82197-5

- Cropper, William H. (2004), *Great Physicists: The Life and Times of Leading Physicists from Galileo to Hawking*, Oxford University Press, ISBN 0-19-517324-4

- Dear, Peter (2001), *Revolutionizing the Sciences: European Knowledge and Its Ambitions, 1500–1700*, Princeton: Princeton University Press, ISBN 0-691-08859-4, OCLC 46622656.

- Gamow, George (1988), *The Great Physicists from Galileo to Einstein*, Dover Publications, ISBN 0-486-25767-3

- Heilbron, John L. (2005), *The Oxford Guide to the History of Physics and Astronomy*, Oxford University Press, ISBN 0-19-517198-5

- Nye, Mary Jo (1996), *Before Big Science: The Pursuit of Modern Chemistry and Physics, 1800–1940*, New York: Twayne, ISBN 0-8057-9512-X, OCLC 185866968 34878783.

- Segrè, Emilio (1984), *From Falling Bodies to Radio Waves: Classical Physicists and Their Discoveries*, New York: W. H. Freeman, ISBN 0-7167-1482-5, OCLC 9943504.

- Segrè, Emilio (1980), *From X-Rays to Quarks: Modern Physicists and Their Discoveries*, San Francisco: W. H. Freeman, ISBN 0-7167-1147-8, OCLC 237246197 56100286 5946636.

- Weaver, Jefferson H. (editor) (1987), *The World of Physics*, Simon and Schuster, ISBN 0-671-49931-9 A selection of 56 articles, written by physicists. Commentaries and notes by Lloyd Motz and Dale McAdoo.

- de Haas, Paul, "Historic Papers in Physics (20th Century)"

1.15 External links

- "Selected Works about Isaac Newton and His Thought" from *The Newton Project*.

Chapter 2

Classical physics

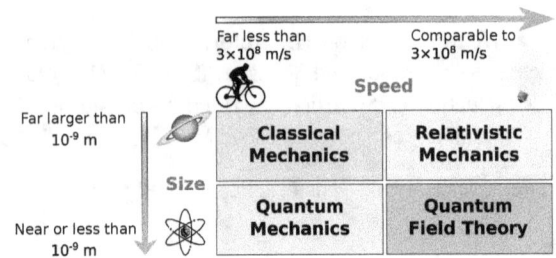

The four major domains of modern physics

Classical physics refers to theories of physics that predate modern, more complete, or more widely applicable theories. If a currently accepted theory is considered to be "modern," and its introduction represented a major paradigm shift, then the previous theories, or new theories based on the older paradigm, will often be referred to as belonging to the realm of "classical" physics.

As such, the definition of a classical theory depends on context. Classical physical concepts are often used when modern theories are unnecessarily complex for a particular situation.

2.1 Overview

Classical theory has at least two distinct meanings in physics. In the context of quantum mechanics, classical theory refers to theories of physics that do not use the quantisation paradigm, which includes classical mechanics and relativity.[*][1] Likewise, classical field theories, such as general relativity and classical electromagnetism, are those that do not use quantum mechanics.[*][2] Classical theories are those that obey Galilean relativity, while general and special relativity use a different framework, but all are part of classical physics.[*][3]

Among the branches of theory included in classical physics are:

- Classical mechanics
  - Newton's laws of motion
  - Classical Lagrangian and Hamiltonian formalisms
- Classical electrodynamics (Maxwell's Equations)
- Classical thermodynamics
- Special relativity and general relativity
- Classical chaos theory and nonlinear dynamics

2.2 Comparison with modern physics

In contrast to classical physics, "modern physics" is a slightly looser term which may refer to just quantum physics or to 20th and 21st century physics in general. Modern physics includes quantum theory and relativity, when applicable.

A physical system can be described by classical physics when it satisfies conditions such that the laws of classical physics are approximately valid. In practice, physical objects ranging from those larger than atoms and molecules, to objects in the macroscopic and astronomical realm, can be well-described (understood) with classical mechanics. Beginning at the atomic level and lower, the laws of classical physics break down and generally do not provide a correct description of nature. Electromagnetic fields and forces can be described well by classical electrodynamics at length scales and field strengths large enough that quantum mechanical effects are negligible. Unlike quantum physics, classical physics is generally characterized by the principle of complete determinism, although deterministic interpretations of quantum mechanics do exist.

From the point of view of classical physics as being non-relativistic physics, the predictions of general and special relativity are significantly different than those of classical theories, particularly concerning the passage of time, the geometry of space, the motion of bodies in free fall, and the propagation of light. Traditionally, light was reconciled with classical mechanics by assuming the existence of a stationary medium through which light propagated, the luminiferous aether, which was later shown not to exist.

Mathematically, classical physics equations are those in which Planck's constant does not appear. According to the correspondence principle and Ehrenfest's theorem, as a system becomes larger or more massive the classical dynamics tends to emerge, with some exceptions, such as superfluidity. This is why we can usually ignore quantum mechanics when dealing with everyday objects and the classical description will suffice. However, one of the most vigorous on-going fields of research in physics is classical-quantum correspondence. This field of research is concerned with the discovery of how the laws of quantum physics give rise to classical physics found at the limit of the large scales of the classical level.

2.3 Computer modeling and manual calculation, modern and classic comparison

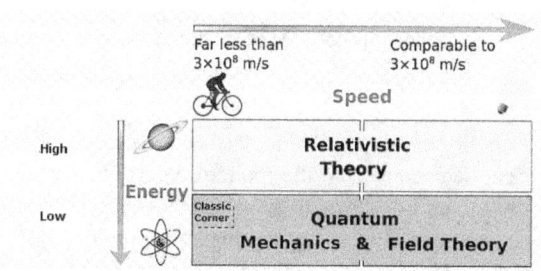

A computer model would use quantum theory and relativistic theory only

Today a computer performs millions of arithmetic operations in seconds to solve a classical differential equation, while Newton (one of the fathers of the differential calculus) would take hours to solve the same equation by manual calculation, even if he were the discoverer of that particular equation.

Computer modeling is essential for quantum and relativistic physics. Classic physics is considered the limit of quantum mechanics for large number of particles. On the other hand, classic mechanics is derived from relativistic mechanics. For example, in many formulations from special relativity, a correction factor $(v/c)^2$ appears, where v is the velocity of the object and c is the speed of light. For ve-

locities much smaller than that of light, one can neglect the terms with c^2 and higher that appear. These formulas then reduce to the standard definitions of Newtonian kinetic energy and momentum. This is as it should be, for special relativity must agree with Newtonian mechanics at low velocities. Computer modeling has to be as real as possible. Classical physics would introduce an error as in the superfluidity case. In order to produce reliable models of the world, we can not use classic physics. It is true that quantum theories consume time and computer resources, and the equations of classical physics could be resorted to provide a quick solution, but such a solution would lack reliability.

Computer modeling would use only the energy criteria to determine which theory to use: relativity or quantum theory, when attempting to describe the behavior of an object. A physicist would use a classical model to provide an approximation before more exacting models are applied and those calculations proceed.

In a computer model, there is no need to use the speed of the object if classical physics is excluded. Low energy objects would be handled by quantum theory and high energy objects by relativity theory.*[4]*[5]*[6] .

2.4 References

[1] Morin, David (2008). *Introduction to Classical Mechanics.* New York: Cambridge University Press. ISBN 9780521876223.

[2] Barut, Asim O. (1980) [1964]. *Introduction to Classical Mechanics.* New York: Dover Publications. ISBN 9780486640389.

[3] Einstein, Albert (2004) [1920]. *Relativity.* Translated by Robert W. Lawson. New York: Barnes & Noble. ISBN 9780760759219.

[4] [Wojciech H. Zurek, Decoherence, einselection, and the quantum origins of the classical, Reviews of Modern Physics 2003, 75, 715 or http://arxiv.org/abs/quant-ph/0105127/>

[5] Wojciech H. Zurek, Decoherence and the transition from quantum to classical, *Physics Today*, 44, pp 36–44 (1991)

[6] Wojciech H. Zurek: *Decoherence and the Transition from Quantum to Classical—Revisited* Los Alamos Science Number 27 2002

2.5 See also

- Glossary of classical physics
- Semiclassical physics

Chapter 3

Modern physics

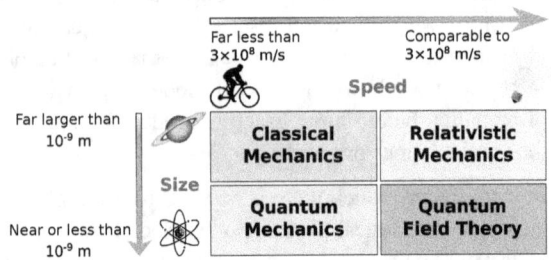

Classical physics is usually concerned with everyday conditions: speeds much lower than the speed of light, and sizes much greater than that of atoms. Modern physics is usually concerned with high velocities and small distances.

Modern physics is an effort to understand the underlying processes of the interactions of matter utilizing the tools of science & engineering. It implies that nineteenth century descriptions of phenomena are not sufficient to describe nature as observed with modern instruments. It is generally assumed that a consistent description of these observations will incorporate elements of quantum mechanics & relativity.

Small velocities and large distances is usually the realm of classical physics. Modern physics often involves extreme conditions; in practice, quantum effects typically involve distances comparable to atoms (roughly $10^{*}-9$ m), while relativistic effects typically involve velocities comparable to the speed of light (roughly 10^8 m/s).

3.1   Overview

In a literal sense, the term *modern physics*, means up-to-date physics. In this sense, a significant portion of so-called *classical physics* is modern. However, since roughly 1890, new discoveries have caused significant paradigm shifts: the advent of quantum mechanics (QM), and of Einsteinian relativity (ER). Physics that incorporates elements of either QM or ER (or both) is said to be *modern physics*. It is in this latter sense that the term is generally used.

Modern physics is often encountered when dealing with extreme conditions. Quantum mechanical effects tend to appear when dealing with "lows" (low temperatures, small distances), while relativistic effects tend to appear when dealing with "highs" (high velocities, large distances), the "middles" being classical behaviour. For example, when analysing the behaviour of a gas at room temperature, most phenomena will involve the (classical) Maxwell–Boltzmann distribution. However near absolute zero, the Maxwell–Boltzmann distribution fails to account for the observed behaviour of the gas, and the (modern) Fermi–Dirac or Bose–Einstein distributions have to be used instead.

- German physicist Albert Einstein, founder of the theory of relativity.

- German physicist Max Planck, founder of quantum theory.

Very often, it is possible to find – or "retrieve" – the classical behaviour from the modern description by analysing the modern description at low speeds and large distances (by taking a limit, or by making an approximation). When doing so, the result is called the *classical limit*.

3.2   Hallmarks of modern physics

Main articles: History of quantum mechanics and History of relativity

These are generally considered to be the topics regarded as the "core" of the foundation of modern physics:

- Atomic theory and the evolution of the atomic model in general

- Black-body radiation

- Franck–Hertz experiment

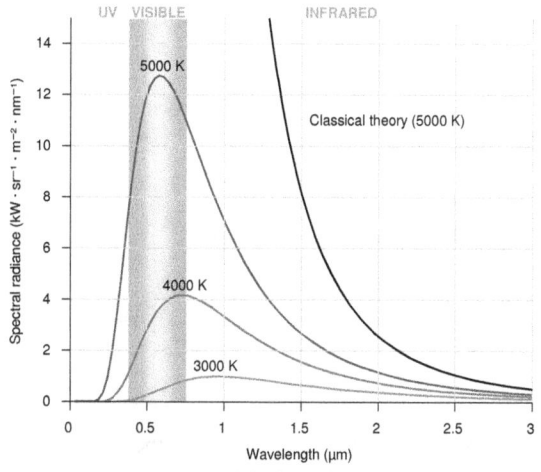

Classical physics (Rayleigh–Jeans law, black line) failed to explain black body radiation – the so-called ultraviolet catastrophe. The quantum description (Planck's law, colored lines) is said to be modern physics.

- Geiger–Marsden experiment (Rutherford's experiment)
- Gravitational lensing
- Michelson–Morley experiment
- Photoelectric effect
- Quantum thermodynamics
- Radioactive phenomena in general
- Perihelion precession of Mercury
- Stern–Gerlach experiment
- Wave–particle duality

3.3 See also

- History of physics
- Classical physics
- Quantum mechanics
- Theory of relativity
- Quantum field theory

3.4 References

[1] F.K. Richtmyer; E.H. Kennard; T. Lauristen (1955). *Introduction to Modern Physics* (5th ed.). New York: McGraw-Hill. p. 1. LCCN 55006862.

3.5 Further reading

- A. Beiser (2003). *Concepts of Modern Physics* (6th ed.). McGraw-Hill. ISBN 0-07-123460-8.

- John D. Walecka (2008). *Introduction to Modern Physics*. World Scientific. ISBN 978-981-2812-25-4.

- John D. Walecka (2010). *Advanced Modern Physics*. World Scientific. ISBN 978-981-4291-52-1.

- John D. Walecka (2013). *Topics in Modern Physics*. World Scientific. ISBN 978-981-4436-89-2.

- P. Tipler, R. Llewellyn (2002). *Modern Physics* (4th ed.). W. H. Freeman. ISBN 0-7167-4345-0.

Chapter 4

Aristotelian physics

Aristotelian physics is a form of natural science described in the works of the Greek philosopher Aristotle (384–322 BCE). In the *Physics*, Aristotle established general principles of change that govern all natural bodies, both living and inanimate, celestial and terrestrial – including all motion, change with respect to place, change with respect to size or number, qualitative change of any kind; and "coming to be" (coming into existence, "generation") and "passing away" (no longer existing, "corruption").

To Aristotle, "physics" was a broad field that included subjects such as the philosophy of mind, sensory experience, memory, anatomy and biology. It constitutes the foundation of the thought underlying many of his works.

4.1   Concepts

> nature is everywhere the cause of order.[*][1]
> —Aristotle, *Physics* VIII.1

While consistent with common human experience, Aristotle's principles were not based on controlled, quantitative experiments, so, while they account for many broad features of nature, they do not describe our universe in the precise, quantitative way now expected of science. Contemporaries of Aristotle like Aristarchus rejected these principles in favor of heliocentrism, but their ideas were not widely accepted. Aristotle's principles were difficult to disprove merely through casual everyday observation, but later development of the scientific method challenged his views with experiments and careful measurement, using increasingly advanced technology such as the telescope and vacuum pump.

In claiming novelty for their doctrines, those natural philosophers who developed the "new science" of the seventeenth century frequently contrasted "Aristotelian" physics with their own. Physics of the former sort, so they claimed,

emphasized the qualitative at the expense of the quantitative, neglected mathematics and its proper role in physics (particularly in the analysis of local motion), and relied on such suspect explanatory principles as final causes and "occult" essences. Yet in his *Physics* Aristotle characterizes physics or the "science of nature" as pertaining to magnitudes (*megethê*), motion (or "process" or "gradual change" – *kinêsis*), and time (*chronon*) (*Phys* III.4 202b30–1). Indeed, the *Physics* is largely concerned with an analysis of motion, particularly local motion, and the other concepts that Aristotle believes are requisite to that analysis.[*][2]
> —Michael J. White, "Aristotle on the Infinite, Space, and Time" in *Blackwell Companion to Aristotle*

4.1.1   Terrestrial change

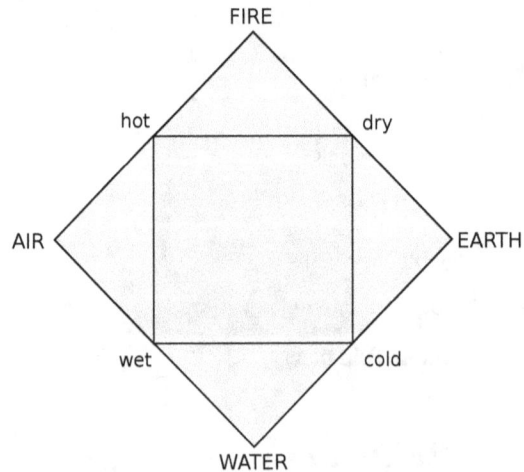

The four terrestrial elements

Unlike the eternal and unchanging celestial aether, each of

the four terrestrial elements are capable of changing into either of the two elements they share a property with: e.g. the cold and wet (water) can transform into the hot and wet (air) or the cold and dry (earth) and any apparent change into the hot and dry (fire) is actually a two-step process. These properties are predicated of an actual substance relative to the work it is able to do; that of heating or chilling and of desiccating or moistening. The four elements exist *only* with regard to this capacity and relative to some potential work. The celestial element is eternal and unchanging, so only the four terrestrial elements account for "coming to be" and "passing away" – or, in the terms of Aristotle's *De Generatione et Corruptione* (Περὶ γενέσεως καὶ φθορᾶς), "generation" and "corruption".

4.1.2 Elements

According to Aristotle, the elements which compose the terrestrial spheres are different from that constituting the celestial spheres.[*][3] He believed that four elements make up everything under the Moon, i.e. everything terrestrial: earth, air, fire and water.[*][a][*][4] He also held that the heavens are made of a special weightless and incorruptible (i.e. unchangeable) fifth element called "aether".[*][4] Aether also has the name "quintessence", meaning, literally, "fifth substance".[*][5]

Aristotle considered heavy substances such as iron and other metals to consist primarily of the element earth, with a smaller amount of the other three terrestrial elements. Other, lighter objects, he believed, have less earth, relative to the other three elements in their composition.[*][5]

A page from an 1837 edition of the ancient Greek philosopher Aristotle's Physica, *a book addressing a variety of subjects including the philosophy of nature and topics now part of its modern-day namesake: physics.*

Aether

Main articles: Aether (classical element) and Dynamics of the celestial spheres

The Sun, Moon, planets and stars – are embedded in perfectly concentric "crystal spheres" that rotate eternally at fixed rates. Because the celestial spheres are incapable of any change except rotation, the terrestrial sphere of fire must account for the heat, starlight and occasional meteorites.[*][6] The lunar sphere is the only celestial sphere that actually comes in contact with the sublunary orb's changeable, terrestrial matter, dragging the rarefied fire and air along underneath as it rotates.[*][7] Like Homer's *æthere* (αἰθήρ) – the "pure air" of Mount Olympus – was the divine counterpart of the air breathed by mortal beings (ἀήρ, *aer*). The celestial spheres are composed of the special element *aether*, eternal and unchanging, the sole capability of which is a uniform circular motion at a given rate (relative to the diurnal motion of the outermost sphere of fixed stars).

The concentric, aetherial, cheek-by-jowl "crystal spheres" that carry the Sun, Moon and stars move eternally with unchanging circular motion. Spheres are embedded within spheres to account for the "wandering stars" (i.e. the planets, which, in comparison with the Sun, Moon and stars, appear to move erratically). Later, the belief that all spheres are concentric was forsaken in favor of Ptolemy's deferent and epicycle model. Aristotle submits to the calculations of astronomers regarding the total number of spheres and various accounts give a number in the neighborhood of fifty spheres. An unmoved mover is assumed for each sphere, including a "prime mover" for the sphere of fixed stars. The unmoved movers do not push the spheres (nor could they, being immaterial and dimensionless) but are the final cause of the spheres' motion, i.e. they explain it in a way that's similar to the explanation "the soul is moved by beauty".

4.1.3   Four causes

Main articles: Four causes and Teleology

According to Aristotle, there are four ways to explain the *aitia* or causes of change. He writes that "we do not have knowledge of a thing until we have grasped its why, that is to say, its cause." *[8]*[9]

Aristotle held that there were four kinds of causes.*[9]*[10]

Material

The material cause of a thing is that of which it is made. For a table, that might be wood; for a statue, that might be bronze or marble.

> "In one way we say that the *aition* is that out of which. as existing, something comes to be, like the bronze for the statue, the silver for the phial, and their genera" (194b2 3—6). By "genera," Aristotle means more general ways of classifying the matter (e.g. "metal"; "material"); and that will become important. A little later on. he broadens the range of the material cause to include letters (of syllables), fire and the other elements (of physical bodies), parts (of wholes), and even premises (of conclusions: Aristotle re-iterates this claim, in slightly different terms, in *An. Post* II. 11).*[11]
> —R.J. Hankinson, "The Theory of the Physics" in *Blackwell Companion to Aristotle*

Formal

The formal cause of a thing is the essential property that makes it the kind of thing it is. In *Metaphysics* Book A Aristotle emphasizes that form is closely related to essence and definition. He says for example that the ratio 2:1, and number in general, is the cause of the octave.

> "Another [cause] is the form and the exemplar: this is the formula (logos) of the essence *(to ti en einai)*, and its genera, for instance the ratio 2:1 of the octave" (*Phys* 11.3 194b26—8)... Form is not just shape... We are asking (and this is the connection with essence, particularly in its canonical Aristotelian formulation) what it is to be some thing. And it is a feature of musical harmonics (first noted and wondered at by the Pythagoreans) that intervals of this type do indeed exhibit this ratio in some form in the instruments used to create them (the length of pipes, of strings, etc.). In some sense, the ratio explains what all the intervals have in common, why they turn out the same.*[12]
> —R.J. Hankinson, "Cause" in *Blackwell Companion to Aristotle*

Efficient

The efficient cause of a thing is the primary agency by which its matter took its form. For example, the efficient cause of a baby is a parent of the same species and that of a table is a carpenter, who knows the form of the table. In his *Physics* II, 194b29—32, Aristotle writes: "there is that which is the primary originator of the change and of its cessation, such as the deliberator who is responsible [sc. for the action] and the father of the child, and in general the producer of the thing produced and the changer of the thing changed".

> Aristotle's examples here are instructive: one case of mental and one of physical causation, followed by a perfectly general characterization. But they conceal (or at any rate fail to make patent) a crucial feature of Aristotle's concept of efficient causation, and one which serves to distinguish it from most modern homonyms. For Aristotle, any process requires a constantly operative efficient cause as long as it continues. This commitment appears most starkly to modern eyes in Aristotle's discussion of projectile motion: what keeps the projectile moving after it leaves the hand? "Impetus," "momentum," much less "inertia," are not possible answers. There must be a mover, distinct (at least in some sense) from the thing moved, which is exercising its motive capacity at every moment of the projectile's flight (see *Phys* VIII. 10 266b29—267a11). Similarly, in every case of animal generation, there is always some thing responsible for the continuity of that generation, although it may do so by way of some intervening instrument (*Phys* II.3 194b35—195a3).*[12]
> —R.J. Hankinson, "Causes" in *Blackwell Companion to Aristotle*

Final

The final cause is that for the sake of which something takes place, its aim or teleological purpose: for a germinating seed, it is the adult plant,*[13] for a ball at the top of a ramp, it is coming to rest at the bottom, for an eye, it is seeing, for a knife, it is cutting.

> Goals have an explanatory function: that is a commonplace, at least in the context of action-ascriptions. Less of a commonplace is the view espoused by Aristotle, that finality and purpose are to be found throughout nature. which is for him the realm of those things which contain within themselves principles of movement and rest (i.e. efficient
>
> causes); thus it makes sense to attribute purposes not only to natural things themselves, but also to their parts: the parts of a natural whole exist for the sake of the whole. As Aristotle himself notes, "for the sake of" locutions are ambiguous: "A is for the sake of B" may mean that A exists or is undertaken in order to bring B about; or it may mean that A is for B's benefit (An II.4 415b2–3, 20–1); but both types of finality have, he thinks. a crucial role to play in natural. as well as deliberative, contexts. Thus a man may exercise for the sake of his health: and so "health," and not just the hope of achieving it, is the cause of his action (this distinction is not trivial). But the eyelids are for the sake of the eye (to protect it: *PA* II.1 3) and the eye for the sake of the animal as a whole (to help it function properly: cf. *An* II.7).[14]
>
> —R.J. Hankinson, "Causes" in *Blackwell Companion to Aristotle*

4.1.4 Biology

According to Aristotle, the science of living things proceeds by gathering observations about each natural kind of animal, organizing them into genera and species (the *differentiae* in *History of Animals*) and then going on to study the causes (in *Parts of Animals* and *Generation of Animals*, his three main biological works).[15]

> The four causes of animal generation can be summarized as follows. The mother and father represent the material and efficient causes, respectively. The mother provides the matter out of which the embryo is formed, while the father provides the agency that informs that material and triggers its development. The formal cause is the definition of the animal's substantial being (*GA* I.1 715a4: *ho logos tês ousias*). The final cause is the adult form, which is the end for the sake of which development takes place.[15]

> —Devin M. Henry, "Generation of Animals"
> in *Blackwell Companion to Aristotle*

Organism and mechanism

Main articles: Organism (philosophy) and Mechanism (philosophy)

The four elements make up the uniform materials such as blood, flesh and bone, which are themselves the matter out of which are created the non-uniform organs of the body (e.g. the heart, liver and hands) "which in turn, as parts, are matter for the functioning body as a whole (*PA* II. 1 646a 13—24)".[11]

> [There] is a certain obvious conceptual economy about the view that in natural processes naturally constituted things simply seek to realize in full actuality the potentials contained within them (indeed, this is what *is* for them to be natural); on the other hand, as the detractors of Aristotelianism from the seventeenth century on were not slow to point out, this economy is won at the expense of any serious empirical content. Mechanism, at least as practiced by Aristotle's contemporaries and predecessors, may have been explanatorily inadequate —but at least it was an *attempt* at a general account given in reductive terms of the lawlike connections between things. Simply introducing what later reductionists were to scoff at as "occult qualities" does not explain —it merely, in the manner of Molière's famous satirical joke, serves to re-describe the effect. Formal talk, or so it is said, is vacuous.
>
> Things are not however quite as bleak as this. For one thing, there's no point in trying to engage in reductionist science if you don't have the wherewithal, empirical and conceptual, to do so successfully: science shouldn't be simply unsubstantiated speculative metaphysics. But more than that. there is a point to describing the world in such teleologically loaded terms: it makes sense of things in a way that atomist speculations do not. And further. Aristotle's talk of species-forms is not as empty as his opponents would insinuate. He doesn't simply say that things do what they do because that's the sort of thing they do: the whole point of his classificatory biology, most clearly exemplified in *PA*, is to show what sorts of function go with what, which presuppose which and which are subservient to which. And in this sense, formal

or functional biology is susceptible of a type of reductionism. We start, he tells us, with the basic animal kinds which we all pre-theoretically (although not indefeasibly) recognize (cf. *PA* I.4): but we then go on to show how their parts relate to one another: why it is, for instance that only blooded creatures have lungs, and how certain structures in one species are analogous or homologous to those in another (such as scales in fish, feathers in birds, hair in mammals). And the answers, for Aristotle, are to be found in the economy of functions, and how they all contribute to the overall well-being (the final cause in this sense) of the animal.[*16]
—R.J. Hankinson, "The Relations between the Causes" in *Blackwell Companion to Aristotle*

See also Organic Form.

Psychology

According to Aristotle, perception and thought are similar, though not exactly alike in that perception is concerned only with the external objects that are acting on our sense organs at any given time, whereas we can think about anything we choose. Thought is about universal forms, in so far as they've been successfully understood, based on our memory of having encountered instances of those forms directly.[*17]

Aristotle's theory of cognition rests on two central pillars: his account of perception and his account of thought. Together, they make up a significant portion of his psychological writings, and his discussion of other mental states depends critically on them. These two activities, moreover, are conceived of in an analogous manner, at least with regard to their most basic forms. Each activity is triggered by its object – each, that is, is about the very thing that brings it about. This simple causal account explains the reliability of cognition: perception and thought are, in effect, transducers, bringing information about the world into our cognitive systems, because, at least in their most basic forms, they are infallibly about the causes that bring them about (*An* III.4 429a13–18). Other, more complex mental states are far from infallible. But they are still tethered to the world, in so far as they rest on the unambiguous and direct contact perception and thought enjoy with their objects.[*17]

—Victor Caston, "Phantasia and Thought" in *Blackwell Companion to Aristotle*

4.1.5   Natural place

The Aristotelian explanation of gravity is that all bodies move toward their natural place. For the elements earth and water, that place is the center of the (geocentric) universe;[*18] the natural place of water is a concentric shell around the earth because earth is heavier; it sinks in water. The natural place of air is likewise a concentric shell surrounding that of water; bubbles rise in water. Finally, the natural place of fire is higher than that of air but below the innermost celestial sphere (carrying the Moon).

In Book *Delta* of his *Physics* (IV.5), Aristotle defines *topos* (place) in terms of two bodies, one of which contains the other: a "place" is where the inner surface of the former (the containing body) touches the outer surface of the other (the contained body). This definition remained dominant until the beginning of the 17th century, even though it had been questioned and debated by philosophers since antiquity.[*19] The most significant early critique was made in terms of geometry by the 11th-century Arab polymath al-Hasan Ibn al-Haytham (Alhazen) in his *Discourse on Place*.[*20]

4.1.6   Natural motion

Terrestrial objects rise or fall, to a greater or lesser extent, according to the ratio of the four elements of which they are composed. For example, earth, the heaviest element, and water, fall toward the center of the cosmos; hence the Earth and for the most part its oceans, will have already come to rest there. At the opposite extreme, the lightest elements, air and especially fire, rise up and away from the center.[*21]

The elements are not proper *substances* in Aristotelian theory (or the modern sense of the word). Instead, they are abstractions used to explain the varying natures and behaviors of actual materials in terms of ratios between them.

Motion and change are closely related in Aristotelian physics. Motion, according to Aristotle, involved a change from potentiality to actuality.[*22] He gave example of four types of change.

Aristotle proposed that the speed at which two identically shaped objects sink or fall is directly proportional to their weights and inversely proportional to the density of the medium through which they move.[*23] While describing their terminal velocity, Aristotle must stipulate that there would be no limit at which to compare the speed of atoms falling through a vacuum, (they could move indefinitely fast because there would be no particular place for them to come to rest in the void). Now however it is understood that at any time prior to achieving terminal velocity in a relatively resistance-free medium like air, two such objects are expected to have nearly identical speeds because both are experiencing a force of gravity proportional to their masses and have thus been accelerating at nearly the same rate. This became especially apparent from the eighteenth century when partial vacuum experiments began to be made, but some two hundred years earlier Galileo had already demonstrated that objects of different weights reach the ground in similar times.*[24]

4.1.7 Unnatural motion

Apart from the natural tendency of terrestrial exhalations to rise and objects to fall, unnatural or forced motion from side to side results from the turbulent collision and sliding of the objects as well as transmutation between the elements (On Generation and Corruption).

Chance

In his *Physics* Aristotle examines accidents (συμβεβηκός, *sumbebekos*) that have no cause but chance. "Nor is there any definite cause for an accident, but only chance (τύχη, *tukhe*), namely an indefinite (ἀόριστον) cause" (*Metaphysics* V, 1025a25).

> It is obvious that there are principles and causes which are generable and destructible apart from the actual processes of generation and destruction; for if this is not true, everything will be of necessity: that is, if there must necessarily be some cause, other than accidental, of that which is generated and destroyed. Will this be, or not? Yes, if this happens; otherwise not (*Metaphysics* VI, 1027a29).

4.1.8 Continuum and vacuum

Aristotle argues against the indivisibles of Democritus (which differ considerably from the historical and the modern use of the term "atom"). As a place without anything existing at or within it, Aristotle argued against the possibility of a vacuum or void. Because he believed that the speed of an object's motion is proportional to the force being applied (or, in the case of natural motion, the object's weight) and inversely proportional to the viscosity of the medium, he reasoned that objects moving in a void would move indefinitely fast – and thus any and all objects surrounding the void would immediately fill it. The void, therefore, could never form.*[25]

The "voids" of modern-day astronomy (such as the Local Void adjacent to our own galaxy) have the opposite effect: ultimately, bodies off-center are ejected from the void due to the gravity of the material outside.*[26]

4.1.9 Speed, weight and resistance

The ideal speed of a terrestrial object is directly proportional to its weight. In nature however, vacuum does not occur, the matter obstructing an object's path is a limiting factor that is inversely proportional to the viscosity of the medium.

4.2 Medieval commentary

Main article: Theory of impetus

The Aristotelian theory of motion came under criticism and modification during the Middle Ages. Modifications began with John Philoponus in the 6th century, who partly accepted Aristotle's theory that "continuation of motion depends on continued action of a force" but modified it to include his idea that a hurled body also acquires an inclination (or "motive power") for movement away from whatever caused it to move, an inclination that secures its continued motion. This impressed virtue would be temporary and self-expending, meaning that all motion would tend toward the form of Aristotle's natural motion.

In *The Book of Healing* (1027), the 11th-century Persian polymath Avicenna developed Philoponean theory into the first coherent alternative to Aristotelian theory. Inclinations in the Avicennan theory of motion were not self-consuming but permanent forces whose effects were dissipated only as a result of external agents such as air resistance, making him "the first to conceive such a permanent type of impressed virtue for non-natural motion". Such a self-motion (*mayl*) is "almost the opposite of the Aristotelian conception of violent motion of the projectile type, and it is rather reminiscent of the principle of inertia, i.e. Newton's first law of motion." *[27]

The eldest Banū Mūsā brother, Ja'far Muhammad ibn Mūsā ibn Shākir (800-873), wrote the *Astral Motion* and *The Force of Attraction*. The Persian physicist, Ibn al-Haytham (965-1039) discussed the theory of attraction between bodies. It seems that he was aware of the magnitude of acceleration due to gravity and he discovered that the heavenly bodies "were accountable to the laws of physics".*[28]

The Persian polymath Abū Rayhān al-Bīrūnī (973-1048) was the first to realize that acceleration is connected with non-uniform motion (as later expressed by Newton's second law of motion).*[29] During his debate with Avicenna, al-Biruni also criticized the Aristotelian theory of gravity firstly for denying the existence of levity or gravity in the celestial spheres; and, secondly, for its notion of circular motion being an innate property of the heavenly bodies.*[30]

In 1121, al-Khazini, in *The Book of the Balance of Wisdom*, proposed that the gravity and gravitational potential energy of a body varies depending on its distance from the centre of the Earth.*[31] Hibat Allah Abu'l-Barakat al-Baghdaadi (1080–1165) wrote *al-Mu'tabar*, a critique of Aristotelian physics where he negated Aristotle's idea that a constant force produces uniform motion, as he realized that a force applied continuously produces acceleration, a fundamental law of classical mechanics and an early foreshadowing of Newton's second law of motion.*[32] Like Newton, he described acceleration as the rate of change of speed.*[33]

In the 14th century, Jean Buridan developed the theory of impetus as an alternative to the Aristotelian theory of motion. The theory of impetus was a precursor to the concepts of inertia and momentum in classical mechanics.*[34] Buridan and Albert of Saxony also refer to Abu'l-Barakat in explaining that the acceleration of a falling body is a result of its increasing impetus.*[35] In the 16th century, Al-Birjandi discussed the possibility of the Earth's rotation and, in his analysis of what might occur if the Earth were rotating, developed a hypothesis similar to Galileo's notion of "circular inertia" .*[36] He described it in terms of the following observational test:

> "The small or large rock will fall to the Earth along the path of a line that is perpendicular to the plane (*sath*) of the horizon; this is witnessed by experience (*tajriba*). And this perpendicular is away from the tangent point of the Earth's sphere and the plane of the perceived (*hissi*) horizon. This point moves with the motion of the Earth and thus there will be no difference in place of fall of the two rocks." *[37]

4.3   Life and death of Aristotelian physics

The reign of Aristotelian physics, the earliest known speculative theory of physics, lasted almost two millennia. After the work of many pioneers such as Copernicus, Galileo, Descartes and Newton, it became generally accepted that Aristotelian physics was neither correct nor viable.*[5] De-

Aristotle depicted by Rembrandt.

spite this, it survived as a scholastic pursuit well into the seventeenth century, until universities amended their curricula.

In Europe, Aristotle's theory was first convincingly discredited by Galileo's studies. Using a telescope, Galileo observed that the Moon was not entirely smooth, but had craters and mountains, contradicting the Aristotelian idea of the incorruptibly perfect smooth Moon. Galileo also criticized this notion theoretically; a perfectly smooth Moon would reflect light unevenly like a shiny billiard ball, so that the edges of the moon's disk would have a different brightness than the point where a tangent plane reflects sunlight directly to the eye. A rough moon reflects in all directions equally, leading to a disk of approximately equal brightness which is what is observed.*[38] Galileo also observed that Jupiter has moons – i.e. objects revolving around a body other than the Earth – and noted the phases of Venus, which demonstrated that Venus (and, by implication, Mercury) traveled around the Sun, not the Earth.

According to legend, Galileo dropped balls of various densities from the Tower of Pisa and found that lighter and heavier ones fell at almost the same speed. His experiments actually took place using balls rolling down inclined planes, a form of falling sufficiently slow to be measured without advanced instruments.

In a relatively dense medium such as water, a heavier body falls faster than a lighter one. This led Aristotle to speculate that the rate of falling is proportional to the weight and inversely proportional to the density of the medium. From his experience with objects falling in water, he concluded that water is approximately ten times denser than air. By weighing a volume of compressed air, Galileo showed that this overestimates the density of air by a factor of forty.[*][39] From his experiments with inclined planes, he concluded that if friction is neglected, all bodies fall at the same rate.

Galileo also advanced a theoretical argument to support his conclusion. He asked if two bodies of different weights and different rates of fall are tied by a string, does the combined system fall faster because it is now more massive, or does the lighter body in its slower fall hold back the heavier body? The only convincing answer is neither: all the systems fall at the same rate.[*][38]

Followers of Aristotle were aware that the motion of falling bodies was not uniform, but picked up speed with time. Since time is an abstract quantity, the peripatetics postulated that the speed was proportional to the distance. Galileo established experimentally that the speed is proportional to the time, but he also gave a theoretical argument that the speed could not possibly be proportional to the distance. In modern terms, if the rate of fall is proportional to the distance, the differential equation for the distance y travelled after time t is:

$$\frac{dy}{dt} = y$$

...with the condition that $y(0) = 0$. Galileo demonstrated that this system would stay at $y = 0$ for all time. If a perturbation set the system into motion somehow, the object would pick up speed exponentially in time, not quadratically.[*][39]

Standing on the surface of the Moon in 1971, David Scott famously repeated Galileo's experiment by dropping a feather and a hammer from each hand at the same time. In the absence of a substantial atmosphere, the two objects fell and hit the Moon's surface at the same time.

The first convincing mathematical theory of gravity – in which two masses are attracted toward each other by a force whose effect decreases according to the inverse square of the distance between them – was Newton's law of universal gravitation. This, in turn, was replaced by the General theory of relativity due to Albert Einstein.

Further information: Gravity

4.4 See also

Minima naturalia, a hylomorphic concept suggested by Aristotle broadly analogous in Peripatetic and Scholastic physical speculation to the atoms of Epicureanism.

4.5 Works

4.6 Notes

a [*][^] Here, the term "Earth" does not refer to planet Earth, known by modern science to be composed of a large number of chemical elements. Modern chemical elements are not conceptually similar to Aristotle's elements; the term "air", for instance, does not refer to breathable air.

4.7 References

[1] Lang, H.S. (2007). *The Order of Nature in Aristotle's Physics: Place and the Elements.* Cambridge University Press. p. 290. ISBN 9780521042291.

[2] White, Michael J. (2009). "Aristotle on the Infinite, Space, and Time". *Blackwell Companion to Aristotle.* p. 260.

[3] "Physics of Aristotle vs. The Physics of Galileo". Archived from the original on 11 April 2009. Retrieved 6 April 2009.

[4] "www.hep.fsu.edu" (PDF). Retrieved 26 March 2007.

[5] "Aristotle's physics". Retrieved 6 April 2009.

[6] Aristotle, meteorology.

[7] Sorabji, R. (2005). *The Philosophy of the Commentators, 200-600 AD: Physics.* G - Reference, Information and Interdisciplinary Subjects Series. Cornell University Press. p. 352. ISBN 978-0-8014-8988-4. LCCN 2004063547.

[8] Aristotle, *Physics* 194 b17–20; see also: *Posterior Analytics* 71 b9–11; 94 a20.

[9] "Four Causes". Falcon, Andrea. Aristotle on Causality. *Stanford Encyclopedia of Philosophy* 2008.

[10] Aristotle, "Book 5, section 1013a", *Metaphysics*, Hugh Tredennick (trans.) Aristotle in 23 Volumes, Vols. 17, 18, Cambridge, MA, Harvard University Press; London, William Heinemann Ltd. 1933, 1989; (hosted at perseus.tufts.edu.) Aristotle also discusses the four causes in his Physics, Book B, chapter 3.

[11] Hankinson, R.J. "The Theory of the Physics". *Blackwell Companion to Aristotle.* p. 216.

[12] Hankinson, R.J. "Causes". *Blackwell Companion to Aristotle.* p. 217.

[13] Aristotle. *Parts of Animals I.1.*

[14] Hankinson, R.J. "Causes". *Blackwell Companion to Aristotle.* p. 218.

[15] Henry, Devin M. (2009). "Generation of Animals". *Blackwell Companion to Aristotle.* p. 368.

[16] Hankinson, R.J. "Causes". *Blackwell Companion to Aristotle.* p. 222.

[17] Caston, Victor (2009). "Phantasia and Thought". *Blackwell Companion to Aristotle.* pp. 322–2233.

[18] *De Caelo* II. 13-14.

[19] For instance, by Simplicius in his *Corollaries on Place.*

[20] El-Bizri, Nader (2007). "In Defence of the Sovereignty of Philosophy: al-Baghdadi's Critique of Ibn al-Haytham's Geometrisation of Place". *Arabic Sciences and Philosophy (Cambridge University Press)* **17**: 57–80. doi:10.1017/s0957423907000367.

[21] Tim Maudlin (2012-07-22). *Philosophy of Physics: Space and Time: Space and Time* (Princeton Foundations of Contemporary Philosophy) (p. 2). Princeton University Press. Kindle Edition. "The element earth's natural motion is to fall—that is, to move downward. Water also strives to move downward but with less initiative than earth: a stone will sink though water, demonstrating its overpowering natural tendency to descend. Fire naturally rises, as anyone who has watched a bonfire can attest, as does air, but with less vigor."

[22] Bodnar, Istvan, "Aristotle's Natural Philosophy" in *The Stanford Encyclopedia of Philosophy* (Spring 2012 Edition, ed. Edward N. Zalta).

[23] Gindikin, S.G. (1988). *Tales of Physicists and Mathematicians.* Birkh. p. 29. ISBN 9780817633172. LCCN 87024971.

[24] Lindberg, D. (2008), *The beginnings of western science: The European scientific tradition in philosophical, religious, and institutional context, prehistory to AD 1450* (2nd ed.), University of Chicago Press.

[25] Land, Helen, *The Order of Nature in Aristotle's Physics: Place and the Elements* (1998).

[26] Tully; Shaya; Karachentsev; Courtois; Kocevski; Rizzi; Peel (2008). "Our Peculiar Motion Away From the Local Void". *The Astrophysical Journal* **676** (1): 184. arXiv:0705.4139. Bibcode:2008ApJ...676..184T. doi:10.1086/527428.

[27] Aydin Sayili (1987), "Ibn Sīnā and Buridan on the Motion of the Projectile", *Annals of the New York Academy of Sciences* 500 (1): 477–482 [477]:

> According to Aristotle, continuation of motion depends on continued action of a force. The motion of a hurled body, therefore, requires elucidation. Aristotle maintained that the air of the atmosphere was responsible for the continuation of such motion. John Philoponos of the 6th century rejected this Aristotelian view. He claimed that the hurled body acquires a motive power or an inclination for forced movement from the agent producing the initial motion and that this power or condition and not

the ambient medium secures the continuation of such motion. According to Philoponos this impressed virtue was temporary. It was a self-expending inclination, and thus the violent motion thus produced comes to an end and changes into natural motion. Ibn Sina adopted this idea in its rough outline, but the violent inclination as he conceived it was a non-self-consuming one. It was a permanent force whose effect got dissipated only as a result of external agents such as air resistance. He is apparently the first to conceive such a permanent type of impressed virtue for non-natural motion. [...] Indeed, self-motion of the type conceived by Ibn Sina is almost the opposite of the Aristotelian conception of violent motion of the projectile type, and it is rather reminiscent of the principle of inertia, i.e., Newton's first law of motion.

[28] Duhem, Pierre (1908, 1969). *To Save the Phenomena: An Essay on the Idea of Physical theory from Plato to Galileo,* University of Chicago Press, Chicago, p. 28.

[29] O'Connor, John J.; Robertson, Edmund F., "Al-Biruni", *MacTutor History of Mathematics archive*, University of St Andrews.

[30] Rafik Berjak and Muzaffar Iqbal, "Ibn Sina--Al-Biruni correspondence", *Islam & Science*, June 2003.

[31] Mariam Rozhanskaya and I. S. Levinova (1996), "Statics", in Roshdi Rashed, ed., *Encyclopedia of the History of Arabic Science*, vol. 2, pp. 614–642 [621-622]. (Routledge, London and New York.)

[32] Shlomo Pines (1970). "Abu'l-Barakāt al-Baghdādī, Hibat Allah". *Dictionary of Scientific Biography* **1**. New York: Charles Scribner's Sons. pp. 26–28. ISBN 0-684-10114-9. (cf. Abel B. Franco (October 2003). "Avempace, Projectile Motion, and Impetus Theory", *Journal of the History of Ideas* **64** (4), pp. 521–546 [528].)

[33] A. C. Crombie, *Augustine to Galileo* 2, p. 67.

[34] Aydin Sayili (1987), "Ibn Sīnā and Buridan on the Motion of the Projectile", *Annals of the New York Academy of Sciences* 500 (1): 477–482

[35] Gutman, Oliver (2003). *Pseudo-Avicenna, Liber Celi Et Mundi: A Critical Edition.* Brill Publishers. p. 193. ISBN 90-04-13228-7.

[36] (Ragep 2001b, pp. 63–4)

[37] (Ragep 2001a, pp. 152–3)

[38] Galileo Galilei, *Dialogue Concerning the Two Chief World Systems.*

[39] Galileo Galilei, *Two New Sciences.*

4.8 Sources

- Ragep, F. Jamil (2001a). "Tusi and Copernicus: The Earth's Motion in Context". *Science in Context* (Cambridge University Press) **14** (1–2): 145–163. doi:10.1017/s0269889701000060.

- Ragep, F. Jamil; Al-Qushji, Ali (2001b). "Freeing Astronomy from Philosophy: An Aspect of Islamic Influence on Science". *Osiris, 2nd Series* **16** (Science in Theistic Contexts: Cognitive Dimensions): 49–64 and 66–71. Bibcode:2001Osir...16...49R. doi:10.1086/649338.

- H. Carteron (1965) "Does Aristotle Have a Mechanics?" in *Articles on Aristotle 1. Science* eds. Jonathan Barnes, Malcolm Schofield, Richard Sorabji (London: General Duckworth and Company Limited), 161-174.

4.9 Further reading

- Katalin Martinás, "Aristotelian Thermodynamics" in *Thermodynamics: history and philosophy: facts, trends, debates* (Veszprém, Hungary 23–28 July 1990), pp. 285–303.

Chapter 5

History of science and technology in China

For the history of science and technology of modern China, see History of science and technology in the People's Republic of China. For the science and technology of modern China, see Science and technology in the People's Republic of China. For the science and technology of modern Taiwan, see Ministry of Science and Technology (Republic of China).

+ Also, for the history of science and technology in the Republic of China (1912–49), a period of tremendous growth, see the brief section below.

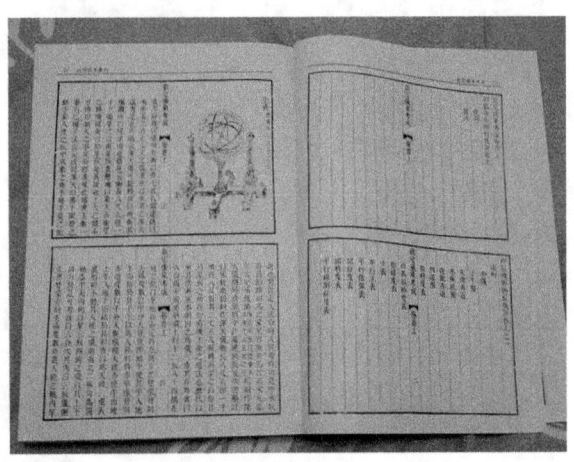

Instructions for making astronomical instruments from the time of the Qing Dynasty.

Ancient Han Chinese scientists, engineers, astronomers, philosophers, mathematicians and medical doctors made significant innovations, scientific discoveries and technological advances in science, technology, engineering, medicine, military technology, mathematics, geology and astronomy. Traditional Chinese medicine, acupuncture and herbal medicine were also developed through empirical observation and scientific experimentation.

Among the earliest inventions were the abacus, the "shadow clock," and the first items such as Kongming lanterns.[1] The *Four Great Inventions*: the compass, gunpowder, papermaking, and printing, were among the most important technological advances, only known to Europe by the end of the Middle Ages 1000 years later. The Tang Dynasty (AD 618 - 906) in particular, was a time of great innovation.[1] A good deal of exchange occurred between Western and Chinese discoveries up to the Qing Dynasty.

The Jesuit China missions of the 16th and 17th centuries introduced Western science and astronomy, then undergoing its own revolution, to China, and knowledge of Chinese technology was brought to Europe.[2][3] In the 19th and 20th century the introduction of Western technology was a major factor in the modernization of China. Much of the early Western work in the history of science in China was done by Joseph Needham.

5.1 Mo Di and the School of Names

The Warring States period began 2500 years ago at the time of the invention of the crossbow.[4] Needham notes that the invention of the crossbow "far outstripped the progress in defensive armor", which made the wearing of armor useless to the princes and dukes of the states.[5] At this time, there were also many nascent schools of thought in China —the Hundred Schools of Thought (諸子百家), scattered among many polities. The schools served as communities which advised the rulers of these states. Mo Di (墨翟 Mozi, 470 BCE–ca. 391 BCE) introduced concepts useful to one of those rulers, such as defensive fortification. One of these concepts, *fa* (法 principle or method)[6] was extended by the School of Names (名家 *Ming jia*, *ming*=name), which began a systematic exploration of logic. The development of a school of logic was cut short by the defeat of Mohism's political sponsors by the Qin Dynasty, and the subsumption of *fa* as law rather than method by the Legalists (法家 *Fa jia*).

Needham further notes that the Han Dynasty, which conquered the short-lived Qin, were made aware of the need for law by Lu Chia and by Shu-Sun Thung, as defined by the scholars, rather than the generals.[*][5]

> You conquered the empire on horseback, but from horseback you will never succeed in ruling it.
>
> —Lu Chia, [*][7]

Derived from Taoist philosophy, one of the newest long-standing contributions of the ancient Chinese are in Traditional Chinese medicine, including acupuncture and herbal medicine. The practice of acupuncture can be traced back as far as the 1st millennium BC and some scientists believe that there is evidence that practices similar to acupuncture were used in Eurasia during the early Bronze Age.[*][8]

Using shadow clocks and the abacus (both invented in the ancient Near East before spreading to China), the Chinese were able to record observations, documenting the first recorded solar eclipse in 2137 BC, and making the first recording of any planetary grouping in 500 BC.[*][9] These claims, however, are highly disputed and rely on much supposition.[*][10][*][11] The Book of Silk was the first definitive atlas of comets, written *c.*400 BC. It listed 29 comets (referred to as *sweeping stars*) that appeared over a period of about 300 years, with renderings of comets describing an event its appearance corresponded to.[*][9]

In architecture, the pinnacle of Chinese technology manifested itself in the Great Wall of China, under the first Chinese Emperor Qin Shi Huang between 220 and 200 BC. Typical Chinese architecture changed little from the succeeding Han Dynasty until the 19th century.[*][12] The Qin Dynasty also developed the crossbow, which later became the mainstream weapon in Europe. Several remains of crossbows have been found among the soldiers of the Terracotta Army in the tomb of Qin Shi Huang.[*][13]

5.2 Han Dynasty

Main article: Science and technology of the Han Dynasty

The Eastern Han Dynasty scholar and astronomer Zhang Heng (78-139 AD) invented the first water-powered rotating armillary sphere (the first armillary sphere having been invented by the Greek Eratosthenes), and catalogued 2500 stars and over 100 constellations. In 132, he invented the first seismological detector, called the "*Houfeng Didong Yi*" ("Instrument for inquiring into the wind and the shaking of the earth").[*][14] According to the *History of Later Han Dynasty* (25-220 AD), this seismograph was an urn-like instrument, which would drop one of eight balls to indicate when and in which direction an earthquake had occurred.[*][14]

Remains of a Chinese crossbow, 2nd century BC.

On June 13, 2005, Chinese seismologists announced that they had created a replica of the instrument.[*][14]

The mechanical engineer Ma Jun (c.200-265 AD) was another impressive figure from ancient China. Ma Jun improved the design of the silk loom,[*][15] designed mechanical chain pumps to irrigate palatial gardens,[*][15] and created a large and intricate mechanical puppet theatre for Emperor Ming of Wei, which was operated by a large hidden waterwheel.[*][16] However, Ma Jun's most impressive invention was the south-pointing chariot, a complex mechanical device that acted as a mechanical compass vehicle. It incorporated the use of a differential gear in order to apply equal amount of torque to wheels rotating at different speeds, a device that is found in all modern automobiles.[*][17]

Sliding calipers were invented in China almost 2,000 years ago.[*][1] The Chinese civilization was the earliest civilization to experiment successfully with aviation, with the kite and Kongming lantern (proto Hot air balloon) being the first flying machines.

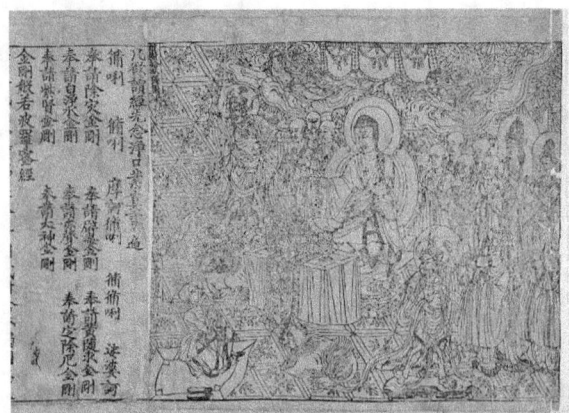

The intricate frontispiece of the Diamond Sutra from Tang Dynasty China, 868 AD (British Library)

5.3 "Four Great Inventions"

Main article: Four Great Inventions

The "Four Great Inventions" (simplified Chinese: 四大发明; traditional Chinese: 四大發明; pinyin: *sì dà fāmíng*) are the compass, gunpowder, papermaking and printing. Paper and printing were developed first. Printing was recorded in China in the Tang Dynasty, although the earliest surviving examples of printed cloth patterns date to before 220.[18] Pin-pointing the development of the compass can be difficult: the magnetic attraction of a needle is attested by the *Louen-heng*, composed between AD 20 and 100,[19] although the first undisputed magnetized needles in Chinese literature appear in 1086.[20]

By AD 300, Ge Hong, an alchemist of the Jin Dynasty, conclusively recorded the chemical reactions caused when saltpetre, pine resin and charcoal were heated together, in *Book of the Master of the Preservations of Solidarity*.[21] Another early record of gunpowder, a Chinese book from *c.* 850 AD, indicates:

> "Some have heated together sulfur, realgar and saltpeter with honey; smoke and flames result, so that their hands and faces have been burnt, and even the whole house where they were working burned down." [22]

These four discoveries had an enormous impact on the development of Chinese civilization and a far-ranging global impact. Gunpowder, for example, spread to the Arabs in the 13th century and thence to Europe.[23] According to English philosopher Francis Bacon, writing in *Novum Organum*:

> Printing, gunpowder and the compass: These

three have changed the whole face and state of things throughout the world; the first in literature, the second in warfare, the third in navigation; whence have followed innumerable changes, in so much that no empire, no sect, no star seems to have exerted greater power and influence in human affairs than these mechanical discoveries.

—[24]

One of the most important military treatises of all Chinese history was the *Huo Long Jing* written by Jiao Yu in the 14th century. For gunpowder weapons, it outlined the use of fire arrows and rockets, fire lances and firearms, land mines and naval mines, bombards and cannons, along with different compositions of gunpowder, including 'magic gunpowder', 'poisonous gunpowder', and 'blinding and burning gunpowder' (refer to his article).

For the 11th century invention of ceramic movable type printing by Bi Sheng (990-1051), it was enhanced by the wooden movable type of Wang Zhen in 1298 and the bronze metal movable type of Hua Sui in 1490.

5.4 China's scientific revolution

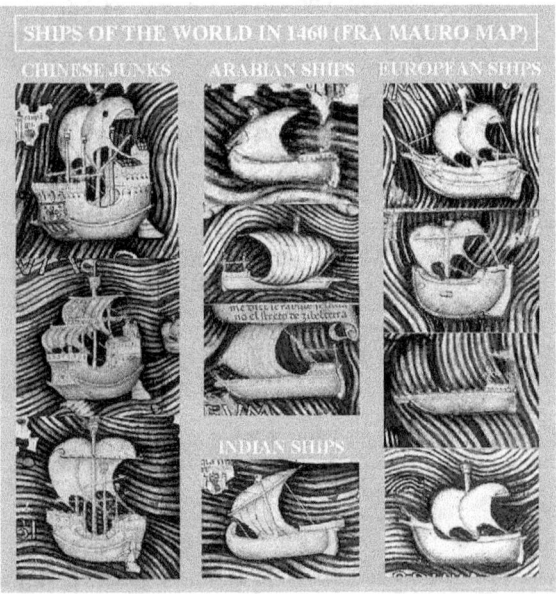

Ships of the world in 1460 (Fra Mauro map). Chinese junks are described as very large, three or four-masted ships.

Among the engineering accomplishments of early China were matches, dry docks, the double-action piston pump, cast iron, the iron plough, the horse collar, the multi-tube seed drill, the wheelbarrow, the suspension bridge, the parachute, natural gas as fuel, the raised-relief map, the propeller, the sluice gate, and the pound lock. The Tang Dynasty (618–906 AD) in particular was a time of great innovation.[*][1]

In the 7th century, book-printing was developed in China, Korea and Japan, using delicate hand-carved wooden blocks to print individual pages.[*][1] The 9th century *Diamond Sutra* is the earliest known printed document.[*][1] Movable type was also used in China for a time, but was abandoned because of the number of characters needed; it would not be until Johannes Gutenberg that the technique was reinvented in a suitable environment.[*][1]

In addition to gunpowder, the Chinese also developed improved delivery systems for the Byzantine weapon of Greek fire, Meng Huo You and Pen Huo Qi first used in China *c.* 900.[*][25] Chinese illustrations were more realistic than in Byzantine manuscripts,[*][25] and detailed accounts from 1044 recommending its use on city walls and ramparts show the brass container as fitted with a horizontal pump, and a nozzle of small diameter.[*][25] The records of a battle on the Yangtze near Nanjing in 975 offer an insight into the dangers of the weapon, as a change of wind direction blew the fire back onto the Song forces.[*][25]

5.4.1 Song Dynasty

Main article: Technology of the Song Dynasty

The Song Dynasty (960–1279) brought a new stability for China after a century of civil war, and started a new area of modernisation by encouraging examinations and meritocracy. The first Song Emperor created political institutions that allowed a great deal of freedom of discourse and thought, which facilitated the growth of scientific advance, economic reforms, and achievements in arts and literature.[*][26] Trade flourished both within China and overseas, and the encouragement of technology allowed the mints at Kaifeng and Hangzhou to gradually increase in production.[*][26] In 1080, the mints of Emperor Shenzong had produced 5 billion coins (roughly 50 per Chinese citizen), and the first banknotes were produced in 1023.[*][26] These coins were so durable that they would still be in use 700 years later, in the 18th century.[*][26]

There were many famous inventors and early scientists in the Song Dynasty period. The statesman Shen Kuo is best known for his book known as the *Dream Pool Essays* (1088 AD). In it, he wrote of use for a drydock to repair boats, the navigational magnetic compass, and the discovery of the concept of true north (with magnetic declination towards the North Pole). Shen Kuo also devised a geological theory for land formation, or geomorphology, and theorized

that there was climate change in geological regions over an enormous span of time.

The equally talented statesman Su Song was best known for his engineering project of the Astronomical Clock Tower of Kaifeng, by 1088 AD. The clock tower was driven by a rotating waterwheel and escapement mechanism. Crowning the top of the clock tower was the large bronze, mechanically-driven, rotating armillary sphere. In 1070, Su Song also compiled the *Ben Cao Tu Jing* (Illustrated Pharmacopoeia, original source material from 1058–1061 AD) with a team of scholars. This pharmaceutical treatise covered a wide range of other related subjects, including botany, zoology, mineralogy, and metallurgy.

Chinese astronomers were the first to record observations of a supernova, the first being the SN 185, recorded during the Han Dynasty. Chinese astronomers made two more notable supernova observations during the Song Dynasty: the SN 1006, the brightest recorded supernova in history; and the SN 1054, making the Crab Nebula the first astronomical object recognized as being connected to a supernova explosion.[*][27]

Archaeology

During the early half of the Song Dynasty (960–1279), the study of archaeology developed out of the antiquarian interests of the educated gentry and their desire to revive the use of ancient vessels in state rituals and ceremonies.[*][28] This and the belief that ancient vessels were products of 'sages' and not common people was criticized by Shen Kuo, who took an interdisciplinary approach to archaeology, incorporating his archaeological findings into studies on metallurgy, optics, astronomy, geometry, and ancient music measures.[*][28] His contemporary Ouyang Xiu (1007–1072) compiled an analytical catalogue of ancient rubbings on stone and bronze, which Patricia B. Ebrey says pioneered ideas in early epigraphy and archaeology.[*][29] In accordance with the beliefs of the later Leopold von Ranke (1795–1886), some Song gentry—such as Zhao Mingcheng (1081–1129)—supported the primacy of contemporaneous archaeological finds of ancient inscriptions over historical works written after the fact, which they contested to be unreliable in regard to the former evidence.[*][30] Hong Mai (1123–1202) used ancient Han Dynasty era vessels to debunk what he found to be fallacious descriptions of Han vessels in the *Bogutu* archaeological catalogue compiled during the latter half of Huizong's reign (1100–1125).[*][30]

Geology and climatology

In addition to his studies in meteorology, astronomy, and archaeology mentioned above, Shen Kuo also made hypotheses in regards to geology and climatology in his *Dream Pool Essays* of 1088, specifically his claims regarding geomorphology and climate change. Shen believed that land was reshaped over time due to perpetual erosion, uplift, and deposition of silt, and cited his observance of horizontal strata of fossils embedded in a cliffside at Taihang as evidence that the area was once the location of an ancient seashore that had shifted hundreds of miles east over an enormous span of time.*[31]*[32]*[33] Shen also wrote that since petrified bamboos were found underground in a dry northern climate zone where they had never been known to grow, climates naturally shifted geographically over time.*[33]*[34]

5.4.2 Mongol transmission

Mongol rule under the Yuan Dynasty saw technological advances from an economic perspective, with the first mass production of paper banknotes by Kublai Khan in the 13th century.*[1] Numerous contacts between Europe and the Mongols occurred in the 13th century, particularly through the unstable Franco-Mongol alliance. Chinese corps, expert in siege warfare, formed an integral part of the Mongol armies campaigning in the West. In 1259-1260 military alliance of the Franks knights of the ruler of Antioch, Bohemond VI and his father-in-law Hetoum I with the Mongols under Hulagu, in which they fought together for the conquests of Muslim Syria, taking together the city of Aleppo, and later Damascus.*[35] William of Rubruck, an ambassador to the Mongols in 1254-1255, a personal friend of Roger Bacon, is also often designated as a possible intermediary in the transmission of gunpowder know-how between the East and the West.*[36] The compass is often said to have been introduced by the Master of the Knights Templar Pierre de Montaigu between 1219 to 1223, from one of his travels to visit the Mongols in Persia.*[37]

Chinese and Arabic astronomy intermingled under Mongol rule. Muslim astronomers worked in the Chinese Astronomical Bureau established by Kublai Khan, while some Chinese astronomers also worked at the Persian Maragha observatory.*[38] Before this, in ancient times, Indian astronomers had lent their expertise to the Chinese court.*[39]

5.4.3 Theory and hypothesis

As Toby E. Huff notes, pre-modern Chinese science developed precariously without solid scientific theory, while there was a lacking of consistent systemic treatment in comparison to contemporaneous European works such as the *Concordance and Discordant Canons* by Gratian of Bologna (fl. 12th century).*[40] This drawback to Chinese science

A 1726 illustration of The Sea Island Mathematical Manual, *written by Liu Hui in the 3rd century.*

was lamented even by the mathematician Yang Hui (1238–1298), who criticized earlier mathematicians such as Li Chunfeng (602–670) who were content with using methods without working out their theoretical origins or principle, stating:

> The men of old changed the name of their methods from problem to problem, so that as no specific explanation was given, there is no way of telling their theoretical origin or basis.
> —*[41]

Despite this, Chinese thinkers of the Middle Ages proposed some hypotheses which are in accordance with modern principles of science. Yang Hui provided theoretical proof for the proposition that the complements of the parallelograms which are about the diameter of any given parallelogram are equal to one another.*[41] Sun Sikong (1015–1076) proposed the idea that rainbows were the result of the contact between sunlight and moisture in the air, while Shen Kuo (1031–1095) expanded upon this with description of atmospheric refraction.*[42]*[43]*[44] Shen believed that rays of sunlight refracted before reaching the surface of the earth, hence the appearance of the observed sun from earth did not match its exact location.*[44] Coinciding with the astronomical work of his colleague Wei Pu, Shen and Wei realized that the old calculation technique for the mean sun was inaccurate compared to the apparent sun, since the latter was ahead of it in the accelerated phase of motion, and behind it in the retarded phase.*[45] Shen supported and expanded upon beliefs earlier proposed by Han Dynasty (202 BCE–202 CE) scholars such as Jing Fang (78–37 BCE) and Zhang Heng (78–139 CE) that lunar eclipse occurs when the earth obstructs the sunlight traveling towards the moon, a solar eclipse is the moon's obstruction of sunlight reaching earth, the moon is spherical like a ball and not flat like a disc, and moonlight is merely sunlight reflected from the moon's surface.*[46] Shen also explained that the observance of a full moon occurred when the sun's light was slanting at a certain degree and that crescent phases of the moon proved that the moon was spherical, using a metaphor of observing different angles of a silver ball with white powder thrown onto one side.*[47]*[48] It should be noted that, although the Chinese accepted the idea of spherical-shaped heavenly bodies, the concept of a spherical earth (as opposed to a flat earth) was not accepted in Chinese thought until the works of Italian Jesuit Matteo Ricci (1552–1610) and Chinese astronomer Xu Guangqi (1562–1633) in the early 17th century.*[49]

5.4.4 Pharmacology

Main article: Traditional Chinese medicine

There were noted advances in traditional Chinese medicine during the Middle Ages. Emperor Gaozong (reigned 649–683) of the Tang Dynasty (618–907) commissioned the scholarly compilation of a *materia medica* in 657 that documented 833 medicinal substances taken from stones, minerals, metals, plants, herbs, animals, vegetables, fruits, and cereal crops.*[50] In his *Bencao Tujing* ('Illustrated Pharmacopoeia'), the scholar-official Su Song (1020–1101) not only systematically categorized herbs and minerals according to their pharmaceutical uses, but he also took an interest in zoology.*[51]*[52]*[53]*[54] For example, Su made systematic descriptions of animal species and the environmental regions they could be found, such as the freshwater crab *Eriocher sinensis* found in the Huai River running through Anhui, in waterways near the capital city, as well as reservoirs and marshes of Hebei.*[55]

Muhammad ibn Zakariya al-Razi in 896, mentions the popular introduction of various Chinese herbs and aloes in Baghdad.

5.4.5 Horology and clockworks

Although the *Bencao Tujing* was an important pharmaceutical work of the age, Su Song is perhaps better known for his work in horology. His book *Xinyi Xiangfayao* (新儀象法要; lit. 'Essentials of a New Method for Mechanizing the Rotation of an Armillary Sphere and a Celestial Globe') documented the intricate mechanics of his astronomical clock tower in Kaifeng. This included the use of an escapement mechanism and world's first known chain drive to power the rotating armillary sphere crowning the top as well as the 133 clock jack figurines positioned on a rotating wheel that sounded the hours by banging drums, clashing gongs, striking bells, and holding plaques with special announcements appearing from open-and-close shutter windows.*[56]*[57]*[58]*[59] While it had been Zhang Heng who applied the first motive power to the armillary sphere via hydraulics in 125 CE,*[60]*[61] it was Yi Xing (683–727) in 725 CE who first applied an escapement mechanism to a water-powered celestial globe and striking clock.*[62] The early Song Dynasty horologist Zhang Sixun (fl. late 10th century) employed liquid mercury in his astronomical clock because there were complaints that water would freeze too easily in the clepsydra tanks during winter.*[63]

Al-Jazari (1136–1206), a Muslim engineer and inventor of various clocks, including the Elephant clock, wrote: "[T]he elephant represents the Indian and African cultures, the two dragons represents Chinese culture, the phoenix represents Persian culture, the water work represents ancient Greek culture, and the turban represents Islamic culture" .

5.4.6 Magnetism and metallurgy

Shen Kuo's written work of 1088 also contains the first written description of the magnetic needle compass, the first description in China of experiments with camera obscura, the invention of movable type printing by the artisan Bi Sheng (990–1051), a method of repeated forging of cast iron under a cold blast similar to the modern Bessemer process, and the mathematical basis for spherical trigonometry that would later be mastered by the astronomer and engineer Guo Shoujing (1231–1316).*[65]*[66]*[67]*[68]*[69]*[70]*[71] While using a sighting tube of improved width to correct the position of the pole star (which had shifted over the centuries), Shen discovered the concept of true north and magnetic declination towards the North Magnetic Pole, a concept which would aid navigators in the years to come.*[72]*[73]

In addition to the method similar to the Bessemer process mentioned above, there were other notable advancements in Chinese metallurgy during the Middle Ages. During the

The elephant clock in a manuscript by Al-Jazari (1206 AD) from The Book of Knowledge of Ingenious Mechanical Devices.*[64]*

11th century, the growth of the iron industry caused vast deforestation due to the use of charcoal in the smelting process.*[74]*[75] To remedy the problem of deforestation, the Song Chinese discovered how to produce coke from bituminous coal as a substitute for charcoal.*[74]*[75] Although hydraulic-powered bellows for heating the blast furnace had been written of since Du Shi's (d. 38) invention of the 1st century CE, the first known drawn and printed illustration of it in operation is found in a book written in 1313 by Wang Zhen (fl. 1290–1333).*[76]

5.4.7   Mathematics

Main article: Chinese mathematics

Qin Jiushao (c. 1202–1261) was the first to introduce the zero symbol into Chinese mathematics.*[77] Before this innovation, blank spaces were used instead of zeros in the system of counting rods.*[78] Pascal's triangle was first illustrated in China by Yang Hui in his book *Xiangjie Jiuzhang*

Suanfa (详解九章算法), although it was described earlier around 1100 by Jia Xian.*[79] Although the *Introduction to Computational Studies* (算学启蒙) written by Zhu Shijie (fl. 13th century) in 1299 contained nothing new in Chinese algebra, it had a great impact on the development of Japanese mathematics.*[80]

5.4.8   Alchemy and Taoism

Main article: Chinese alchemy

In their pursuit for an elixir of life and desire to create gold from various mixtures of materials, Taoists became heavily associated with alchemy.*[81] Joseph Needham labeled their pursuits as proto-scientific rather than merely pseudoscience.*[81] Fairbank and Goldman write that the futile experiments of Chinese alchemists did lead to the discovery of new metal alloys, porcelain types, and dyes.*[81] However, Nathan Sivin discounts such a close connection between Taoism and alchemy, which some sinologists have asserted, stating that alchemy was more prevalent in the secular sphere and practiced by laymen.*[82]

Experimentation with various materials and ingredients in China during the middle period led to the discovery of many ointments, creams, and other mixtures with practical uses. In a 9th-century Arab work *Kitāb al-Khawāss al Kabīr*, there are numerous products listed that were native to China, including waterproof and dust-repelling cream or varnish for clothes and weapons, a Chinese lacquer, varnish, or cream that protected leather items, a completely fireproof cement for glass and porcelain, recipes for Chinese and Indian ink, a waterproof cream for the silk garments of underwater divers, and a cream specifically used for polishing mirrors.*[83]

5.4.9   Gunpowder warfare

The significant change that distinguished Medieval warfare to early Modern warfare was the use of gunpowder weaponry in battle. A 10th-century silken banner from Dunhuang portrays the first artistic depiction of a fire lance, a prototype of the gun.*[84] The *Wujing Zongyao* military manuscript of 1044 listed the first known written formulas for gunpowder, meant for light-weight bombs lobbed from catapults or thrown down from defenders behind city walls.*[85] By the 13th century, the iron-cased bomb shell, hand cannon, land mine, and rocket were developed.*[86]*[87] As evidenced by the *Huolongjing* of Jiao Yu and Liu Bowen, by the 14th century the Chinese had developed the heavy cannon, hollow and gunpowder-packed exploding cannonballs, the two-stage rocket with a booster rocket, the naval mine and wheellock mechanism to ignite trains of fuses.*[88]*[89]

5.5   Jesuit activity in China

Jesuits in China.

The Jesuit China missions of the 16th and 17th centuries introduced Western science and astronomy, then undergoing its own revolution, to China. One modern historian writes that in late Ming courts, the Jesuits were "regarded as impressive especially for their knowledge of astronomy, calendar-making, mathematics, hydraulics, and geography."*[90] The Society of Jesus introduced, according to Thomas Woods, "a substantial body of scientific knowledge and a vast array of mental tools for understanding the physical universe, including the Euclidean geometry that made planetary motion comprehensible." *[2] Another expert quoted by Woods said the scientific revolution brought by the Jesuits coincided with a time when science was at a very low level in China:

> [The Jesuits] made efforts to translate western mathematical and astronomical works into Chinese and aroused the interest of Chinese scholars in these sciences. They made very extensive astronomical observation and carried out the first modern cartographic work in China. They also learned to appreciate the scientific achievements of this ancient culture and made them known in Europe. Through their correspondence European scientists first learned about the Chinese science and culture.
>
> —*[3]

Conversely, the Jesuits were very active in transmitting Chinese knowledge to Europe. Confucius's works were translated into European languages through the agency of Jesuit scholars stationned in China. Matteo Ricci started to report on the thoughts of Confucius, and Father Prospero Intorcetta published the life and works of Confucius into Latin in 1687.*[91] It is thought that such works had considerable importance on European thinkers of the period, particularly among the Deists and other philosophical groups of the Enlightenment who were interested by the integration of the system of morality of Confucius into Christianity.*[92]*[93]

The followers of the French physiocrat François Quesnay habitually referred to him as "the Confucius of Europe", and he personally identified himself with the Chinese sage.*[94] The doctrine and even the name of "Laissez-faire" may have been inspired by the Chinese concept of Wu wei.*[95]*[96] However, the economic insights of ancient Chinese political thought had otherwise little impact outside China in later centuries.*[97] Goethe, was known as "the Confucius of Weimar".*[98]

5.6   Scientific and technological stagnation

Further information: Great Divergence

One question that has been the subject of debate among historians has been why China did not develop a scientific revolution and why Chinese technology fell behind that of Europe. Many hypotheses have been proposed ranging from the cultural to the political and economic. John K. Fairbank, for example, argued that the Chinese political system was hostile to scientific progress. As for Needham, he wrote that cultural factors prevented traditional Chinese achievements from developing into what could be called "science." It was the religious and philosophical framework of the Chinese intellectuals which made them unable to believe in the ideas of laws of nature:

> It was not that there was no order in nature for the Chinese, but rather that it was not an order ordained by a rational personal being, and hence there was no conviction that rational personal beings would be able to spell out in their lesser earthly languages the divine code of laws which he had decreed aforetime. The Taoists, indeed, would have scorned such an idea as being too naïve for the subtlety and complexity of the universe as they intuited it.
>
> —*[99]

Another prominent historian of science, Nathan Sivin, has argued that China indeed had a scientific revolution in the 17th century but it's just that we are still not able to really understand the scientific revolution that took place in China. Sivin suggests that we need to look at the scientific development in China on its own terms.[*][100]

There are also questions about the philosophy behind traditional Chinese medicine, which, derived partly from Taoist philosophy, reflects the classical Chinese belief that individual human experiences express causative principles effective in the environment at all scales. Because its theory predates use of the scientific method, it has received various criticisms based on scientific thinking. Philosopher Robert Todd Carroll, a member of the Skeptics Society, deemed acupuncture a pseudoscience because it "confuse(s) metaphysical claims with empirical claims" ..[*][101]

More recent historians have questioned political and cultural explanations and have put greater focus on economic causes. Mark Elvin's high level equilibrium trap is one well-known example of this line of thought. It argues that the Chinese population was large enough, workers cheap enough, and agrarian productivity high enough to not require mechanization: thousands of Chinese workers were perfectly able to quickly perform any needed task. Other events such as Haijin, the Opium Wars and the resulting hate of European influence prevented China from undergoing an Industrial Revolution; copying Europe's progress on a large scale would be impossible for a lengthy period of time. Political instability under Cixi rule (opposition and frequent oscillation between modernists and conservatives), the Republican wars (1911–1933), the Sino-Japanese War (1933–1945), the Communist/Nationalist War (1945–1949) as well as the later Cultural Revolution isolated China at the most critical times. Kenneth Pomeranz has made the argument that the substantial resources taken from the New World to Europe made the crucial difference between European and Chinese development.

In his book *Guns, Germs, and Steel*, Jared Diamond postulates that the lack of geographic barriers within much of China – essentially a wide plain with two large navigable rivers and a relatively smooth coastline – led to a single government without competition. At the whim of a ruler who disliked new inventions, technology could be stifled for half a century or more. In contrast, Europe's barriers of the Pyrennes, the Alps, and the various defensible peninsulas (Denmark, Scandinavia, Italy, Greece, etc.) and islands (Britain, Ireland, Sicily, etc.) led to smaller countries in constant competition with each other. If a ruler chose to ignore a scientific advancement (especially a military or economic one), his more-advanced neighbors would soon usurp his throne. This explanation, however, ignores the fact that

China had been politically fragmented in the past, and was thus not inherently disposed to political unification.[*][102]

5.7   The Republic of China (1912–49)

The Republican period (1911-1949) saw the introduction in earnest of modern science to China. Large numbers of Chinese students studied abroad in Japan and in Europe and the US. Many returned to help teach and to found numerous schools and universities. Among them were numerous outstanding figures, including Cai Yuanpei, Hu Shi, Weng Wenhao, Ding Wenjiang, Fu Ssu-nien, and many others. As a result, there was a tremendous growth of modern science in China. As the Communist Party took over China's mainland in 1949, some of these Chinese scientists and institutions moved to Taiwan. The central science academy, Academia Sinica, also moved there.

5.8   People's Republic of China

Main article: History of science and technology in the People's Republic of China
See also: Science and technology in the People's Republic of China

After the establishment of the People's Republic in 1949, China reorganized its science establishment along Soviet lines. From 1975, science and technology was one of the Four Modernizations, and its high-speed development was declared essential to all national economic development by Deng Xiaoping. Scientific research in nuclear weapons, satellite launching and recovery, superconductivity, high-yield hybrid rice led to new developments due to the application of science to industry and foreign technology transfer.

As the People's Republic of China becomes better connected to the global economy, the government has placed more emphasis on science and technology. This has led to increases in funding, improved scientific structure, and more money for research. These factors have led to advancements in agriculture, medicine, genetics, and global change.

In 2003, China became the third country to send humans into space.

5.9   See also

- Chinese astronomy

- Chinese mathematics
- List of Chinese inventions
- Military history of China
- Science and Civilization in China
- Traditional Chinese medicine
- Yongle Encyclopedia

5.10 Notes

[1] *Inventions* (Pocket Guides).

[2] Thomas Woods, *How the Catholic Church Built Western Civilization* (Washington, DC: Regenery, 2005)

[3] Agustín Udías, p.53

[4] Needham, Robinson & Huang 2004, p.218

[5] Needham, Robinson & Huang 2004, p.10

[6] Needham 1956 p. 185

[7] Lu Chia (196 BCE, 前漢書 *(Chi'en Han Shu)* (History of the former Han dynasty) ch. 43, p. 6b and *Tung Chien Kang Mu* (Essential Mirror of Universal History) ch. 3, p. 46b) as referenced in Needham, Robinson & Huang 2004, p. 10.

[8] ,

[9] Ancient Chinese Astronomy

[10] F. Espenak. "Solar Eclipses of Historical Interest".

[11] F.R. Stephenson (1997). *Historical Eclipses and Earth's Rotation*. Cambridge University Press.

[12] *Buildings* (Pocket Guides).

[13] Weapons of the terracotta army

[14] People's Daily Online

[15] Needham, Volume 4, Part 2, 39.

[16] Needham, Volume 4, Part 2, 158.

[17] Needham, Volume 4, Part 2, 40.

[18] Shelagh Vainker

[19] "A lodestone attracts a needle." Li Shu-hua, p.176

[20] Li Shu-hua, p.182f.

[21] Liang, pp. Appendix C VII

[22] Kelly, p. 4

[23] Kelly, p. 22. "Around 1240 the Arabs acquired knowledge of saltpeter ("Chinese snow") from the East, perhaps through India. They knew of gunpowder soon afterward. They also learned about fireworks ("Chinese flowers") and rockets ("Chinese arrows")."

[24] Novum Organum, Liber I, CXXIX - Adapted from the 1863 translation

[25] Turnbull, p. 43

[26] *Money of the World* Special Christmas Edition, Orbis Publishing Ltd, 1998.

[27] Mayall N.U. (1939), *The Crab Nebula, a Probable Supernova*, Astronomical Society of the Pacific Leaflets, v. 3, p.145

[28] Julius Thomas Fraser and Francis C. Haber, *Time, Science, and Society in China and the West* (Amherst: University of Massachusetts Press, ISBN 0-87023-495-1, 1986), pp. 227.

[29] Patricia B. Ebrey, The Cambridge Illustrated History of China (Cambridge: Cambridge University Press, 1999, ISBN 0-521-66991-X), pp. 148.

[30] Rudolph, R.C. "Preliminary Notes on Sung Archaeology," *The Journal of Asian Studies* (Volume 22, Number 2, 1963): 169–177.

[31] Joseph Needham, *Science and Civilization in China: Volume 3, Mathematics and the Sciences of the Heavens and the Earth* (Taipei: Caves Books, Ltd., 1986) pp. 603–604, 618.

[32] Nathan Sivin, *Science in Ancient China: Researches and Reflections.* (Brookfield, Vermont: VARIORUM, Ashgate Publishing, 1995), Chapter III, pp. 23.

[33] Alan Kam-leung Chan, Gregory K. Clancey, and Hui-Chieh Loy, *Historical Perspectives on East Asian Science, Technology and Medicine* (Singapore: Singapore University Press, 2002, ISBN 9971-69-259-7) pp. 15.

[34] Joseph Needham, *Science and Civilization in China: Volume 3, Mathematics and the Sciences of the Heavens and the Earth* (Taipei: Caves Books, Ltd., 1986) pp. 618.

[35] "Histoire des Croisades", René Grousset, p581, ISBN 2-262-02569-X

[36] "The Eastern Origins of Western Civilization", John M.Hobson, p186, ISBN 0-521-54724-5

[37] Source

[38] Abstracta Iranica

[39] http://www.nybooks.com/articles/article-preview?article_id=17608

[40] Toby E. Huff, *The Rise of Early Modern Science: Islam, China, and the West* (Cambridge: Cambridge University Press, 2003, ISBN 0-521-52994-8) pp 303.

[41] Joseph Needham, *Science and Civilization in China: Volume 3, Mathematics and the Sciences of the Heavens and the Earth* (Taipei: Caves Books, Ltd., 1986) pp. 104.

[42] Nathan Sivin, *Science in Ancient China: Researches and Reflections.* (Brookfield, Vermont: VARIORUM, Ashgate Publishing, 1995), Chapter III, pp. 24.

[43] Yung Sik Kim, *The Natural Philosophy of Chu Hsi (1130-1200)* (DIANE Publishing, 2002, ISBN 0-87169-235-X), pp. 171.

[44] Paul Dong, *China's Major Mysteries: Paranormal Phenomena and the Unexplained in the People's Republic* (San Francisco: China Books and Periodicals, Inc., 2000, ISBN 0-8351-2676-5), pp. 72.

[45] Nathan Sivin, *Science in Ancient China: Researches and Reflections.* (Brookfield, Vermont: VARIORUM, Ashgate Publishing, 1995), Chapter III, pp. 16–19.

[46] Joseph Needham, *Science and Civilization in China: Volume 3, Mathematics and the Sciences of the Heavens and the Earth* (Taipei: Caves Books, Ltd., 1986) pp. 227 & 414–416

[47] "Joseph Needham, *Science and Civilization in China: Volume 3, Mathematics and the Sciences of the Heavens and the Earth* (Taipei: Caves Books, Ltd., 1986) pp. 415–416.

[48] Paul Dong, *China's Major Mysteries: Paranormal Phenomena and the Unexplained in the People's Republic* (San Francisco: China Books and Periodicals, Inc., 2000, ISBN 0-8351-2676-5), pp. 71–72.

[49] Dainian Fan and Robert Sonné Cohen, *Chinese Studies in the History and Philosophy of Science and Technology* (Dordrecht: Kluwer Academic Publishers, 1996, ISBN 0-7923-3463-9), pp. 431–432.

[50] Charles Benn, *China's Golden Age: Everyday Life in the Tang Dynasty.* Oxford University Press, 2002, ISBN 0-19-517665-0), pp. 235.

[51] Wu Jing-nuan, *An Illustrated Chinese Materia Medica.* (New York: Oxford University Press, 2005), pp. 5.

[52] Joseph Needham, *Science and Civilization in China: Volume 3, Mathematics and the Sciences of the Heavens and the Earth* (Taipei: Caves Books, Ltd., 1986) pp. 648–649.

[53] Joseph Needham, *Science and Civilization in China: Volume 6, Biology and Biological Technology, Part 1, Botany.* (Taipei: Caves Books Ltd., 1986), pp. 174–175.

[54] Schafer, Edward H. "Orpiment and Realgar in Chinese Technology and Tradition," *Journal of the American Oriental Society* (Volume 75, Number 2, 1955): 73–89.

[55] West, Stephen H. "Cilia, Scale and Bristle: The Consumption of Fish and Shellfish in The Eastern Capital of The Northern Song," *Harvard Journal of Asiatic Studies* (Volume 47, Number 2, 1987): 595–634.

[56] Joseph Needham, *Science and Civilization in China: Volume 4, Physics and Physical Technology, Part 2: Mechanical Engineering* (Taipei: Caves Books, Ltd. 1986) pp. 111 & 165 & 445–448.

[57] Liu, Heping. ""The Water Mill"and Northern Song Imperial Patronage of Art, Commerce, and Science,"The Art Bulletin (Volume 84, Number 4, 2002): 566–595.

[58] Tony Fry, *The Architectural Theory Review: Archineering in Chinatime* (Sydney: University of Sydney, 2001), pp. 10–11.

[59] Derk Bodde, *Chinese Thought, Society, and Science* (Honolulu: University of Hawaii Press, 1991), pp. 140.

[60] Joseph Needham, *Science and Civilization in China: Volume 4, Physics and Physical Technology, Part 2: Mechanical Engineering* (Taipei: Caves Books, Ltd. 1986), pp. 30.

[61] W. Scott Morton and Charlton M. Lewis, China: Its History and Culture. (New York: McGraw-Hill, Inc., 2005), pp. 70.

[62] Joseph Needham, *Science and Civilization in China: Volume 4, Physics and Physical Technology, Part 2: Mechanical Engineering* (Taipei: Caves Books, Ltd. 1986) pp. 470–475.

[63] Joseph Needham, *Science and Civilization in China: Volume 4, Physics and Physical Technology, Part 2: Mechanical Engineering* (Taipei: Caves Books, Ltd. 1986), pp. 469–471.

[64] Ibn al-Razzaz Al-Jazari (ed. 1974), *The Book of Knowledge of Ingenious Mechanical Devices.* Translated and annotated by Donald Routledge Hill, Dordrecht/D. Reidel.

[65] Sal Restivo, *Mathematics in Society and History: Sociological Inquiries* (Dordrecht: Kluwer Academic Publishers, 1992, ISBN 1-4020-0039-1), pp 32.

[66] Nathan Sivin, *Science in Ancient China: Researches and Reflections.* (Brookfield, Vermont: VARIORUM, Ashgate Publishing, 1995), Chapter III, pp. 21, 27, & 34.

[67] Joseph Needham, *Science and Civilization in China: Volume 4, Physics and Physical Technology, Part 1, Physics* (Taipei: Caves Books Ltd., 1986), pp. 98 & 252.

[68] Hsu, Mei-ling. "Chinese Marine Cartography: Sea Charts of Pre-Modern China," *Imago Mundi* (Volume 40, 1988): 96–112.

[69] Jacques Gernet, *A History of Chinese Civilization* (Cambridge: Cambridge University Press, 1996, ISBN 0-521-49781-7), pp. 335.

[70] Joseph Needham, *Science and Civilization in China: Volume 5, Chemistry and Chemical Technology, Part 1: Paper and Printing* (Taipei: Caves Books, Ltd, 1986), pp 201.

[71] Hartwell, Robert. "Markets, Technology, and the Structure of Enterprise in the Development of the Eleventh-Century Chinese Iron and Steel Industry," The Journal of Economic History (Volume 26, Number 1, 1966): 29–58.

[72] Nathan Sivin, *Science in Ancient China: Researches and Reflections.* (Brookfield, Vermont: VARIORUM, Ashgate Publishing, 1995), Chapter III, pp. 22.

[73] Peter Mohn, *Magnetism in the Solid State: An Introduction* (New York: Springer-Verlag Inc., 2003, ISBN 3-540-43183-7), pp. 1.

[74] Wagner, Donald B. "The Administration of the Iron Industry in Eleventh-Century China," Journal of the Economic and Social History of the Orient (Volume 44 2001): 175-197.

[75] Patricia B. Ebrey, Anne Walthall, and James B. Palais, *East Asia: A Cultural, Social, and Political History* (Boston: Houghton Mifflin Company, 2006, ISBN 0-618-13384-4), pp. 158.

[76] Joseph Needham, *Science and Civilization in China: Volume 4, Physics and Physical Technology, Part 2, Mechanical Engineering* (Taipei: Caves Books, Ltd., 1986), pp. 376.

[77] Joseph Needham, *Science and Civilization in China: Volume 3, Mathematics and the Sciences of the Heavens and the Earth* (Taipei: Caves Books, Ltd., 1986) pp. 43.

[78] Joseph Needham, *Science and Civilization in China: Volume 3, Mathematics and the Sciences of the Heavens and the Earth* (Taipei: Caves Books, Ltd., 1986) pp. 62–63.

[79] Needham, *Science and Civilization in China: Volume 3, Mathematics and the Sciences of the Heavens and the Earth* (Taipei: Caves Books, Ltd., 1986) pp. 134–137.

[80] Joseph Needham, *Science and Civilization in China: Volume 3, Mathematics and the Sciences of the Heavens and the Earth* (Taipei: Caves Books, Ltd., 1986) pp. 46.

[81] John King Fairbank and Merle Goldman, *China: A New History* (Cambridge: MA; London: The Belknap Press of Harvard University Press, 2nd ed., 2006, ISBN 0-674-01828-1), pp. 81.

[82] Nathan Sivin, "Taoism and Science" in *Medicine, Philosophy and Religion in Ancient China* (Variorum, 1995). Retrieved on 2008-08-13.

[83] Joseph Needham, *Science and Civilization in China: Volume 5, Chemistry and Chemical Technology, Part 4, Spagyrical Discovery and Invention: Apparatus, Theories and Gifts* (Taipei: Caves Books Ltd., 1986), pp. 452.

[84] Joseph Needham, *Science and Civilization in China: Volume 5, Chemistry and Chemical Technology, Part 7, Military Technology; the Gunpowder Epic* (Taipei: Caves Books, Ltd., 1986), pp. 220–262.

[85] Joseph Needham, *Science and Civilization in China: Volume 5, Chemistry and Chemical Technology, Part 7, Military Technology; the Gunpowder Epic* (Taipei: Caves Books, Ltd., 1986), pp. 70–73 & 117–124.

[86] Joseph Needham, *Science and Civilization in China: Volume 5, Chemistry and Chemical Technology, Part 7, Military Technology; the Gunpowder Epic* (Taipei: Caves Books, Ltd., 1986), pp. 173–174, 192, 290, & 477.

[87] Alfred W. Crosby, *Throwing Fire: Projectile Technology Through History* (Cambridge: Cambridge University Press, 2002, ISBN 0-521-79158-8), pp. 100–103.

[88] Joseph Needham, *Science and Civilization in China: Volume 5, Chemistry and Chemical Technology, Part 7, Military Technology; the Gunpowder Epic* (Taipei: Caves Books, Ltd., 1986), pp. 203–205, 264, 508.

[89] John Norris, *Early Gunpowder Artillery: 1300–1600* (Marlborough: The Crowood Press, Ltd., 2003), pp. 11.

[90] Patricia Buckley Ebrey, p. 212.

[91] "Windows into China", John Parker, p.25

[92] "Windows into China", John Parker, p.25, ISBN 0-89073-050-4

[93] "The Eastern origins of Western civilization", John Hobson, p194-195, ISBN 0-521-54724-5

[94] Rothbard, p 366

[95] 1 Source

[96] "The Eastern Origins of Western Civilization", John M. Hobson, p.196

[97] Rothbard, p 23

[98] "Confucius (K'ung Tzu)" (PDF). *Prospects: the quarterly review of comparative education* (Paris: UNESCO: International Bureau of Education). XXIII (1/2): 211–19. 1993. doi:10.1007/bf02195036.

[99] Needham & Wang 1954, p. 581.

[100]

[101] http://skepdic.com/pseudosc.html

[102] James M. Blaut, "Environmentalism and Eurocentrism," Geographical Review 89.3 (1999): 391-408

5.11 References

- *Inventions* (Pocket Guides). Publisher: DK CHILDREN; Pocket edition (March 15, 1995). ISBN 1-56458-889-0. ISBN 978-1-56458-889-0

- *Buildings* (Pocket Guides). Publisher: DK CHILDREN; Pocket edition (March 15, 1995). ISBN 1-56458-885-8. ISBN 978-1-56458-885-2

- Patricia Buckley Ebrey, *The Cambridge Illustrated History of China*. Cambridge, New York and Melbourne: Cambridge University Press, 1996.   ISBN 0-521-43519-6.

- Mark Elvin, "The high-level equilibrium trap: the causes of the decline of invention in the traditional Chinese textile industries" in W. E. Willmott, *Economic Organization in Chinese Society*, (Stanford, Calif., Stanford University Press, 1972) pp. 137–172.

    - The High-level Equilibrium Trap PDF (64.3 KB)

- Kelly, Jack (2004).   *Gunpowder: Alchemy, Bombards, & Pyrotechnics: The History of the Explosive that Changed the World*. Basic Books. ISBN 0-465-03718-6.

- Liang, Jieming (2006). *Chinese Siege Warfare: Mechanical Artillery & Siege Weapons of Antiquity*. Singapore, Republic of Singapore: Leong Kit Meng. ISBN 981-05-5380-3.

- Needham, Joseph; Wang, Ling (王 玲) (1954). "Science and Civilisation in China". 1 *Introductory Orientations*. Cambridge University Press.

- Needham, Joseph (1956). *Science and Civilisation in China*. 2 *History of Scientific Thought*. p. 697. ISBN 0-521-05800-7.

- Joseph Needham (1986). *Science and Civilization in China*, Volume **4, Part 2**: *Mechanical Engineering*. Taipei: Caves Books Pty. Ltd.

- Needham, Joseph; Robinson, Kenneth G.; Huang, Jen-Yü (2004). "Science and Civilisation in China". 7, part II *General Conclusions and Reflections*. Cambridge University Press.

- Li Shu-hua, "Origine de la Boussole 11. Aimant et Boussole," *Isis*, Vol. 45, No. 2. (Jul., 1954)

- Rothbard, Murray N. (2006). *Economic thought before Adam Smith: An Austrian Perspective on the History of Economic Thought*. Cheltnam, UK: Edward Elgar. ISBN 0-945466-48-X.

- Stephen Turnbull, *The Walls of Constantinople, AD 324–1453*, Osprey Publishing, ISBN 1-84176-759-X

- Agustín Udías, *Searching the Heavens and the Earth: The History of Jesuit Observatories* (Dordrecht, The Netherlands: Kluwer Academic Publishers, 2003)

- Shelagh Vainker in Anne Farrer (ed), "Caves of the Thousand Buddhas", 1990, British Museum publications, ISBN 0-7141-1447-2

- Thomas Woods, *How the Catholic Church Built Western Civilization*, (Washington, DC: Regenery, 2005), ISBN 0-89526-038-7

5.12   External links

- Institute for the History of Natural Science, Chinese Academy of Sciences

- Chinese Society for the History of Science and Technology

- Popular Science Alliance Network, Internet Society of China

- China Association for Science and Technology

- China International Association for Promotion of Science and Technology (CIAPST)

- China Popular Science Network

- China Research Institute for Science Popularization

- Science Education Network

- China Association of Children's Science Instructors

- China Science

Chapter 6

History of science and technology in the Indian subcontinent

The **history of science and technology in the Indian Subcontinent** begins with prehistoric human activity at Mehrgarh, in present-day Pakistan, and continues through the Indus Valley Civilization to early states and empires. Following independence science and technology in the Republic of India has included automobile engineering, information technology, communications as well as space, polar, and nuclear sciences.

6.1 Prehistory

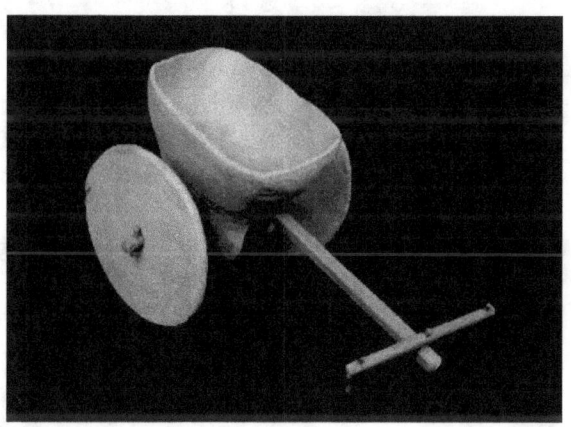

Hand-propelled wheel cart, Indus Valley Civilization (3300–1300 BCE). Housed at the National Museum, New Delhi.

By 5500 BCE a number of sites similar to Mehrgarh had appeared, forming the basis of later chalcolithic cultures.*[1] The inhabitants of these sites maintained trading relations with Near East and Central Asia.*[1]

This was developed in the Indus Valley Civilization by around 4500 BCE.*[2] The size and prosperity of the Indus civilization grew as a result of this innovation, which eventually led to more planned settlements making use of drainage and sewerage.*[2] Sophisticated irrigation and wa-

ter storage systems were developed by the Indus Valley Civilization, including artificial reservoirs at Girnar dated to 3000 BCE, and an early canal irrigation system from c. 2600 BCE.*[3] Cotton was cultivated in the region by the 5th–4th millennia BCE.*[4] Sugarcane was originally from tropical South and Southeast Asia.*[5] Different species likely originated in different locations with *S. barberi* originating in India, and *S. edule* and *S. officinarum* coming from New Guinea.*[5]

The inhabitants of the Indus valley developed a system of standardization, using weights and measures, evident by the excavations made at the Indus valley sites.*[6] This technical standardization enabled gauging devices to be effectively used in angular measurement and measurement for construction.*[6] Calibration was also found in measuring devices along with multiple subdivisions in case of some devices.*[6] One of the earliest known docks is at Lothal (2400 BCE), located away from the main current to avoid deposition of silt.*[7] Modern oceanographers have observed that the Harappans must have possessed knowledge relating to tides in order to build such a dock on the ever-shifting course of the Sabarmati, as well as exemplary hydrography and maritime engineering.*[7]

Excavations at Balakot (c. 2500–1900 BCE), present day Pakistan, have yielded evidence of an early furnace.*[8] The furnace was most likely used for the manufacturing of ceramic objects.*[8] Ovens, dating back to the civilization's mature phase (c. 2500–1900 BCE), were also excavated at Balakot.*[8] The Kalibangan archeological site further yields evidence of potshaped hearths, which at one site have been found both on ground and underground.*[9] Kilns with fire and kiln chambers have also been found at the Kalibangan site.*[9]

Based on archaeological and textual evidence, Joseph E. Schwartzberg (2008)—a University of Minnesota professor emeritus of geography—traces the origins of Indian cartography to the Indus Valley Civilization (c. 2500–1900

View of the Ashokan Pillar at Vaishali. One of the edicts of Ashoka (272−231 BCE) reads: "Everywhere King Piyadasi (Ashoka) erected two kinds of hospitals, hospitals for people and hospitals for animals. Where there were no healing herbs for people and animals, he ordered that they be bought and planted." *[10]

Ink drawing of Ganesha under an umbrella (early 19th century). Ink, called masi, *an admixture of several chemical components, has been used in India since at least the 4th century BCE.* *[16] The practice of writing with ink and a sharp pointed needle was common in early South India.* *[17] Several Jain sutras in India were compiled in ink.* *[18]

BCE).*[11] The use of large scale constructional plans, cosmological drawings, and cartographic material was known in India with some regularity since the Vedic period (2nd - 1st millennium BCE).*[11] Climatic conditions were responsible for the destruction of most of the evidence, however, a number of excavated surveying instruments and measuring rods have yielded convincing evidence of early cartographic activity.*[12] Schwartzberg (2008)—on the subject of surviving maps—further holds that: 'Though not numerous, a number of map-like graffiti appear among the thousands of Stone Age Indian cave paintings; and at least one complex Mesolithic diagram is believed to be a representation of the cosmos.'*[13]

Archeological evidence of an animal-drawn plough dates back to 2500 BCE in the Indus Valley Civilization.*[14] The earliest available swords of copper discovered from the Harappan sites date back to 2300 BCE.*[15] Swords have been recovered in archaeological findings throughout the Ganges–Jamuna Doab region of India, consisting of bronze but more commonly copper.*[15]

6.2 Early kingdoms

The religious texts of the Vedic Period provide evidence for the use of large numbers.*[19] By the time of the last Veda, the *Yajurvedasaṃhitā* (1200-900 BCE), numbers as high as 10^{12} were being included in the texts.*[19] For example, the *mantra* (sacrificial formula) at the end of the *annahoma* ("food-oblation rite") performed during the *aśvamedha* ("an allegory for a horse sacrifice"), and uttered just before-, during-, and just after sunrise, invokes powers of ten from

Value	0	1	2	3	4	5	6	7	8	9
Western Arabic	٠	١	٢	٣	٤	٥	٦	٧	٨	٩
Eastern Arabic	٠	١	٢	٣	۴	۵	۶	٧	٨	٩
Devanagari	०	१	२	३	४	५	६	७	८	९
Gujarati	૦	૧	૨	૩	૪	૫	૬	૭	૮	૯
Gurmukhi	੦	੧	੨	੩	੪	੫	੬	੭	੮	੯
Limbu	᥆	᥇	᥈	᥉	᥊	᥋	᥌	᥍	᥎	᥏
Bengali	০	১	২	৩	৪	৫	৬	৭	৮	৯
Oriya	୦	୧	୨	୩	୪	୫	୬	୭	୮	୯
Telugu	౦	౧	౨	౩	౪	౫	౬	౭	౮	౯
Kannada	೦	೧	೨	೩	೪	೫	೬	೭	೮	೯
Malayalam	൦	൧	൨	൩	൪	൫	൬	൭	൮	൯
Tamil (Grantha)	௦	௧	௨	௩	௪	௫	௬	௭	௮	௯
Tibetan	༠	༡	༢	༣	༤	༥	༦	༧	༨	༩
Burmese	၀	၁	၂	၃	၄	၅	၆	၇	၈	၉
Thai	๐	๑	๒	๓	๔	๕	๖	๗	๘	๙
Khmer	០	១	២	៣	៤	៥	៦	៧	៨	៩
Lao	໐	໑	໒	໓	໔	໕	໖	໗	໘	໙

The Hindu-Arabic numeral *system. The inscriptions on the edicts of Ashoka (1st millennium BCE) display this number system being used by the Imperial Mauryas.*

a hundred to a trillion.*[19] The Satapatha Brahmana (9th century BCE) contains rules for ritual geometric constructions that are similar to the Sulba Sutras.*[20]

Baudhayana (c. 8th century BCE) composed the *Baudhayana Sulba Sutra*, which contains examples of simple Pythagorean triples, such as: $(3, 4, 5)$, $(5, 12, 13)$, $(8, 15, 17)$, $(7, 24, 25)$, and $(12, 35, 37)$ *[21] as well as a statement of the Pythagorean theorem for the sides of a square: "The rope which is stretched across the diagonal of a square produces an area double the size of the original square." *[21] It also contains the general statement of the Pythagorean theorem (for the sides of a rectangle): "The rope stretched along the length of the diagonal of a rectangle makes an area which the vertical and horizontal sides make together." *[21] Baudhayana gives a formula for the square root of two.*[22] Mesopotamian influence at this stage is considered likely.*[23]

The earliest Indian astronomical text—named *Vedānga Jyotiṣa*—attributed to *Lagadha*, is considered one of the oldest astronomical texts, dating from 1400–1200 BCE (with the extant form possibly from 700–600 BCE),*[24] it details several astronomical attributes generally applied for timing social and religious events. It also details astronomical calculations, calendrical studies, and establishes rules for empirical observation.*[25] Since the *Vedānga Jyotiṣa* is a religious text, it has connections with Indian astrology and details several important aspects of the time and seasons, including lunar months, solar months, and their adjustment by a lunar leap month of *Adhimāsa*.*[26] *Ritus* and *Yugas* are also described.*[26] Tripathi (2008) holds that "Twenty-seven constellations, eclipses, seven planets, and twelve signs of the zodiac were also known at that time." *[26]

The Egyptian *Papyrus of Kahun* (1900 BCE) and literature of the Vedic period in India offer early records of veterinary medicine.*[27] Kearns & Nash (2008) state that mention of leprosy is described in the medical treatise *Sushruta Samhita* (6th century BCE). The Sushruta Samhita an Ayurvedic text contains 184 chapters and description of 1120 illnesses, 700 medicinal plants, a detailed study on Anatomy, 64 preparations from mineral sources and 57 preparations based on animal sources.*[28]*[29] However, *The Oxford Illustrated Companion to Medicine* holds that the mention of leprosy, as well as ritualistic cures for it, were described in the Hindu religious book *Atharva-veda*, written in 1500–1200 BCE.*[30]

Cataract surgery was known to the physician Sushruta (6th century BCE).*[31] Traditional cataract surgery was performed with a special tool called the *Jabamukhi Salaka*, a curved needle used to loosen the lens and push the cataract out of the field of vision.*[31] The eye would later be soaked with warm butter and then bandaged.*[31] Though this method was successful, Susruta cautioned that it should only be used when necessary.*[31] The removal of cataract by surgery was also introduced into China from India.*[32]

During the 5th century BCE, the scholar Pāṇini had made several discoveries in the fields of phonetics, phonology, and morphology.*[33] Pāṇini's morphological analysis remained more advanced than any equivalent Western theory until the mid-20th century.*[34] Metal currency was minted in India before the 5th century BCE,*[35]*[36] with coinage (400 BCE—100 CE) being made of silver and copper, bearing animal and plant symbols on them.*[37]

Zinc mines of Zawar, near Udaipur, Rajasthan, were active during 400 BCE.*[38] Diverse specimens of swords have been discovered in Fatehgarh, where there are several varieties of hilt.*[39] These swords have been variously dated to periods between 1700–1400 BCE, but were probably used more extensively during the opening centuries of the 1st millennium BCE.*[40] Archaeological sites in such as Malhar, Dadupur, Raja Nala Ka Tila and Lahuradewa in present-day Uttar Pradesh show iron implements from the period between 1800 BCE and 1200 BCE.*[41] Early iron objects found in India can be dated to 1400 BCE by employing the method of radio carbon dating.*[42] Some scholars believe that by the early 13th century BCE iron smelting was practiced on a bigger scale in India, suggesting that the date of the technology's inception may be placed earlier.*[41] In Southern India (present day Mysore) iron appeared as early as 11th to 12th centuries BCE.*[43] These developments were too early for any significant close contact with the northwest of the country.*[43]

6.3 Post Maha Janapadas —High Middle Ages

The *Arthashastra* of Kautilya mentions the construction of dams and bridges.*[44] The use of suspension bridges using plaited bamboo and iron chain was visible by about the 4th century.*[45] The *stupa*, the precursor of the pagoda and torii, was constructed by the 3rd century BCE.*[46]*[47] Rock-cut step wells in the region date from 200-400 CE.*[48] Subsequently, the construction of wells at Dhank (550-625 CE) and stepped ponds at Bhinmal (850-950 CE) took place.*[48]

During the 1st millennium BCE, the Vaisheshika school of atomism was founded. The most important proponent of this school was Kanada, an Indian philosopher who lived around 200 BCE.*[49] The school proposed that atoms are indivisible and eternal, can neither be created nor destroyed,*[50] and that each one possesses its own distinct viśeṣa (individuality).*[51] It was further elaborated on by the Buddhist school of atomism, of which the philosophers Dharmakirti and Dignāga in the 7th century CE were the most important proponents. They considered atoms to be point-sized, durationless, and made of energy.*[52]

By the beginning of the Common Era glass was being used for ornaments and casing in the region.*[53] Contact with the Greco-Roman world added newer techniques, and local artisans learnt methods of glass molding, decorating and coloring by the early centuries of the Common Era.*[53] The Satavahana period further reveals short cylinders of

The iron pillar of Delhi (375–413 CE). The first iron pillar was the Iron pillar of Delhi, erected at the times of Chandragupta II Vikramaditya.

Model of a Chola (200–848) ship's hull, built by the ASI, based on a wreck 19 miles off the coast of Poombuhar, displayed in a Museum in Tirunelveli.

composite glass, including those displaying a lemon yellow matrix covered with green glass.*[54] Wootz originated in the region before the beginning of the common era.*[55] Wootz was exported and traded throughout Europe, China, the Arab world, and became particularly famous in the Middle East, where it became known as Damascus steel. Archaeological evidence suggests that manufacturing process for Wootz was also in existence in South India before the Christian era.*[56]*[57]

Evidence for using bow-instruments for carding comes from India (2nd century CE).*[58] The mining of diamonds and its early use as gemstones originated in India.*[59] Golconda served as an important early center for diamond mining and processing.*[59] Diamonds were then exported to other parts of the world.*[59] Early reference to diamonds comes from Sanskrit texts.*[60] The *Arthashastra* also mentions diamond trade in the region.*[61] The Iron pillar of Delhi was erected at the times of Chandragupta II Vikramaditya (375–413).*[62] The Rasaratna Samuccaya (800) explains the existence of two types of ores for zinc metal, one of which is ideal for metal extraction while the other is used for medicinal purpose.*[38]

The origins of the spinning wheel are unclear but India is one of the probable places of its origin.*[63]*[64] The device certainly reached Europe from India by the 14th century.*[65] The cotton gin was invented in India as a mechanical device known as *charkhi*, the "wooden-worm-worked roller" .*[58] This mechanical device was, in some parts of the region, driven by water power.*[58] The Ajanta caves yield evidence of a single roller cotton gin in use by the 5th century.*[66] This cotton gin was used until further innovations were made in form of foot powered gins.*[66] Chinese documents confirm at least two missions to India, initiated in 647, for obtaining technology for sugar-refining.*[67] Each mission returned with different results on refining sugar.*[67] (300-200 BCE) was a musical theorist who authored a Sanskrit treatise on prosody. There is evidence that in his work on the enumeration of syllabic combinations, Pingala stumbled upon both the Pascal triangle and Binomial coefficients, although he did not have knowledge of the Binomial theorem itself.*[68]*[69] A description of binary numbers is also found in the works of Pingala.*[70] The Indians also developed the use of the law of signs in multiplication. Negative numbers and the subtrahend had been used in East Asia since the 2nd century BCE, and Indian mathematicians were aware of negative numbers by the 7th century CE,*[71] and their role in mathematical problems of debt was understood.*[72] Although the Indians were not the first to use the subtrahend, they were the first to establish the "law of signs" with regards to the multiplication of positive and negative numbers, which did not appear in East Asian texts until 1299.*[71] Mostly consistent and correct rules for working with negative numbers were formulated,*[73] and the diffusion of these rules led the Arab intermediaries to pass it on to Europe.*[72]

A decimal number system using hieroglyphics dates back to 3000 BC in Egypt,*[74] and was later in use in ancient India where the modern numeration system was developed.[*][75] By the 9th century CE, the Hindu–Arabic numeral system was transmitted from India through the Middle East and to the rest of the world.[*][76] The concept of 0 as a number, and not merely a symbol for separation is attributed to India.[*][73] In India, practical calculations were carried out using zero, which was treated like any other number by the 9th century CE, even in case of division.[*][73][*][77] Brahmagupta (598–668) was able to find (integral) solutions of Pell's equation.[*][78] Conceptual design for a perpetual motion machine by Bhaskara II dates to 1150. He described a wheel that he claimed would run forever.[*][79]

The trigonometric functions of sine and versine, from which it was trivial to derive the cosine, were used by the mathematician, Aryabhata, in the late 5th century.[*][80][*][81] The calculus theorem now known as "Rolle's theorem" was stated by mathematician, Bhāskara II, in the 12th century.[*][82]

Akbarnama—written by August 12, 1602—depicts the defeat of Baz Bahadur of Malwa by the Mughal troops, 1561. The Mughals extensively improved metal weapons and armor used by the armies of India.

Indigo was used as a dye in India, which was also a major center for its production and processing.[*][83] The *Indigofera tinctoria* variety of Indigo was domesticated in India.[*][83] Indigo, used as a dye, made its way to the Greeks and the Romans via various trade routes, and was valued as a luxury product.[*][83] The cashmere wool fiber, also known as *pashm* or *pashmina*, was used in the handmade shawls of Kashmir.[*][84] The woolen shawls from Kashmir region find written mention between 3rd century BCE and the 11th century CE.[*][85] Crystallized sugar was discovered by the time of the Gupta dynasty,[*][86] and the earliest reference to candied sugar comes from India.[*][87] Jute was also cultivated in India.[*][88] Muslin was named after the city where Europeans first encountered it, Mosul, in what is now Iraq, but the fabric actually originated from Dhaka in what is now Bangladesh.[*][89][*][90] In the 9th century, an Arab merchant named Sulaiman makes note of the material's origin in Bengal (known as *Ruhml* in Arabic).[*][90]

European scholar Francesco I reproduced a number of Indian maps in his magnum opus *La Cartografia Antica dell India*.[*][91] Out of these maps, two have been reproduced using a manuscript of *Lokaprakasa*, originally compiled by the polymath Ksemendra (Kashmir, 11th century CE), as a source.[*][91] The other manuscript, used as a source by Francesco I, is titled *Samgraha'.[*][91]*

6.4   Late Middle Ages

Jantar Mantar, Delhi—consisting of 13 architectural astronomy instruments, built by Jai Singh II of Jaipur, from 1724 onwards.

Madhava of Sangamagrama (c. 1340 – 1425) and his Kerala school of astronomy and mathematics developed and founded mathematical analysis.[*][92] The infinite series for π was stated by him and he made use of the series expansion of arctan x to obtain an infinite series expression, now known as the *Madhava-Gregory series*, for π . Their rational approximation of the *error* for the finite sum of their series are of particular interest. They manipulated the error term to derive a faster converging series for π. They used the improved series to derive a rational expression,[93] 104348/33215 for π correct up to nine decimal places, *i.e.* 3.141592653 .[93] The development of the series expansions for trigonometric functions (sine, cosine, and arc tangent) was carried out by mathematicians of the Kerala School in the 15th century CE.[94] Their work, completed two centuries before the invention of calculus in Europe, provided what is now considered the first example of a power series (apart from geometric series).[94]

Shēr Shāh of northern India issued silver currency bearing Islamic motifs, later imitated by the Mughal empire.[37] The Chinese merchant Ma Huan (1413–51) noted that gold coins, known as *fanam*, were issued in Cochin and weighed a total of one *fen* and one *li* according to the Chinese standards.[95] They were of fine quality and could be exchanged in China for 15 silver coins of four-*li* weight each.[95]

In 1500, Nilakantha Somayaji of the Kerala school of astronomy and mathematics, in his Tantrasangraha, revised Aryabhata's elliptical model for the planets Mercury and Venus. His equation of the centre for these planets remained the most accurate until the time of Johannes Kepler in the 17th century.[96]

The seamless celestial globe was invented in Kashmir by Ali Kashmiri ibn Luqman in 998 AH (1589-90 CE), and twenty other such globes were later produced in Lahore and Kashmir during the Mughal Empire.[97] Before they were rediscovered in the 1980s, it was believed by modern metallurgists to be technically impossible to produce metal globes without any seams, even with modern technology.[97] These Mughal metallurgists pioneered the method of lost-wax casting in order to produce these globes.[97]

Gunpowder and gunpowder weapons were transmitted to India through the Mongol invasions of India.[98][99] The Mongols were defeated by Alauddin Khilji of the Delhi Sultanate, and some of the Mongol soldiers remained in northern India after their conversion to Islam.[99] It was written in the *Tarikh-i Firishta* (1606–1607) that the envoy of the Mongol ruler Hulagu Khan was presented with a pyrotechnics display upon his arrival in Delhi in 1258 CE.[100] As a part of an embassy to India by Timurid leader Shah Rukh (1405–1447), 'Abd al-Razzaq mentioned naphtha-throwers mounted on elephants and a variety of pyrotechnics put on display.[101] Firearms known as *top-o-tufak* also existed in the Vijayanagara Empire by as early as 1366 CE.[100] From then on the employment of gunpowder warfare in the region was prevalent, with events such as the siege of Belgaum in 1473 CE by the Sultan Muhammad Shah Bahmani.[102]

In *A History of Greek Fire and Gunpowder*, James Rid-

Portrait of a young Indian scholar, Mughal miniature by Mir Sayyid Ali, c. 1550.

dick Partington describes the gunpowder warfare of 16th and 17th century Mughal India, and writes that "Indian war rockets were formidable weapons before such rockets were used in Europe. They had bamboo rods, a rocket-body lashed to the rod, and iron points. They were directed at the target and fired by lighting the fuse, but the trajectory was rather erratic... The use of mines and counter-mines with explosive charges of gunpowder is mentioned for the times of Akbar and Jahāngir." [103]

By the 16th century, Indians were manufacturing a diverse variety of firearms; large guns in particular, became visible in Tanjore, Dacca, Bijapur and Murshidabad.[104] Guns made of bronze were recovered from Calicut (1504) and Diu (1533).[103] Gujarāt supplied Europe saltpeter for use in gunpowder warfare during the 17th century.[105] Bengal and Mālwa participated in saltpeter production.[105] The Dutch, French, Portuguese, and English used Chhapra as a center of saltpeter refining.[106]

The construction of water works and aspects of water technology in India is described in Arabic and Persian works.*[107] During medieval times, the diffusion of Indian and Persian irrigation technologies gave rise to an advanced irrigation system which bought about economic growth and also helped in the growth of material culture.*[107] The founder of the cashmere wool industry is traditionally held to be the 15th-century ruler of Kashmir, Zayn-ul-Abidin, who introduced weavers from Central Asia.*[85]

The scholar Sadiq Isfahani of Jaunpur compiled an atlas of the parts of the world which he held to be 'suitable for human life'.*[108] The 32 sheet atlas —with maps oriented towards the south as was the case with Islamic works of the era—is part of a larger scholarly work compiled by Isfahani during 1647 CE.*[108] According to Joseph E. Schwartzberg (2008): 'The largest known Indian map, depicting the former Rajput capital at Amber in remarkable house-by-house detail, measures 661 × 645 cm. (260 × 254 in., or approximately 22 × 21 ft).'*[109]

6.5 Colonial era

- The armies of Sultan Hyder Ali of Mysore employed rockets whose gunpowder was packed in metal cylinders instead of paper ones.

- Extent of the railway network in India in 1871; construction had begun in 1856.

- The Indian railways network in 1909.

- Physicist Satyendra Nath Bose is known for his work on the Bose–Einstein statistics during the 1920s.

Early volumes of the *Encyclopædia Britannica* described cartographic charts made by the seafaring Dravidian people.*[110] In *Encyclopædia Britannica (2008)*, Stephen Oliver Fought & John F. Guilmartin, Jr. describe the gunpowder technology in 18th-century Mysore:*[111]

Hyder Ali, prince of Mysore, developed war rockets with an important change: the use of metal cylinders to contain the combustion powder. Although the hammered soft iron he used was crude, the bursting strength of the container of black powder was much higher than the earlier paper construction. Thus a greater internal pressure was possible, with a resultant greater thrust of the propulsive jet. The rocket body was lashed with leather thongs to a long bamboo stick. Range was perhaps up to three-quarters of a mile (more than a kilometre). Although individually these rockets were not accurate, dispersion error became less important when large numbers were fired rapidly in mass attacks. They were particularly effective against cavalry and were hurled into the air, after lighting, or skimmed along the hard dry ground. Hyder Ali's son, Tippu Sultan, continued

to develop and expand the use of rocket weapons, reportedly increasing the number of rocket troops from 1,200 to a corps of 5,000. In battles at Seringapatam in 1792 and 1799 these rockets were used with considerable effect against the British.

By the end of the 18th century the postal system in the region had reached high levels of efficiency.*[112] According to Thomas Broughton, the Maharaja of Jodhpur sent daily offerings of fresh flowers from his capital to Nathadvara (320 km) and they arrived in time for the first religious Darshan at sunrise.*[112] Later this system underwent modernization with the establishment of the British Raj.*[113] The Post Office Act XVII of 1837 enabled the Governor-General of India to convey messages by post within the territories of the East India Company.*[113] Mail was available to some officials without charge, which became a controversial privilege as the years passed.*[113] The Indian Post Office service was established on October 1, 1837.*[113] The British also constructed a vast railway network in the region for both strategic and commercial reasons.*[114]

The British education system, aimed at producing able civil and administrative services candidates, exposed a number of Indians to foreign institutions.*[115] Sir Jagadis Chandra Bose (1858–1937), Prafulla Chandra Ray (1861-1944), Satyendra Nath Bose (1894–1974), Meghnad Saha (1893–1956), P. C. Mahalanobis (1893–1972), Sir C. V. Raman (1888–1970), Subrahmanyan Chandrasekhar (1910–1995), Homi Bhabha (1909–1966), Srinivasa Ramanujan (1887–1920), Vikram Sarabhai (1919–1971), Har Gobind Khorana (1922–2011), and Harish Chandra (1923–1983) were among the notable scholars of this period.*[115]

Extensive interaction between colonial and native sciences was seen during most of the colonial era.*[116] Western science came to be associated with the requirements of nation building rather than being viewed entirely as a colonial entity,*[117] especially as it continued to fuel necessities from agriculture to commerce.*[116] Scientists from India also appeared throughout Europe.*[117] By the time of India's independence colonial science had assumed importance within the westernized intelligentsia and establishment.*[117]

Further information: For science and technology in the Republic of India refer to Science and technology in the Republic of India.

Further information: For science and technology in Pakistan refer to Science and technology in Pakistan.

6.6   See also

- Science and technology in India

- List of Indian inventions

- Information technology in India

- Project of History of Indian Science, Philosophy and Culture

- List of Indian engineering colleges before 1947

- Digit (magazine)

6.7   Notes

[1] Kenoyer, 230

[2] Rodda & Ubertini, 279

[3] Rodda & Ubertini, 161

[4] Stein, 47

[5] Sharpe (1998)

[6] Baber, 23

[7] Rao, 27–28

[8] Dales, 3–22 [10]

[9] Baber, 20

[10] Finger, 12

[11] "We now believe that some form of mapping was practiced in what is now India as early as the Mesolithic period, that surveying dates as far back as the Indus Civilization (ca. 2500–1900 BCE), and that the construction of large-scale plans, cosmographic maps, and other cartographic works has occurred continuously at least since the late Vedic age (first millennium BCE)" —*Joseph E. Schwartzberg, 1301.*

[12] Schwartzberg, 1301-1302

[13] Schwartzberg, 1301

[14] Lal (2001)

[15] Allchin, 111-112

[16] Banerji, 673

[17] Sircar, 62

[18] Sircar, 67

[19] Hayashi, 360-361

[20] Seidenberg, 301-342

[21] Joseph, 229

[22] Cooke, 200

[23] (Boyer 1991, "China and India" p. 207)

[24] Subbarayappa, B. V. (14 September 1989). "Indian astronomy: An historical perspective". In Biswas, S. K.; Mallik, D. C. V.; Vishveshwara, C. V. *Cosmic Perspectives.* Cambridge University Press. pp. 25–40. ISBN 978-0-521-34354-1.

[25] Subbaarayappa, 25-41

[26] Tripathi, 264-267

[27] Thrusfield, 2

[28] Dwivedi & Dwivedi (2007)

[29] Kearns & Nash (2008)

[30] Lock etc., 420

[31] Finger, 66

[32] Lade & Svoboda, 85

[33] Encyclopædia Britannica (2008), *Linguistics.*

[34] Staal, Frits (1988). *Universals: studies in Indian logic and linguistics.* University of Chicago Press. p. 47.

[35] Dhavalikar, 330-338

[36] Sellwood (2008)

[37] Allan & Stern (2008)

[38] Craddock (1983)

[39] F.R. Allchin, 111-112

[40] Allchin, 114

[41] Tewari (2003)

[42] Ceccarelli, 218

[43] Drakonoff, 372

[44] Dikshitar, pg. 332

[45] Encyclopædia Britannica (2008), *suspension bridge.*

[46] Encyclopædia Britannica (2008), *Pagoda.*

[47] Japanese Architecture and Art Net Users System (2001), *torii.*

[48] Livingston & Beach, xxiii

[49] Oliver Leaman, *Key Concepts in Eastern Philosophy.* Routledge, 1999, page 269.

[50] Chattopadhyaya 1986, pp. 169–70

[51] Radhakrishnan 2006, p. 202

[52] (Stcherbatsky 1962 (1930). Vol. 1. P. 19)

[53] Ghosh, 219

[54] "Ornaments, Gems etc." (Ch. 10) in Ghosh 1990.

[55] Srinivasan & Ranganathan

[56] Srinivasan (1994)

[57] Srinivasan & Griffiths

[58] Baber, 57

[59] Wenk, 535-539

[60] MSN Encarta (2007), *Diamond*. Archived 2009-10-31.

[61] Lee, 685

[62] Balasubramaniam, R., 2002

[63] Britannica Concise Encyclopedia (2007), *spinning wheel*.

[64] Encyclopeedia Britnnica (2008). *spinning*.

[65] MSN Encarta (2008), *Spinning*. Archived 2009-10-31.

[66] Baber, 56

[67] Kieschnick, 258

[68] Fowler, 11

[69] Singh, 623-624

[70] Sanchez & Canton, 37

[71] Smith (1958), page 258

[72] Bourbaki (1998), page 49

[73] Bourbaki 1998, p. 46

[74] Georges Ifrah: *From One to Zero. A Universal History of Numbers*, Penguin Books, 1988, ISBN 0-14-009919-0, pp. 200-213 (Egyptian Numerals)

[75] Ifrah, 346

[76] Jeffrey Wigelsworth (1 January 2006). *Science And Technology in Medieval European Life*. Greenwood Publishing Group. p. 18. ISBN 978-0-313-33754-3.

[77] Britannica Concise Encyclopedia (2007). *algebra*

[78] Stillwell, 72-73

[79] Lynn Townsend White, Jr.

[80] O'Connor, J. J. & Robertson, E.F. (1996)

[81] "Geometry, and its branch trigonometry, was the mathematics Indian astronomers used most frequently. In fact, the Indian astronomers in the third or fourth century, using a pre-Ptolemaic Greek table of chords, produced tables of sines and versines, from which it was trivial to derive cosines. This new system of trigonometry, produced in India, was transmitted to the Arabs in the late eighth century and by them, in an expanded form, to the Latin West and the Byzantine East in the twelfth century" - Pingree (2003).

[82] Broadbent, 307–308

[83] Kriger & Connah, 120

[84] Encyclopædia Britannica (2008), *cashmere*.

[85] Encyclopædia Britannica (2008), *Kashmir shawl*.

[86] Shaffer, 311

[87] Kieschnick (2003)

[88] Encyclopædia Britannica (2008), *jute*.

[89] Banglapedia (2008), *Muslin*, Asiatic Society of Bangladesh.

[90] Ahmad, 5–26

[91] Sircar 328

[92] J J O'Connor and E F Robertson. "Mādhava of Sangamagrāma". School of Mathematics and Statistics University of St Andrews, Scotland. Retrieved 2007-09-08.

[93] Roy, 291-306

[94] Stillwell, 173

[95] Chaudhuri, 223

[96] Joseph, George G. (2000), *The Crest of the Peacock: Non-European Roots of Mathematics*, Penguin Books, ISBN 0-691-00659-8.

[97] Savage-Smith (1985)

[98] Iqtidar Alam Khan (2004). *Gunpowder And Firearms: Warfare In Medieval India*. Oxford University Press. ISBN 978-0-19-566526-0.

[99] Iqtidar Alam Khan (25 April 2008). *Historical Dictionary of Medieval India*. Scarecrow Press. p. 157. ISBN 978-0-8108-5503-8.

[100] Khan, 9-10

[101] Partington, 217

[102] Khan, 10

[103] Partington, 226

[104] Partington, 225

[105] Encyclopædia Britannica (2008), *India*.

[106] Encyclopædia Britannica (2008), *Chāpra*.

[107] Siddiqui, 52–77

[108] Schwartzberg, 1302

[109] Schwartzberg, 1303

[110] Sircar 330

[111] Encyclopædia Britannica (2008), *rocket and missile system*.

[112] Peabody, 71

[113] Lowe, 134

[114] Seaman, 348

[115] Raja (2006)

[116] Arnold, 211

[117] Arnold, 212

6.8　References

- Allan, J. & Stern, S. M. (2008), *coin*, Encyclopædia Britannica.

- Allchin, F.R. (1979), *South Asian Archaeology 1975: Papers from the Third International Conference of the Association of South Asian Archaeologists in Western Europe, Held in Paris* edited by J.E. van Lohuizen-de Leeuw, Brill Academic Publishers, ISBN 90-04-05996-2.

- Ahmad, S. (2005), "Rise and Decline of the Economy of Bengal", *Asian Affairs*, **27** (3): 5–26.

- Arnold, David (2004), *The New Cambridge History of India: Science, Technology and Medicine in Colonial India*, Cambridge University Press, ISBN 0-521-56319-4.

- Baber, Zaheer (1996), *The Science of Empire: Scientific Knowledge, Civilization, and Colonial Rule in India*, State University of New York Press, ISBN 0-7914-2919-9.

- Balasubramaniam, R. (2002), *Delhi Iron Pillar: New Insights*, Indian Institute of Advanced Studies, ISBN 81-7305-223-9.

- BBC (2006), "Stone age man used dentist drill".

- Bourbaki, Nicolas (1998), *Elements of the History of Mathematics*, Springer, ISBN 3-540-64767-8.

- Broadbent, T. A. A. (1968), "Reviewed work(s): The History of Ancient Indian Mathematics by C. N. Srinivasiengar", *The Mathematical Gazette*, **52** (381): 307–308.

- Ceccarelli, Marco (2000), *International Symposium on History of Machines and Mechanisms: Proceedings HMM Symposium*, Springer, ISBN 0-7923-6372-8.

- Chaudhuri, K. N. (1985), *Trade and Civilisation in the Indian Ocean*, Cambridge University Press, ISBN 0-521-28542-9.

- Craddock, P.T. etc. (1983), *Zinc production in medieval India*, World Archaeology, **15** (2), Industrial Archaeology.

- Cooke, Roger (2005), *The History of Mathematics: A Brief Course*, Wiley-Interscience, ISBN 0-471-44459-6.

- Coppa, A. etc. (2006), "Early neolithic tradition of dentistry", *Nature*, **440**: 755-756.

- Dales, George (1974), "Excavations at Balakot, Pakistan, 1973", *Journal of Field Archaeology*, **1** (1-2): 3–22 [10].

- Dhavalikar, M. K. (1975), "The beginning of coinage in India", *World Archaeology*, **6** (3): 330-338, Taylor & Francis.

- Dikshitar, V. R. R. (1993), *The Mauryan Polity*, Motilal Banarsidass, ISBN 81-208-1023-6.

- Drakonoff, I. M. (1991), *Early Antiquity*, University of Chicago Press, ISBN 0-226-14465-8.

- Fowler, David (1996), "Binomial Coefficient Function", *The American Mathematical Monthly*, **103** (1): 1-17.

- Finger, Stanley (2001), *Origins of Neuroscience: A History of Explorations Into Brain Function*, Oxford University Press, ISBN 0-19-514694-8.

- Ghosh, Amalananda (1990), *An Encyclopaedia of Indian Archaeology*, Brill Academic Publishers, ISBN 90-04-09262-5.

- Hayashi, Takao (2005), "Indian Mathematics", *The Blackwell Companion to Hinduism* edited by Gavin Flood, pp. 360–375, Basil Blackwell, ISBN 978-1-4051-3251-0.

- Hopkins, Donald R. (2002), *The Greatest Killer: Smallpox in history*, University of Chicago Press, ISBN 0-226-35168-8.

- Ifrah, Georges (2000), *A Universal History of Numbers: From Prehistory to Computers*, Wiley, ISBN 0-471-39340-1.

- Joseph, G. G. (2000), *The Crest of the Peacock: The Non-European Roots of Mathematics*, Princeton University Press, ISBN 0-691-00659-8.

- Kearns, Susannah C.J. & Nash, June E. (2008), *leprosy*, Encyclopædia Britannica.

- Kenoyer, J.M. (2006), "Neolithic Period", *Encyclopedia of India (vol. 3)* edited by Stanley Wolpert, Thomson Gale, ISBN 0-684-31352-9.

- Khan, Iqtidar Alam (1996), *Coming of Gunpowder to the Islamic World and North India: Spotlight on the Role of the Mongols*, Journal of Asian History **30**: 41–5 .

- Kieschnick, John (2003), *The Impact of Buddhism on Chinese Material Culture*, Princeton University Press, ISBN 0-691-09676-7.

- Kriger, Colleen E. & Connah, Graham (2006), *Cloth in West African History*, Rowman Altamira, ISBN 0-7591-0422-0.

- Lade, Arnie & Svoboda, Robert (2000), *Chinese Medicine and Ayurveda*, Motilal Banarsidass, ISBN 81-208-1472-X.

- Lal, R. (2001), "Thematic evolution of ISTRO: transition in scientific issues and research focus from 1955 to 2000" , *Soil and Tillage Research*, **61** (1-2): 3–12 [3].

- Lee, Sunggyu (2006), *Encyclopedia of Chemical Processing*, CRC Press, ISBN 0-8247-5563-4.

- Livingston, Morna & Beach, Milo (2002), *Steps to Water: The Ancient Stepwells of India*, Princeton Architectural Press, ISBN 1-56898-324-7.

- Lock, Stephen etc. (2001), *The Oxford Illustrated Companion to Medicine*, Oxford University Press, ISBN 0-19-262950-6.

- Lowe, Robson (1951), *The Encyclopedia of British Empire Postage Stamps, 1661–1951 (vol. 3)*.

- MSNBC (2008), "Dig uncovers ancient roots of dentistry" .

- Nair, C.G.R. (2004), "Science and technology in free India" , *Government of Kerala—Kerala Call*, Retrieved on 2006-07-09.

- O'Connor, J. J. & Robertson, E.F. (1996), "Trigonometric functions" , *MacTutor History of Mathematics Archive*.

- O'Connor, J. J. & Robertson, E. F. (2000), "Paramesvara" , *MacTutor History of Mathematics archive*.

- Partington, James Riddick & Hall, Bert S. (1999), *A History of Greek Fire and Gunpowder*, Johns Hopkins University Press, ISBN 0-8018-5954-9.

- Peabody, Norman (2003), *Hindu Kingship and Polity in Precolonial India*, Cambridge University Press, ISBN 0-521-46548-6.

- Peele, Stanton & Marcus Grant (1999), *Alcohol and Pleasure: A Health Perspective*, Psychology Press, ISBN 1-58391-015-8.

- Piercey, W. Douglas & Scarborough, Harold (2008), *hospital*, Encyclopædia Britannica.

- Pingree, David (2003), "The logic of non-Western science: mathematical discoveries in medieval India" , *Daedalus*, **132** (4): 45-54.

- Raja, Rajendran (2006), "Scientists of Indian origin and their contributions" , *Encyclopedia of India (Vol 4.)* edited by Stanley Wolpert, ISBN 0-684-31512-2.

- Rao, S. R. (1985), *Lothal*, Archaeological Survey of India.

- Rodda & Ubertini (2004), *The Basis of Civilization—Water Science?*, International Association of Hydrological Science, ISBN 1-901502-57-0.

- Roy, Ranjan (1990), "Discovery of the Series Formula for π by Leibniz, Gregory, and Nilakantha" , *Mathematics Magazine*, Mathematical Association of America, **63** (5): 291-306.

- Sanchez & Canton (2006), *Microcontroller Programming: The Microchip PIC*, CRC Press, ISBN 0-8493-7189-9.

- Savage-Smith, Emilie (1985), *Islamicate Celestial Globes: Their History, Construction, and Use*, Smithsonian Institution Press, Washington, D.C.

- Schwartzberg, Joseph E. (2008), "Maps and Mapmaking in India" , *Encyclopaedia of the History of Science, Technology, and Medicine in Non-Western Cultures (2nd edition)* edited by Helaine Selin, pp. 1301–1303, Springer, ISBN 978-1-4020-4559-2.

- Seaman, Lewis Charles Bernard (1973), *Victorian England: Aspects of English and Imperial History 1837-1901*, Routledge, ISBN 0-415-04576-2.

- Seidenberg, A. (1978), *The origin of mathematics*, Archive for the history of Exact Sciences, **18**: 301-342.

- Sellwood, D. G. J. (2008), *coin*, Encyclopædia Britannica.

- Shaffer, Lynda N., "Southernization" , *Agricultural and Pastoral Societies in Ancient and Classical History* edited by Michael Adas, pp. 308–324, Temple University Press, ISBN 1-56639-832-0.

- Sharpe, Peter (1998), *Sugar Cane: Past and Present*, Southern Illinois University.

- Siddiqui, I. H. (1986), "Water Works and Irrigation System in India during Pre-Mughal Times", *Journal of the Economic and Social History of the Orient*, **29** (1): 52–77.

- Singh, A. N. (1936), "On the Use of Series in Hindu Mathematics", *Osiris*, **1**: 606-628.

- Sircar, D.C.C. (1990), *Studies in the Geography of Ancient and Medieval India*, Motilal Banarsidass Publishers, ISBN 81-208-0690-5.

- Smith, David E. (1958). *History of Mathematics*. Courier Dover Publications. ISBN 0-486-20430-8.

- Srinivasan, S. & Griffiths, D., "South Indian wootz: evidence for high-carbon steel from crucibles from a newly identified site and preliminary comparisons with related finds", *Material Issues in Art and Archaeology-V*, Materials Research Society Symposium Proceedings Series Vol. 462.

- Srinivasan, S. & Ranganathan, S., *Wootz Steel: An Advanced Material of the Ancient World*, Bangalore: Indian Institute of Science.

- Srinivasan, S. (1994), "Wootz crucible steel: a newly discovered production site in South India", Institute of Archaeology, University College London, **5**: 49-61.

- Stein, Burton (1998), *A History of India*, Blackwell Publishing, ISBN 0-631-20546-2.

- Stillwell, John (2004), *Mathematics and its History (2 edition)*, Springer, ISBN 0-387-95336-1.

- Subbaarayappa, B.V. (1989), "Indian astronomy: an historical perspective", *Cosmic Perspectives* edited by Biswas etc., pp. 25–41, Cambridge University Press, ISBN 0-521-34354-2.

- Teresi, Dick etc. (2002), *Lost Discoveries: The Ancient Roots of Modern Science—from the Babylonians to the Maya*, Simon & Schuster, ISBN 0-684-83718-8.

- Tewari, Rakesh (2003), "The origins of Iron Working in India: New evidence from the central Ganga plain and the eastern Vindhyas", *Antiquity*, **77** (297): 536–544.

- Thrusfield, Michael (2007), *Veterinary Epidemiology*, Blackwell Publishing, ISBN 1-4051-5627-9.

- Tripathi, V.N. (2008), "Astrology in India", *Encyclopaedia of the History of Science, Technology, and Medicine in Non-Western Cultures (2nd edition)* edited by Helaine Selin, pp. 264–267, Springer, ISBN 978-1-4020-4559-2.

- Wenk, Hans-Rudolf etc. (2003), *Minerals: Their Constitution and Origin*, Cambridge University Press, ISBN 0-521-52958-1.

- White, Lynn Townsend, Jr. (1960), "Tibet, India, and Malaya as Sources of Western Medieval Technology", *The American Historical Review* **65** (3): 522-526.

- Whish, Charles (1835), *Transactions of the Royal Asiatic Society of Great Britain and Ireland*.

6.9 External links

- Our Science and Technology Heritage gallery for the National Science Centre in Delhi

- A brief introduction to technological brilliance of Ancient India (Indian Institute of Scientific Heritage)

- Science and Technology in Ancient India

- *India: Science and technology*, U.S. Library of Congress.

- *Pursuit and promotion of science: The Indian Experience*, Indian National Science Academy.

- *India: Science and technology*, U.S. Library of Congress.

- Indian National Science Academy (2001), *Pursuit and promotion of science: The Indian Experience*, Indian National Science Academy,

- Presenting Indian S&T Heritage in Science Museums, Propagation : a Journal of science communication Vol 1, NO.1, January 2010, National Council of Science Museums, Kolkata, India, by S.M Khened, .

- Presenting Indian S&T Heritage in Science Museums, Propagation : a Journal of science communication Vol 1, NO.2, July, 2010, pages 124-132, National Council of Science Museums, Kolkata, India, by S.M Khened,.

Chapter 7

Physics in the medieval Islamic world

Main article: Science in the medieval Islamic world

The natural sciences saw various advancements during the Golden Age of Islam (from roughly the mid 8th to the mid 13th centuries), adding a number of innovations to the Transmission of the Classics (such as Aristotle, Ptolemy, Euclid, Neoplatonism).*[1] During this period, Islamic theology was encouraging of thinkers to find knowledge, .*[2] Thinkers from this period included Al-Farabi, Abu Bishr Matta, Ibn Sina, al-Hassan Ibn al-Haytham and Ibn Bajjah.*[3] These works and the important commentaries on them were the wellspring of science during the medieval period. They were translated into Arabic, the *lingua franca* of this period.

Islamic scholarship had inherited Aristotelian physics from the Greeks and during the Islamic Golden Age developed it further. However the Islamic world had a greater respect for knowledge gained from empirical observation, and believed that the universe is governed by a single set of laws. Their use of empirical observation led to the formation of crude forms of the scientific method.*[4] The study of physics in the Islamic world started in Iraq and Egypt.*[5] Fields of physics studied in this period include optics, mechanics (including statics, dynamics, kinematics and motion), and astronomy.

7.1   Physics

Islamic scholarship had inherited Aristotelian physics from the Greeks and during the Islamic Golden Age developed it further, especially placing emphasis on observation and *a priori* reasoning, developing early forms of the scientific method. With Aristotelian physics, physics was seen as lower than demonstrative mathematical sciences, but in terms of a larger theory of knowledge, physics was higher than astronomy; many of whose principles derive from physics and metaphysics.*[6] The primary subject of physics, according to Aristotle, was motion or change; there

were three factors involved with this change, underlying thing, privation, and form. In his *Metaphysics*, Aristotle believed that the *Unmoved Mover* was responsible for the movement of the cosmos, which Neoplatonists later generalized as the cosmos were eternal.*[1] Al-Kindi argued against the idea of the cosmos being eternal by claiming that the eternality of the world lands one in a different sort of absurdity involving the infinite; Al-Kindi asserted that the cosmos must have a temporal origin because traversing an infinite was impossible.

One of the first commentaries of Aristotle's *Metaphysics* is by Al-Farabi. In "'The Aims of Aristotle's Metaphysics", Al-Farabi argues that metaphysics is not specific to natural beings, but at the same time, metaphysics is higher in universality than natural beings.*[1]

7.2   Optics

One field in physics, optics, developed rapidly in this period. By the ninth century, there were works on physiological optics as well as mirror reflections, and geometrical and physical optics.*[7] In the eleventh century, Ibn al-Haytham not only rejected the Greek idea about vision, he came up with a new theory.*[8] ibn al-Haytham postulated in his "Book of Optics" that light was reflected upon different surfaces in different directions, thus causing different light signatures for a certain object that we see.*[9] It was a different approach than that which was previously thought by Greek scientists, such as Euclid or Ptolemy, who believed rays were emitted from the eye to an object and back again. Al-Haytham, with this new theory of optics, was able to study the geometric aspects of the visual cone theories without explaining the physiology of perception.*[7] Also in his Book of Optics, Ibn al-Haytham used mechanics to try and understand optics. Using projectiles, he observed that objects that hit a target perpendicularly exert much more force than projectiles that hit at an angle. Al-Haytham applied this discovery to optics and tried to explain why direct light hurts the eye, because direct light approaches perpendicu-

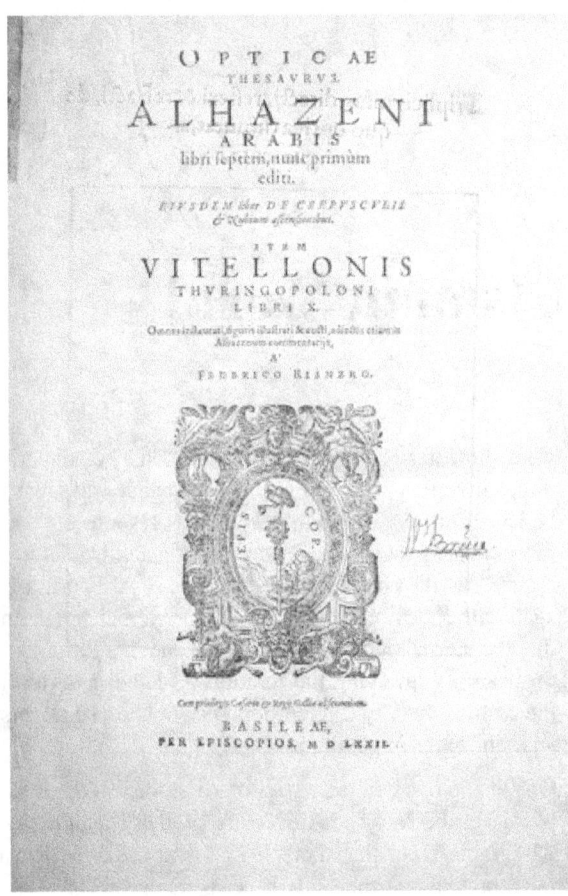

Cover page for Ibn al-Haytham's Book of Optics

larly and not at an oblique angle.[9] Taqī al-Dīn tried to disprove the widely held belief that light is emitted by the eye and not the object that is being observed. He explained that, if light came from our eyes at a constant velocity it would take much too long to illuminate the stars for us to see them while we are still looking at them, because they are so far away. Therefore, the illumination must be coming from the stars so we can see them as soon as we open our eyes.[10]

7.3 Astronomy

The Islamic understanding of the astronomical model was based on the Greek Ptolemaic system. However many early astronomers had started to question the model. It was not always accurate in its predictions and was over complicated because astronomers were trying to mathematically describe the movement of the heavenly bodies. Ibn al-Haytham published *Al-Shukuk ala Batiamyus* ("Doubts on Ptolemy"), which outlined his many criticisms of the Ptolemaic paradigm. This book encouraged other astronomers

to develop new models to explain celestial movement better than Ptolemy.[11] In al-Haytham's Book of Optics he argues that the celestial spheres were not made of solid matter, and that the heavens are less dense that air. [12] Al-Haytham eventually concludes that heavenly bodies follow the same laws of physics as Earthly bodies.[13] Some astronomers theorized about gravity too, al-Khazini suggests that the gravity an object contains varies depending on its distance from the center of the universe. The center of the universe in this case refers to the center of the Earth.[14]

7.4 Mechanics

7.4.1 Inertia

John Philoponus had rejected the Aristotelian view of motion, and argued that an object acquires an inclination to move when it has a motive power impressed on it. In the eleventh century Ibn Sina had roughly adopted this idea, believing that a moving object has force which is dissipated by external agents like air resistance.[15] Ibn Sina made distinction between 'force' and 'inclination' (called "mayl"), he claimed that an objected gained mayl when the object is in opposition to its natural motion. So he concluded that continuation of motion is attributed to the inclination that is transferred to the object, and that object will be in motion until the mayl is spent. He also claimed that projectile in a vacuum would not stop unless it is acted upon. This conception of motion is consistent with Newton's first law of motion, inertia. Which states that an object in motion will stay in motion unless it is acted on by an external force.[16] This idea which dissented from the Aristotelian view was basically abandoned until it was described as "impetus" by John Buridan, who was influenced by Ibn Sina's book Book of Healing.[15]

7.4.2 Acceleration

In Abū Rayḥān al-Bīrūnī text *Shadows*, he recognizes that non-uniform motion is the result of acceleration.[17] Ibn-Sina's theory of mayl tried to relate the velocity and weight of a moving object, this idea closely resembled the concept of momentum[18] Aristotle's theory of motion stated that a constant force produces a uniform motion, Abu'l-Barakāt al-Baghdādī contradicted this and developed his own theory of motion. In his theory he showed that velocity and acceleration are two different things and force is proportional to acceleration and not velocity. [19]

7.4.3 Reaction

Ibn Bajjah proposed that for every force there is always a reaction force. While he did not specify that these forces be equal it is still an early version of the third law of motion which states that for every action there is an equal and opposite reaction.[*][20]

7.5 See also

- Astronomy in medieval Islam
- History of optics
- History of physics
- History of scientific method
- Islamic contributions to Medieval Europe
- Islamic Golden Age
- Science in medieval Islam
- Science in the Middle Ages

7.6 References

[1] *Classical Arabic Philosophy An Anthology of Sources*, Translated by Jon McGinnis and David C. Reisman. Indianapolis: Hackett Publishing Company, 2007. pg. xix

[2] Bakar, Osman. *The History and Philosophy of Islamic Science*. Cambridge: Islamic Texts Society, 1999. pg. 2

[3] Al-Khalili, Jim. "The 'first true scientist'". Archived from the original on 5 January 2009. Retrieved 4 January 2009.

[4] I.A., Ahmad (1995). "The Impact of the Qur'anic Conception of Astronomical Phenomena on Islamic Civilization" (PDF). *Vistas in Astronomy* **39** (4). pp. 395–403.

[5] Thiele, Rüdiger (August 2005), "In Memoriam: Matthias Schramm, 1928–2005", *Historia Mathematica* **32** (3): 271–274, doi:10.1016/j.hm.2005.05.002

[6] . *Islam, Science, and the Challenge of History*. New Haven: Yale University Press. pg 57

[7] Dallal, Ahmad. *Islam, Science, and the Challenge of History*. New Haven: Yale University Press, 2010. pg. 38

[8] Dallal, Ahmad. *Islam, Science, and the Challenge of History*. New Haven:Yale University Press. pg 39

[9] Lindberg, David C. (1976). *Theories of Vision from al-Kindi to Kepler*. University of Chicago Press, Chicago. ISBN 0-226-48234-0. OCLC 1676198 185636643.

[10] Taqī al-Dīn. Kitāb Nūr, Book I, Chapter 5, MS 'O', folio 14b; MS 'S', folio 12a-b

[11] Dallal, Ahmad (1999), "Science, Medicine and Technology", in Esposito, John, The Oxford History of Islam, Oxford University Press, New York

[12] Rosen, Edward. (1985). "The Dissolution of the Solid Celestial Spheres". *Journal of the History of Ideas*. Vol 46(1):13-31.

[13] Duhem, Pierre. (1969). "To Save the Phenomena: An Essay on the Idea of Physical Theory from Plato to Galileo". University of Chicago Press, Chicago.

[14] Mariam Rozhanskaya and I. S. Levinova (1996), "Statics", in Roshdi Rashed, ed., Encyclopedia of the History of Arabic Science, Vol. 2, p. 614-642 Routledge, London and New York

[15] Sayili, Aydin. "Ibn Sina and Buridan on the Motion the Projectile". Annals of the New York Academy of Sciences vol. 500(1). p.477-482.

[16] Espinoza, Fernando. "An Analysis of the Historical Development of Ideas About Motion and its Implications for Teaching". Physics Education. Vol. 40(2).

[17] "Biography of Al-Biruni". University of St. Andrews, Scotland.

[18] Nasr S.H., Razavi M.A.. "The islamic Intellectual Tradition in Persia" (1996). Routledge

[19] Pines, Shlomo (1986), *Studies in Arabic versions of Greek texts and in mediaeval science* **2**, Brill Publishers, p. 203, ISBN 965-223-626-8

[20] Franco, Abel B.. "Avempace, Projectile Motion, and Impetus Theory". *Journal of the History of Ideas*. Vol. 64(4): 543.

Chapter 8

Science in the medieval Islamic world

This article is about the history of science in the Islamic civilization between the 8th and 16th centuries. For information on science in the context of Islam, see Islam and science.

Science in the medieval Islamic world (also known, less accurately, as **Islamic science** or **Arabic science**) was the science developed and practiced in the medieval Islamic world during the Islamic Golden Age (8th century CE – c. 1258 CE, sometimes considered to have extended to the 15th or 16th century). During this time scholars translated Indian, Assyrian, Iranian and Greek knowledge into Arabic. These translations became a wellspring for scientific advances by scientists from Muslim-ruled areas during the Middle Ages.[*][1]

Scientists within the Muslim-ruled areas had diverse ethnic backgrounds they included Persians,[*][2][*][3][*][4][*][5] Arabs, Assyrians, Kurds[*][4] and Egyptians. They also came from diverse religious backgrounds. Most were Muslims,[*][6][*][7][*][8] but their ranks also included some Christians,[*][9] Jews[*][9][*][10] and irreligious.[*][11][*][12]

8.1 Science in the context of Islamic civilization

The term *Islam* refers either to the religion of Islam or to the Islamic civilization that formed around it.[*][13] Islamic civilization is composed of many faiths and cultures, although the proportion of Muslims among its population has increased over time.[*][14]

The religion of Islam was completed during the lifetime of the last Islamic prophet Muhammad. After his death in 632, Islam continued to expand under the leadership of its Muslim rulers, known as Caliphs. Struggles for leadership of the growing religious community began at this time, and continue today. The early periods of Islamic history after the death of Muhammad can be referred to as the Rashidun Caliphate. Then came the period of Umayyad

Caliphate.[*][15]

During the Umayyad Caliphate, the Islamic empire began to consolidate its territorial gains. Arabic became the language of administration. The Arabs became a ruling class assimilated into their new surroundings across the empire, rather than occupiers of conquered territories.[*][16]

8.1.1 The crystallization of Islamic thought and civilization

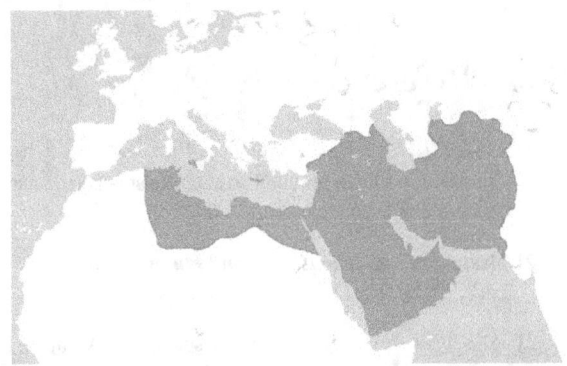

The Abbasid Caliphate at its greatest extent, c. 850

Through the Umayyad and, in particular, the succeeding Abbasid Caliphate's early phase, lies the period of Islamic history known as the Islamic Golden Age. This era can be identified as the years between 692 and 945,[*][15] and ended when the caliphate was marginalized by local Muslim rulers in Baghdad – its traditional seat of power. From 945 onward until the sacking of Baghdad by the Mongols in 1258, the Caliph continued on as a figurehead, with power devolving more to local amirs.[*][17]

During the Islamic Golden Age, stable political structures were established and trade flourished. The Chinese were undergoing a revolution in commerce, and the trade routes between the lands of Islam and China boomed both overland and along the coastal routes between the two civilizations.[*][17] Islamic civilization continued to be primarily based upon agriculture, but commerce began to play a more important role as the caliphate secured peace within the empire. The wars and cultural divisions that had separated peoples before the Arab conquests gradually gave way to a new civilization encompassing diverse ethnic and religious backgrounds. This new Islamic civilization used the Arabic language as transmitters of culture and Arabic increasingly became the language of commerce and government.[18]

Over time, the great religious and cultural works of the empire were translated into Arabic, the population increasingly understood Arabic, and they increasingly professed Islam as their religion. The cultural heritages of the area included strong Greek, Indic, Asyrian and Persian influences. The Greek intellectual traditions were recognized, translated and studied broadly. Through this process, the population of the lands of Islam gained access to all the important works of all the cultures of the empire, and a new common civilization formed in this area of the world, based on the religion of Islam. A new era of high culture and innovation ensued, where these diverse influences were recognized and given their respective places in the social consciousness.[19]

8.1.2 Domains of thought and culture in the High Caliphate

The pious scholars of Islam, men and women collectively known as the ulama, were the most influential element of society in the fields of Sharia law, speculative thought and theology. Their pronouncements defined the external practice of Islam, including prayer, as well as the details of the Islamic way of life. They held strong influence over government, and especially the laws of commerce. They were not rulers themselves, but rather keepers and upholders of the rule of law.[20]

Conversely, among the religious, there were inheritors of the more charismatic expressions of Christianity and Buddhism, in the Sufi orders. These Muslims had a more informal and varied approach to their religion. Islam also expressed itself in other, more esoteric forms that could have significant influence over public discourse during times of social unrest.[21]

Among the more worldly, adab – polite, worldly culture — permeated the lives of the professional, the courtly and genteel classes. Art, literature, poetry, music and even some aspects of religion were among the areas widely appreciated by those of a more refined taste among Muslim and non-Muslim alike. New trends and new topics flowed from the center of the Baghdad courts, to be adopted both quickly and widely across the lands of Islam.[21]

Apart from these other traditions stood *falsafa*; Greek phi-

Illustration of medieval Islamic scholars

losophy, inclusive of the sciences as well as the philosophy of the ancients. This science had been widely known across Mesopotamia and Iran since before the advent of Islam. These "sciences" were in many ways contrary to the teachings of Islam and the ways of the adab, but were nonetheless highly regarded in society. The ulama tolerated these outlooks and practices with reservation. Some *faylasufs* made a good living in the practices of astrology and medicine.[21]

8.2 Medieval Islamic science

8.2.1 Notable fields of inquiry

The roots of Islamic science drew primarily upon Arab, Persian, Indian and Greek learning. The extent of Islamic scientific achievement is not as yet fully understood, but it is extremely vast.[1]

These achievements encompass a wide range of subject areas; most notably[1]

- Mathematics

- Astronomy

- Medicine

Other notable areas, and specialized subjects, of scientific inquiry include

- Physics

- Alchemy and chemistry

- Cosmology

- Ophthalmology

- Geography and cartography

- Sociology

- Psychology

8.2.2   Notable scientists

In medieval Islam, the sciences, which included philosophy, were viewed holistically. The individual scientific disciplines were approached in terms of their relationships to each other and the whole, as if they were branches of a tree. In this regard, the most important scientists of Islamic civilization have been the polymaths, known as *hakim* or sages. Their role in the transmission of the sciences was central.[22]

The *hakim* was most often a poet and a writer, skilled in the practice of medicine as well as astronomy and mathematics. These multi-talented sages, the central figures in Islamic science, elaborated and personified the unity of the sciences. They orchestrated scientific development through their insights, and excelled in their explorations as well.[22]

8.2.3   Arabs

- al-Battani (850–922) was an astronomer who accurately determined the length of the solar year. He contributed to numeric tables, such as the Tables of Toledo, used by astronomers to predict the movements of the sun, moon and planets across the sky. Some of Battani's astronomic tables were later used by Copernicus. Battani also developed numeric tables which could be used to find the direction of Mecca from different locations. Knowing the direction of Mecca is important for Muslims, as this is the direction faced during prayer.[23]

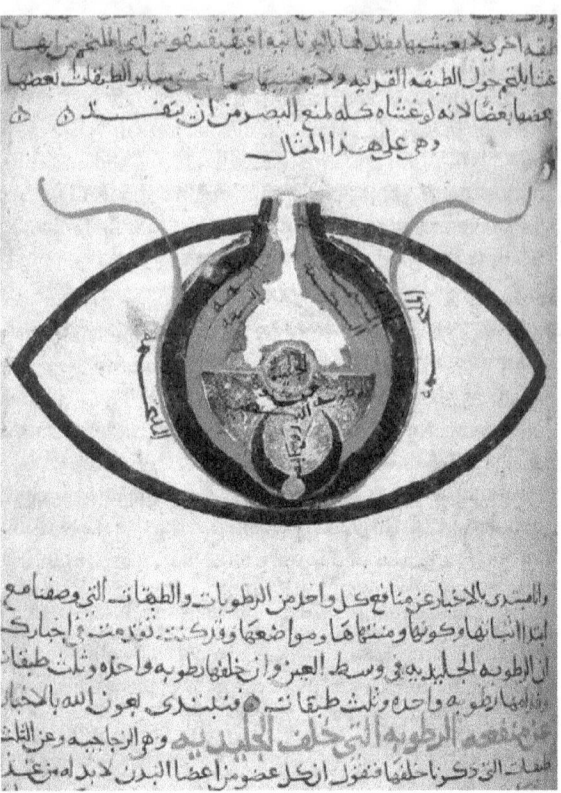

The eye according to Hunain ibn Ishaq. From a manuscript dated circa 1200.

- Ibn Ishaq al-Kindi (801–873) was a philosopher and polymath scientist heavily involved in the translation of Greek classics into Arabic. He worked to reconcile the conflicts between his Islamic faith and his affinity for reason; a conflict that would eventually lead to problems with his rulers. He criticized the basis of alchemy and astrology, and contributed to a wide range of scientific subjects in his writings. He worked on cryptography for the caliphate, and even wrote a piece on the subject of time, space and relative movement.[24]

- ibn al-Haytham (965–1040), also known as Alhazen, was an Arab scientist born in Basra, Iraq. Later, he moved to Egypt as an adult. Hasan Haytham worked in several fields, but is now known primarily for his achievements in astronomy and optics. He was an experimentalist who questioned the ancient Greek works of Ptolemy and Galen. At times, al-Haytham suggested Ptolomey's celestial model, and Galen's explanation of vision, had problems. The prevailing opinion of the time, Galen's opinion, was that vision involved emission of rays from the eye, an explanation al-Haytham cast doubt upon. He also studied the effects of light refraction, and suggested the mathematics of reflection and refraction needed to be consistent with the anatomy of the eye.*[25] He played an important role in the development of optics, experimental physics, theoretical physics, and the scientific method.

- ibn al-Nafis (1213–1288) was a physician who was born in Damascus and practiced medicine as head physician at the al-Mansuri hospital in Cairo. He wrote an influential book on medicine, believed to have replaced ibn-Sina's *Canon* in the Islamic world – if not Europe. He wrote important commentaries on Galen and ibn-Sina's works. One of these commentaries was discovered in 1924, and yielded a description of pulmonary transit, the circulation of blood from the right to left ventricles of the heart through the lungs.*[26]

8.2.4 Moors

- al-Zarqali (1028–1087) was an Andalusian artisan, skilled in working sheet metal, who became a famous maker of astronomical equipment, an astronomer, and a mathematician. He developed a new design for a highly accurate astrolabe which was used for centuries afterwards. He constructed a famous water clock that attracted much attention in Toledo for centuries. He discovered that the Sun's apogee moves slowly relative to the fixed stars, and obtained a very good estimate*[27] for its rate of change.*[28]

- Abbas ibn Firnas (810–887) was an Andalusian scientist, musician and inventor. He developed a clear glass used in drinking vessels, and lenses used for magnification and the improvement of vision. He had a room in his house where the sky was simulated, including the motion of planets, stars and weather complete with clouds, thunder and lightning. He is most well known for reportedly surviving an attempt at controlled flight.*[29]

- al-Zahrawi (936–1013) was an Andalusian surgeon who is known as the greatest surgeon of medieval Islam. His most important surviving work is referred to as al-Tasrif (Medical Knowledge). It is a 30 volume set discussing medical symptoms, treatments, and mostly pharmacology, but it is the last volume of the set which has attracted the most attention over time. This last volume is a surgical manual describing surgical instruments, supplies and procedures. Scholars studying this manual are discovering references to procedures previously believed to belong to more modern times.*[30]

- al-Idrisi (1100–1166) was a Moroccan traveler from Ceuta, cartographer and geographer famous for a map of the world he created for Roger, the Norman King of Sicily. al-Idrisi also wrote the Book of Roger, a geographic study of the peoples, climates, resources and industries of all the world known at that time. In it, he incidentally relates the tale of a Moroccan ship blown west in the Atlantic, and returning with tales of faraway lands.*[31]

8.2.5 Persians

A page from al-Khwārizmī's Algebra

- al-Khwarizmi (ca. 8th–9th centuries) was a Persian mathematician,*[32] geographer and astronomer. He is regarded as the greatest mathematician of Islamic civilization. He was instrumental in the adoption of the Indian numbering system, later known as Arabic numerals. He developed algebra, which also had Indian antecedents, by introducing methods of simplifying the equations. He used Euclidean geometry in his proofs.*[33]

- al-Razi (ca. 854–925/935) was a Persian born in Rey, Iran. He was a polymath who wrote on a variety of topics, but his most important works were in the field of medicine. He identified smallpox and measles, and recognized fever was part of the body's defenses. He wrote a 23-volume compendium of Chinese, Indian, Persian, Syriac and Greek medicine. al-Razi questioned some aspects of the classical Greek medical theory of how the four humors regulate life processes. He challenged Galen's work on several fronts, including the treatment of bloodletting. His trial of bloodletting showed it was effective; a result we now know to be erroneous.*[34]

- al-Farabi (ca. 870–950) was a Persian/Iranian (born in Farab, Iran) rationalist philosopher and mathematician who attempted to describe, geometrically, the repeating patterns popular in Islamic decorative motifs. His book on the subject is titled *Spiritual Crafts and Natural Secrets in the Details of Geometrical Figures.*[35]

- Avicenna (908–946) was a Persian physician, astronomer, physicist and mathematician from Bukhara, Uzbekistan. In addition to his master work, The Canon of Medicine, he also made important astronomical observations, and discussed a variety of topics including the different forms energy can take, and the properties of light. He contributed to the development of mathematical techniques such as Casting out nines.*[36]

- Omar Khayyam (1048–1131) was a Persian poet and mathematician who calculated the length of the year to within 5 decimal places. He found geometric solutions to all 13 forms of cubic equations. He developed some quadratic equations still in use. He is well known in the West for his poetry (rubaiyat).*[37]

- Nasir al-Din al-Tusi (1201–1274) was a Persian astronomer and mathematician whose life was overshadowed by the Mongol invasions of Genghis Khan and his grandson Helagu. al-Tusi wrote an important revision to Ptolemy's celestial model, among other works. When he became Helagu's astrologer, he was furnished with an impressive observatory and gained access to Chinese techniques and observations. He developed trigonometry to the point it became a separate field, and compiled the most accurate astronomical tables available up to that time.*[38]

- The Banu Musa brothers, Jafar-Muhammad, Ahmad and al-Hasan (ca. early 9th century) were three Persian sons of a colorful astronomer and astrologer. They

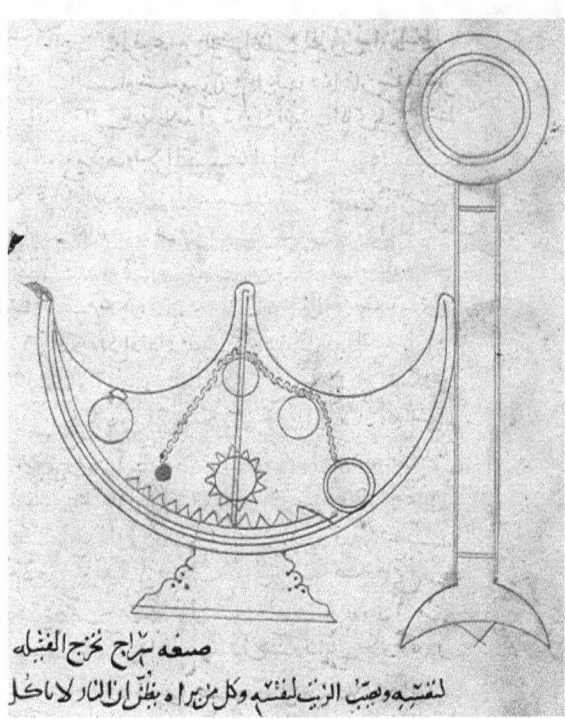

Drawing of Self trimming lamp in Ahmad ibn Mūsā ibn Shākir's treatise on mechanical devices. The manuscript was written in Arabic.

were scholars close to the court of caliph al-Maʾ mun, and contributed greatly to the translation of ancient works into Arabic. They elaborated the mathematics of cones and ellipses, and performed astronomic calculations. Most notably, they contributed to the field of automation with the creations of automated devices such as the ones described in their Book of Ingenious Devices.*[39]*[40]*[41]

- Jabir ibn Hayyan (ca. 8th – 9th centuries) was a Persian*[42] alchemist who used extensive experimentation and produced many works on science and alchemy which have survived to the present day. Jabir described the laboratory techniques and experimental methods of chemistry. He identified many substances including sulfuric and nitric acid. He described processes including sublimation, reduction and distillation. He utilized equipment such as the alembic and the retort. There is considerable uncertainty as to the actual provenance of many works that are ascribed to him.*[43]*[44]

- Jamshid al-Kashi (ca. 1380-1429) is credited with several theorems of trigonometry including the Law of Cosines, also known as Al-Kashi's Theorem. Furthermore he is often credited with the invention of decimal fractions, and a method like Horner's to calculate roots. He calculated π correctly to 17 significant figures.*[45]

- Ibn Sahl (ca. 940–1000) was a Persian physicist and optical engineer who is credited with discovering the law of refraction often referred to as Snell's law. He used the law to produce the first Aspheric lenses that focused light without geometric aberrations.*[46]*[47]

8.2.6 Assyrians

- Hunayn ibn Ishaq (809–873) was an Assyrian Nestorian Christian scholar, physician, and scientist. He was one of the most important translators of the ancient Greek works into Arabic. His translations interpreted, corrected and extended the ancient works. Some of his translations of medical works were used in Europe for centuries. He also wrote on medical subjects, particularly on the human eye. His book *Ten Treatises on the Eye* was influential in the West until the 17th century.*[48]

- Thabit ibn Qurra (835–901) was a Sabian translator and mathematician from Harran, in what is now Turkey. He is known for his translations of Greek mathematics and astronomy, but as was common, he also added his own work to the translations. He is known for having calculated the solution to a chessboard problem involving an exponential series.*[49]

8.3 The views of historians and scholars

8.3.1 On the impact of medieval Islamic science

There are several different views on Islamic science among historians of science.

- The traditionalist view, as exemplified by Bertrand Russell,*[50] holds that Islamic science, while admirable in many technical ways, lacked the intellectual energy required for innovation and was chiefly important as a preserver of ancient knowledge and transmitter to medieval Europe.

- The revisionist view, as exemplified by Abdus Salam,*[51] George Saliba*[52] and John M. Hobson*[53] holds that a Muslim scientific revolution occurred during the Middle Ages,*[54]

- Scholars such as Donald Routledge Hill and Ahmad Y Hassan express the view that Islam was the driving force behind the Muslim achievements,*[55]

- According to Dallal, science in medieval Islam was "practiced on a scale unprecedented in earlier human history or even contemporary human history".*[56]

- Toby E. Huff*[57]*[58] takes the view that, although Islamic science did produce a number of innovations, it did not lead to the Scientific Revolution.

- Will Durant,*[59] Fielding H. Garrison,*[60] Hossein Nasr and Bernard Lewis*[61] held that Muslim scientists helped in laying the foundations for an experimental science with their contributions to the scientific method and their empirical, experimental and quantitative approach to scientific inquiry.

8.3.2 On the historiography of medieval Islamic science

See also: Islam and science and Historiography of early Islam

The history of science in the Islamic world, like all history, is filled with questions of interpretation.

Historians of science generally consider that the study of science in the Islamic world, like all history, must be seen within the particular circumstances of time and place.

- A. I. Sabra opened a recent overview of Arabic science by noting, "I trust no one would wish to contest the proposition that all of history is local history ... and the history of science is no exception." *[62]

Some scholars avoid such local historical approaches and seek to identify essential relations between Islam and science that apply at all times and places.

- The Persian philosopher and historian of science, Seyyed Hossein Nasr saw a more positive connection in "an Islamic science that was spiritual and antisecular" which "point[ed] the way to a new 'Islamic science' that would avoid the dehumanizing and despiritualizing mistakes of Western science." *[63]*[64] Nasr identified a distinctly Muslim approach to science, flowing from Islamic monotheism and the related theological prohibition against portraying graven images. In science, this is reflected in a philosophical disinterest in describing individual material objects, their properties and characteristics and instead a concern with the ideal, the Platonic form, which exists in matter as an expression of the will of the Creator. Thus one can "see why mathematics was to make such a strong appeal to the Muslim: its abstract nature furnished the bridge that Muslims were seeking between multiplicity and unity." *[65]

Some historians of science, however, question the value of drawing boundaries that label the sciences, and the scientists who practice them, in specific cultural, civilizational, or linguistic terms.

- Some scholars consider the practice to be an example of "boosterism" and object that it "defines the achievements of scholars... in terms of their religion rather than their research." *[66]

- While others simply consider it futile. For example, Nasir al-Din Tusi (1201–1274), invented his mathematical theorem, the Tusi Couple, while he was director of Maragheh observatory. Tusi's patron and founder of the observatory was the non-Muslim Mongol conqueror of Baghdad, Hulagu Khan. The Tusi-couple "was first encountered in an Arabic text, written by a man who spoke Persian at home, and used that theorem, like many other astronomers who followed him and were all working in the "Arabic/Islamic" world, in order to reform classical Greek astronomy, and then have his theorem in turn be translated into Byzantine Greek towards the beginning of the 14th century, only to be used later by Copernicus and others in Latin texts of Renaissance Europe." *[67]

8.4   Role of Christians

Christians especially Nestorian contributed to the Arab Islamic Civilization during the Ummayads and the Abbasids by translating works of Greek philosophers to Syriac and afterwards to Arabic.*[68] They also excelled in philosophy, science (such as Hunayn ibn Ishaq, Qusta ibn Luqa, Masawaiyh, Patriarch Eutychius, Jabril ibn Bukhtishu etc) and theology (such as Tatian, Bar Daisan, Babai the Great, Nestorius, Toma bar Yacoub etc.) and the personal physicians of the Abbasid Caliphs were often Assyrian Christians such as the long serving Bukhtishu dynasty.*[69]*[70]

8.5   Role of Persians

As Ibn Khaldun, the fourteenth-century Arab historiographer and sociologist suggests, it is a remarkable fact that with few exceptions, most Muslim scholars in the intellectual sciences were *Ajam*s ("Persians"):

> Thus the founders of grammar were Sibawaih and after him, al-Farisi and Az-Zajjaj. All of them were of Persian descent ···they invented rules of (Arabic) grammar ···great jurists were Persians ···only the Persians engaged in the task of preserving knowledge and writing systematic scholarly works. Thus the truth of the statement of the prophet becomes apparent, 'If learning were suspended in the highest parts of heaven the Persians would attain it' ···The intellectual sciences were also the preserve of the Persians, left alone by the Arabs, who did not cultivate them ···as was the case with all crafts ···This situation continued in the cities as long as the Persians and Persian countries, Iraq, Khorasan and Transoxiana [=modern Central Asia], retained their sedentary culture.
> —Ibn Khaldun, *Muqaddimah*, Translated by Franz Rosenthal (III, pp. 311-15, 271-4 [Arabic]; R.N. Frye. p. 91.

8.6   See also

- Alchemy and chemistry in medieval Islam

- Astronomy in medieval Islam

- Hindu and Buddhist contribution to science in medieval Islam

- History of scientific method

  - Greek contributions to Islamic world

- Islam and science

- Islamic contributions to Medieval Europe

  - Latin translations of the 12th century

- Islamic economics in the world

- Islamic Golden Age

  - Early Islamic philosophy

  - Golden age of Jewish culture in Spain

  - Inventions in the Muslim world

  - Muslim Agricultural Revolution

- Islamic philosophy

  - Logic in Islamic philosophy

- Islamic studies
- List of Muslim scientists
    - Ibn Sina Academy of Medieval Medicine and Sciences
    - List of Arab scientists and scholars
    - List of Persian scientists and scholars
    - List of Turkish Philosophers and scientists
- Mathematics in medieval Islam
- Medicine in the medieval Islamic world
- Medieval Islamic astrology
- Ophthalmology in medieval Islam
- Physics in medieval Islam
- Psychology in medieval Islam
- Qur'an and science
- Scholasticism
- Timeline of Islamic science and technology
- Continuity thesis

8.7 Notes

[1] Robinson (editor), Francis (1996). *The Cambridge Illustrated History of the Islamic World.* Cambridge University Press. pp. 228–229.

[2] William Bayne Fisher, et al, The Cambridge History of Iran 4, Cambridge University Press, 1975, p. 396

[3] Shaikh M. Ghazanfar, Medieval Islamic economic thought: filling the "great gap" in European economics, Psychology Press, 2003 (p. 114-115)

[4] Ibn Khaldun, Franz Rosenthal, N. J. Dawood (1967), *The Muqaddimah: An Introduction to History*, p. 430, Princeton University Press, ISBN 0-691-01754-9.

[5] Joseph A. Schumpeter, Historian of Economics: Selected Papers from the History of Economics Society Conference, 1994, y Laurence S. Moss, Joseph Alois Schumpeter, History of Economics Society. Conference, Published by Routledge, 1996, ISBN 0-415-13353-X, p.64.

[6] Howard R. Turner (1997), *Science in Medieval Islam*, p. 270 (book cover, last page), University of Texas Press, ISBN 0-292-78149-0

[7] Hogendijk, Jan P. (January 1999), *Bibliography of Mathematics in Medieval Islamic Civilization*

[8] A. I. Sabra (1996). "Greek Science in Medieval Islam". In Ragep, F. J.; Ragep, Sally P.; Livesey, Steven John. *Tradition, Transmission, Transformation: Proceedings of Two Conferences on Pre-modern Science held at the University of Oklahoma.* Brill Publishers. p. 20. ISBN 90-04-09126-2.

[9] Bernard Lewis, The Jews of Islam, 1987, p.6

[10] Salah Zaimeche (2003), Introduction to Muslim Science.

[11] Hogendijk 1989

[12] Bernard Lewis, *What Went Wrong? Western Impact and Middle Eastern Response*

[13] Lewis, Brenard (1987). *The Jews of Islam.* Princeton University Press. pp. 5–6.

[14] Courbage, Youssef; Fargues, Phillipe (1995). *Christians and Jews under Islam.* London: I.B. Tauris Publishers. pp. ix–x. ISBN 1-86064-285-3.

[15] Marshall Hodgson, *The Venture of Islam; Conscience and History in a World Civilization Vol 1.* The University of Chicago, 1974, pg. 234.

[16] Marshall Hodgson, *The Venture of Islam Conscience and History in a World Civilization Vol 1.* The University of Chicago, 1974, pg. 230.

[17] Marshall Hodgson, *The Venture of Islam; Conscience and History in a World Civilization Vol 1.* The University of Chicago, 1974, pg. 233.

[18] Marshall Hodgson, *The Venture of Islam; Conscience and History in a World Civilization Vol 1.* The University of Chicago, 1974, pg. 235.

[19] Marshall Hodgson, *The Venture of Islam; Conscience and History in a World Civilization Vol 1.* The University of Chicago, 1974, pg. 236–238.

[20] Marshall Hodgson, *The Venture of Islam; Conscience and History in a World Civilization Vol 1.* The University of Chicago, 1974, pg. 238.

[21] Marshall Hodgson, *The Venture of Islam; Conscience and History in a World Civilization Vol 1.* The University of Chicago, 1974, pg. 238–239.

[22] Nasr, Seyyed Hossein (1968). *Science and Civilization in Islam.* Harvard University Press. p. 41.

[23] Masood (2009, pp.74, 148–50)

[24] Masood (2009, pp.49–52

[25] Masood (2009, pp.173–75)

[26] Masood (2009, pp.110–11)

[27] Linton (2004), p.97. Owing to the unreliability of the data al-Zarqali relied on for this estimate its remarkable accuracy was somewhat fortuitous.

[28] Masood, Ehsan (2009). *Science and Islam A History*. Icon Books Ltd. pp. 73–75.

[29] Masood (2009, pp.71–73)

[30] Masood (2009, pp.108–109)

[31] Masood (2009, pp.79–80)

[32] Toomer, Gerald (1990). "Al-Khwārizmī, Abu Ja'far Muḥammad ibn Mūsā". In Gillispie, Charles Coulston. Dictionary of Scientific Biography. 7. New York: Charles Scribner's Sons. ISBN 0-684-16962-2.

[33] Masood (2009, pp.139–45)

[34] Masood (2009, pp.74, 99–105)

[35] Masood (2009, pp.148–49)

[36] Masood (2009, pp.104–5)

[37] Masood (2009, pp.5, 104, 145–146)

[38] Masood (2009, pp.132–35)

[39] Masood (2009, pp.161–63)

[40] Lindberg, David (1978). *Science in the Middle Ages*. The University of Chicago Press. p. 23,56.

[41] Selin, Helaine, ed. (1997). *Encyclopaedia of the History of Science, Technology, and Medicine in Non-Western Cultures*. Kluwer Academic Publishers. pp. 151, 235, 375.

[42] Henry Corbin, "The Voyage and the Messenger: Iran and Philosophy", Translated by Joseph H. Rowe, North Atlantic Books, 1998. p.45:

[43] Masood (2009, pp.153–55)

[44] Lagerkvist, Urf (2005). *The Enigma of Ferment: from the Philosopher's Stone to the First Biochemical Nobel Prize*. World Scientific Publishing. p. 32.

[45] O'Connor, John J.; Robertson, Edmund F., "Ghiyath al-Din Jamshid Mas'ud al-Kashi", MacTutor History of Mathematics archive, University of St Andrews.

[46] K. B. Wolf, "Geometry and dynamics in refracting systems", European Journal of Physics 16, p. 14-20, 1995.

[47] R. Rashed, "A pioneer in anaclastics: Ibn Sahl on burning mirrors and lenses", Isis 81, p. 464–491, 1990

[48] Masood (2009, pp.47–48, 59, 96–97, 171–72)

[49] Masood (2009, pp.48–49)

[50] Bertrand Russell (1945), *History of Western Philosophy*, book 2, part 2, chapter X

[51] Abdus Salam, H. R. Dalafi, Mohamed Hassan (1994). *Renaissance of Sciences in Islamic Countries*, p. 162. World Scientific, ISBN 9971-5-0713-7.

[52] (Saliba 1994, pp. 245, 250, 256–257)

[53] (Hobson 2004, p. 178)

[54] Abid Ullah Jan (2006), *After Fascism: Muslims and the struggle for self-determination*, "Islam, the West, and the Question of Dominance", Pragmatic Publishings, ISBN 978-0-9733687-5-8.

[55] Ahmad Y Hassan and Donald Routledge Hill (1986), *Islamic Technology: An Illustrated History*, p. 282, Cambridge University Press

[56] Dallal, Ahmad (2010). *Islam, science, and the challenge of history*. Yale University Press. p. 12. ISBN 9780300159110.

[57] (Huff 2003)

[58] Saliba, George (Autumn 1999). "Seeking the Origins of Modern Science? Review of Toby E. Huff, The Rise of Early Modern Science: Islam, China and the West". *Bulletin of the Royal Institute for Inter-Faith Studies* **1** (2). Retrieved 2008-04-10.

[59] Will Durant (1980). *The Age of Faith (The Story of Civilization, Volume 4)*, p. 162–186. Simon & Schuster. ISBN 0-671-01200-2.

[60] Fielding H. Garrison, *An Introduction to the History of Medicine: with Medical Chronology, Suggestions for Study and Biblographic Data*, p. 86

[61] Lewis, Bernard (2001). *What Went Wrong? : Western Impact and Middle Eastern Response*. Oxford University Press. p. 79. ISBN 0-19-514420-1.

[62] Sabra (2000) p. 216.

[63] F. Jamil Ragep, "Freeing Astronomy from Philosophy: An Aspect of Islamic Influence on Science," *Osiris*, topical issue on *Science in Theistic Contexts: Cognitive Dimensions*, n.s. 16(2001): 49–50, note 3

[64] Nasr, Seyyed Hossein (1968). "The Principles of Islam". *Science and Civilization in Islam*. Harvard University Press. ISBN 0-946621-11-X. Retrieved 2008-02-03.

[65] Seyyed Hossein Nasr, *Science and Civilization in Islam*.

[66] Aaen-Stockdale, C.R. (2008). "Ibn al-Haytham and psychophysics". *Perception* **37** (4): 636–638. doi:10.1068/p5940. PMID 18546671.

[67] George Saliba (1999). Whose Science is Arabic Science in Renaissance Europe?

[68] Hill, Donald. *Islamic Science and Engineering*. 1993. Edinburgh Univ. Press. ISBN 0-7486-0455-3, p.4

[69] Rémi Brague, Assyrians contributions to the Islamic civilization

[70] Britannica, Nestorian

8.8 References

- Campbell, Donald (2001). *Arabian Medicine and Its Influence on the Middle Ages*. Routledge. (Reprint of the London, 1926 edition). ISBN 0-415-23188-4.

- d'Alverny, Marie-Thérèse. "Translations and Translators", in Robert L. Benson and Giles Constable, eds., *Renaissance and Renewal in the Twelfth Century*, p. 421–462. Cambridge: Harvard Univ. Pr., 1982.

- Hobson, John M. (2004). *The Eastern Origins of Western Civilisation*. Cambridge University Press. ISBN 0-521-54724-5.

- Hudson, A. (2003). *Equity and Trusts* (3rd ed.). London: Cavendish Publishing. ISBN 1-85941-729-9.

- Huff, Toby E. (2003). "The Rise of Early Modern Science: Islam, China, and the West". Cambridge University Press. ISBN 0-521-52994-8.

- Joseph, George G. (2000). *The Crest of the Peacock*. Princeton University Press. ISBN 0-691-00659-8.

- Katz, Victor J. (1998). *A History of Mathematics: An Introduction*. Addison Wesley. ISBN 0-321-01618-1.

- Levere, Trevor Harvey (2001). *Transforming Matter: A History of Chemistry from Alchemy to the Buckyball*. Johns Hopkins University Press. ISBN 0-8018-6610-3.

- Linton, Christopher M. (2004). *From Eudoxus to Einstein—A History of Mathematical Astronomy*. Cambridge: Cambridge University Press. ISBN 978-0-521-82750-8

- Masood, Ehsan (2009). *Science and Islam A History*. Icon Books Ltd.

- Morelon, Régis; Rashed, Roshdi (1996). *Encyclopedia of the History of Arabic Science* **3**. Routledge. ISBN 0-415-12410-7.

- Phillips, William D.; Carla Rahn Phillips; Jr. Phillips (1992). *The Worlds of Christopher Columbus*. Cambridge University Press. ISBN 0-521-44652-X.

- Sabra, A. I. (2000) "Situating Arab Science: Locality versus Essence," *Isis*, 87(1996):654–70; reprinted in Michael H. Shank, ed., The Scientific Enterprise in Antiquity and the Middle Ages," (Chicago: Univ. of Chicago Pr.), pp. 215–231.

- Saliba, George (1994). *A History of Arabic Astronomy: Planetary Theories During the Golden Age of Islam*. New York University Press. ISBN 0-8147-8023-7.

- Turner, Howard R. (1997). *Science in Medieval Islam: An Illustrated Introduction*. University of Texas Press. ISBN 0-292-78149-0.

8.9 Further reading

- Daffa, Ali Abdullah al-; Stroyls, J.J. (1984). *Studies in the exact sciences in medieval Islam*. New York: Wiley. ISBN 0-471-90320-5.

- Nader El-Bizri, 'A Philosophical Perspective on Alhazen's Optics', *Arabic Sciences and Philosophy* (Cambridge University Press), Vol. 15 (2005), pp. 189–218.

- Nader El-Bizri, 'In Defence of the Sovereignty of Philosophy: al-Baghdadi's Critique of Ibn al-Haytham's Geometrisation of Place', *Arabic Sciences and Philosophy* (Cambridge University Press), Vol. 17 (2007), pp. 57–80.

- Hogendijk, Jan P.; Abdelhamid I. Sabra (2003). *The Enterprise of Science in Islam: New Perspectives*. MIT Press. ISBN 0-262-19482-1. Reviewed by Robert G. Morrison at

- Hogendijk, Jan P.; Berggren, J. L. (1989). "*Episodes in the Mathematics of Medieval Islam* by J. Lennart Berggren". *Journal of the American Oriental Society* **109** (4): 697–698. doi:10.2307/604119. JSTOR 604119.)

- Hill, Donald Routledge, *Islamic Science And Engineering*, Edinburgh University Press (1993), ISBN 0-7486-0455-3

- Huff, Toby E. (1993, 2nd edition 2003), *The Rise of Early Modern Science: Islam, China and the West*. New York: Cambridge University Press. ISBN 0-521-52994-8. Reviewed by George Saliba at Seeking the Origins of Modern Science?

- Huff, Toby E. (2000). "Science and Metaphysics in the Three Religions of the Books" (PDF). *Intellectual Discourse* **8** (2): 173–198.

- Kennedy, Edward S. (1970). "The Arabic Heritage in the Exact Sciences". *Al-Abhath* **23**: 327–344.

- Kennedy, Edward S. (1983). *Studies in the Islamic Exact Sciences*. Syracuse University Press. ISBN 0-8156-6067-7.

- Morelon, Régis; Rashed, Roshdi (1996). *Encyclopedia of the History of Arabic Science* **2–3**. Routledge. ISBN 0-415-02063-8.

- Saliba, George (2007). *Islamic Science and the Making of the European Renaissance*. The MIT Press. ISBN 0-262-19557-7.

- Nasr, Seyyed Hossein (1976). *Islamic Science: An Illustrated Study*. Kazi Publications. ISBN 1-56744-312-5.

- Nasr, Seyyed Hossein (2003). *Science & Civilization in Islam* (2nd ed.). Islamic Texts Society. ISBN 1-903682-40-1.

- Suter, Heinrich (1900). *Die Mathematiker und Astronomen der Araber und ihre Werke*. Abhandlungen zur Geschichte der Mathematischen Wissenschaften Mit Einschluss Ihrer Anwendungen, X Heft. Leipzig.

Popular

- Deen, S M (2007). *Science Under Islam: Rise, Decline, Revival*. LULU. ISBN 978-1-84799-942-9.

Television

- BBC (2010). *Science and Islam*.

8.10   External links

Academic institutes

- Commission on the History of Science and Technology in Islamic Societies at University of Barcelona

Other

- "How Greek Science Passed to the Arabs" by De Lacy O'Leary

- Saliba, George. "Whose Science is Arabic Science in Renaissance Europe?".

- Habibi, Golareh. Review article, *Science Creative Quarterly*.

- Richard Covington, *Rediscovering Arabic Science*, 2007, Saudi Aramco World

Chapter 9

History of classical mechanics

This article deals with the **history of classical mechanics**.

9.1 Antiquity

Main article: Aristotelian physics

The ancient Greek philosophers, Aristotle in particular, were among the first to propose that abstract principles govern nature. Aristotle argued, in *On the Heavens*, that terrestrial bodies rise or fall to their "natural place" and mistakenly claimed that an object twice as heavy as some other would fall to the ground from the same height in half the time. Aristotle believed in logic and observation but it would be more than eighteen hundred years before Francis Bacon would first develop the scientific method of experimentation, which he called a *vexation of nature*.[1]

Aristotle saw a distinction between "natural motion" and "forced motion", and he believed that in a hypothetical vacuum, there would be no reason for a body to move naturally toward one point rather than any other, and so he concluded a body in a vacuum must either stay at rest or else move indefinitely fast. In this way, Aristotle was the first to approach something similar to the law of inertia. However, he believed a vacuum would be impossible because the surrounding air would rush in to fill it immediately. He also believed that an object would stop moving in an unnatural direction once the applied forces were removed. Later Aristotelians developed an elaborate explanation for why an arrow continues to fly through the air after it has left the bow, proposing that an arrow creates a vacuum in its wake, into which air rushes, pushing it from behind. Aristotle's beliefs were influenced by Plato's teachings on the perfection of the circular uniform motions of the heavens. As a result, he conceived of a natural order in which the motions of the heavens were necessarily perfect, in contrast to the terrestrial world of changing elements, where individuals come to be and pass away.

Galileo would later observe "the resistance of the air exhibits itself in two ways: first by offering greater impedance to less dense than to very dense bodies, and secondly by offering greater resistance to a body in rapid motion than to the same body in slow motion".[2]

9.2 Medieval thought

French priest Jean Buridan developed the Theory of impetus. Albert, Bishop of Halberstadt, developed the theory further.

9.3 Modern age —formation of classical mechanics

It wasn't until Galileo Galilei's development of the telescope and his observations that it became clear that the heavens were not made from a perfect, unchanging substance. Adopting Copernicus's heliocentric hypothesis, Galileo believed the Earth was the same as other planets. Galileo may have performed the famous experiment of dropping two cannonballs from the tower of Pisa. (The theory and the practice showed that they both hit the ground at the same time.) Though the reality of this experiment is disputed, he did carry out quantitative experiments by rolling balls on an inclined plane; his correct theory of accelerated motion was apparently derived from the results of the experiments. Galileo also found that a body dropped vertically hits the ground at the same time as a body projected horizontally, so an Earth rotating uniformly will still have objects falling to the ground under gravity. More significantly, it asserted that uniform motion is indistinguishable from rest, and so forms the basics of the theory of relativity.

Sir Isaac Newton was the first to unify the three laws of motion (the law of inertia, his second law mentioned above, and the law of action and reaction), and to prove that these laws govern both earthly and celestial objects. Newton and most of his contemporaries, with the notable exception of

Christiaan Huygens, hoped that classical mechanics would be able to explain all entities, including (in the form of geometric optics) light. Newton's own explanation of Newton's rings avoided wave principles and supposed that the light particles were altered or excited by the glass and resonated.

Newton also developed the calculus which is necessary to perform the mathematical calculations involved in classical mechanics. However it was Gottfried Leibniz who, independently of Newton, developed a calculus with the notation of the derivative and integral which are used to this day. Classical mechanics retains Newton's dot notation for time derivatives.

Leonhard Euler extended Newton's laws of motion from particles to rigid bodies with two additional laws.

After Newton, re-formulations progressively allowed solutions to a far greater number of problems. The first was constructed in 1788 by Joseph Louis Lagrange, an Italian-French mathematician. In Lagrangian mechanics the solution uses the path of least action and follows the calculus of variations. William Rowan Hamilton re-formulated Lagrangian mechanics in 1833. The advantage of Hamiltonian mechanics was that its framework allowed a more in-depth look at the underlying principles. Most of the framework of Hamiltonian mechanics can be seen in quantum mechanics however the exact meanings of the terms differ due to quantum effects.

Although classical mechanics is largely compatible with other "classical physics" theories such as classical electrodynamics and thermodynamics, some difficulties were discovered in the late 19th century that could only be resolved by more modern physics. When combined with classical thermodynamics, classical mechanics leads to the Gibbs paradox in which entropy is not a well-defined quantity. As experiments reached the atomic level, classical mechanics failed to explain, even approximately, such basic things as the energy levels and sizes of atoms. The effort at resolving these problems led to the development of quantum mechanics. Similarly, the different behaviour of classical electromagnetism and classical mechanics under velocity transformations led to the theory of relativity.

9.4 Present

By the end of the 20th century, classical mechanics in physics was no longer an independent theory. Along with classical electromagnetism, it has become imbedded in relativistic quantum mechanics or quantum field theory[*]. It defines the non-relativistic, non-quantum mechanical limit for massive particles.

Classical mechanics has also been a source of inspiration for mathematicians. The realization that the phase space in classical mechanics admits a natural description as a symplectic manifold (indeed a cotangent bundle in most cases of physical interest), and symplectic topology, which can be thought of as the study of global issues of Hamiltonian mechanics, has been a fertile area of mathematics research since the 1980s.

9.5 Notes

[1] Peter Pesic (March 1999). "Wrestling with Proteus: Francis Bacon and the "Torture"of Nature". *Isis* (The University of Chicago Press on behalf of The History of Science Society) **90** (1): 81–94. doi:10.1086/384242. JSTOR 237475.

[2] Galileo Galilei, *Dialogues Concerning Two New Sciences by Galileo Galilei*. Translated from the Italian and Latin into English by Henry Crew and Alfonso de Salvio. With an Introduction by Antonio Favaro (New York: Macmillan, 1914). Chapter: The Motion of Projectiles

9.6 References

- René Dugas *A History of Mechanics* Dover, (1988) ISBN 0-486-65632-2

9.7 See also

- Mechanics

- Classical mechanics

- Timeline of classical mechanics

Chapter 10

Philosophiæ Naturalis Principia Mathematica

For Russell's 1910 book on mathematical logic, see Principia Mathematica.

Philosophiæ Naturalis Principia Mathematica (Latin for "Mathematical Principles of Natural Philosophy"),[1] often referred to as simply the ***Principia***, is a work in three books by Sir Isaac Newton, in Latin, first published 5 July 1687.[2][3] After annotating and correcting his personal copy of the first edition,[4] Newton also published two further editions, in 1713 and 1726.[5] The *Principia* states Newton's laws of motion, forming the foundation of classical mechanics, also Newton's law of universal gravitation, and a derivation of Kepler's laws of planetary motion (which Kepler first obtained empirically). The *Principia* is "justly regarded as one of the most important works in the history of science".[6]

The French mathematical physicist Alexis Clairaut assessed it in 1747: "The famous book of *mathematical Principles of natural Philosophy* marked the epoch of a great revolution in physics. The method followed by its illustrious author Sir Newton ... spread the light of mathematics on a science which up to then had remained in the darkness of conjectures and hypotheses."[7] A more recent assessment has been that while acceptance of Newton's theories was not immediate, by the end of a century after publication in 1687, "no one could deny that" (out of the *Principia*) "a science had emerged that, at least in certain respects, so far exceeded anything that had ever gone before that it stood alone as the ultimate exemplar of science generally."[8]

In formulating his physical theories, Newton developed and used mathematical methods now included in the field of calculus. But the language of calculus as we know it was largely absent from the *Principia*; Newton gave many of his proofs in a geometric form of infinitesimal calculus, based on limits of ratios of vanishing small geometric quantities.[9] In a revised conclusion to the *Principia* (see *General Scholium*), Newton used his expression that

became famous, *Hypotheses non fingo* ("I contrive no hypotheses".[10]).

10.1 Contents

10.1.1 Expressed aim and topics covered

Sir Isaac Newton (1643–1727) author of the Principia

In the preface of the *Principia*, Newton wrote[11]

> [...] Rational Mechanics will be the science of motions resulting from any forces whatsoever,

and of the forces required to produce any motions, accurately proposed and demonstrated [...] And therefore we offer this work as mathematical principles of philosophy. For all the difficulty of philosophy seems to consist in this —from the phenomena of motions to investigate the forces of Nature, and then from these forces to demonstrate the other phenomena [...]

The *Principia* deals primarily with massive bodies in motion, initially under a variety of conditions and hypothetical laws of force in both non-resisting and resisting media, thus offering criteria to decide, by observations, which laws of force are operating in phenomena that may be observed. It attempts to cover hypothetical or possible motions both of celestial bodies and of terrestrial projectiles. It explores difficult problems of motions perturbed by multiple attractive forces. Its third and final book deals with the interpretation of observations about the movements of planets and their satellites. It shows how astronomical observations prove the inverse square law of gravitation (to an accuracy that was high by the standards of Newton's time); offers estimates of relative masses for the known giant planets and for the Earth and the Sun; defines the very slow motion of the Sun relative to the solar-system barycenter; shows how the theory of gravity can account for irregularities in the motion of the Moon; identifies the oblateness of the figure of the Earth; accounts approximately for marine tides including phenomena of spring and neap tides by the perturbing (and varying) gravitational attractions of the Sun and Moon on the Earth's waters; explains the precession of the equinoxes as an effect of the gravitational attraction of the Moon on the Earth's equatorial bulge; and gives theoretical basis for numerous phenomena about comets and their elongated, near-parabolic orbits.

The opening sections of the *Principia* contain, in revised and extended form, nearly[12] all of the content of Newton's 1684 tract *De motu corporum in gyrum*.

The *Principia* begin with "Definitions"[13] and "Axioms or Laws of Motion",[14] and continues in three books:

10.1.2　Book 1, De motu corporum

Book 1, subtitled *De motu corporum* (*On the motion of bodies*) concerns motion in the absence of any resisting medium. It opens with a mathematical exposition of "the method of first and last ratios",[15] a geometrical form of infinitesimal calculus.[9]

The second section establishes relationships between centripetal forces and the law of areas now known as Kepler's second law (Propositions 1–3),[16] and relates

Newton's proof of Kepler's second law, as described in the book. If an instantaneous centripetal force (red arrow) is considered on the planet during its orbit, the area of the triangles defined by the path of the planet will be the same. This is true for any fixed time interval. When the interval tends to zero, the force can be considered continuous. (Click image for a detailed description).

circular velocity and radius of path-curvature to radial force[17] (Proposition 4), and relationships between centripetal forces varying as the inverse-square of the distance to the center and orbits of conic-section form (Propositions 5–10).

Propositions 11–31[18] establish properties of motion in paths of eccentric conic-section form including ellipses, and their relation with inverse-square central forces directed to a focus, and include Newton's theorem about ovals (lemma 28).

Propositions 43–45[19] are demonstration that in an eccentric orbit under centripetal force where the apse may move, a steady non-moving orientation of the line of apses is an indicator of an inverse-square law of force.

Book 1 contains some proofs with little connection to real-world dynamics. But there are also sections with far-reaching application to the solar system and universe:

Propositions 57–69[20] deal with the "motion of bodies drawn to one another by centripetal forces". This section is of primary interest for its application to the solar system, and includes Proposition 66[21] along with its 22 corollaries:[22] here Newton took the first steps in the definition and study of the problem of the movements of three massive bodies subject to their mutually perturbing gravitational attractions, a problem which later gained name and fame (among other reasons, for its great difficulty) as the three-body problem.

Propositions 70–84[23] deal with the attractive forces of spherical bodies. The section contains Newton's proof that a massive spherically symmetrical body attracts other bodies outside itself as if all its mass were concentrated at its centre. This fundamental result, called the Shell theorem, enables the inverse square law of gravitation to be applied to the real solar system to a very close degree of approximation.

10.1.3 Book 2

Part of the contents originally planned for the first book was divided out into a second book, which largely concerns motion through resisting mediums. Just as Newton examined consequences of different conceivable laws of attraction in Book 1, here he examines different conceivable laws of resistance; thus Section 1 discusses resistance in direct proportion to velocity, and Section 2 goes on to examine the implications of resistance in proportion to the square of velocity. Book 2 also discusses (in Section 5) hydrostatics and the properties of compressible fluids. The effects of air resistance on pendulums are studied in Section 6, along with Newton's account of experiments that he carried out, to try to find out some characteristics of air resistance in reality by observing the motions of pendulums under different conditions. Newton compares the resistance offered by a medium against motions of bodies of different shape, attempts to derive the speed of sound, and gives accounts of experimental tests of the result.

Less of Book 2 has stood the test of time than of Books 1 and 3, and it has been said that Book 2 was largely written on purpose to refute a theory of Descartes which had some wide acceptance before Newton's work (and for some time after). According to this Cartesian theory of vortices, planetary motions were produced by the whirling of fluid vortices that filled interplanetary space and carried the planets along with them.[24] Newton wrote at the end of Book 2[25] his conclusion that the hypothesis of vortices was completely at odds with the astronomical phenomena, and served not so much to explain as to confuse them.

10.1.4 Book 3, De mundi systemate

Book 3, subtitled *De mundi systemate* (*On the system of the world*), is an exposition of many consequences of universal gravitation, especially its consequences for astronomy. It builds upon the propositions of the previous books, and applies them with further specificity than in Book 1 to the motions observed in the solar system. Here (introduced by Proposition 22,[26] and continuing in Propositions 25–35[27]) are developed several of the features and irregularities of the orbital motion of the Moon, especially the variation. Newton lists the astronomical observations on which he relies,[28] and establishes in a stepwise manner that the inverse square law of mutual gravitation applies to solar system bodies, starting with the satellites of Jupiter[29] and going on by stages to show that the law is of universal application.[30] He also gives starting at Lemma 4[31] and Proposition 40[32]) the theory of the motions

of comets, for which much data came from John Flamsteed and Edmond Halley, and accounts for the tides,[33] attempting quantitative estimates of the contributions of the Sun[34] and Moon[35] to the tidal motions; and offers the first theory of the precession of the equinoxes.[36] Book 3 also considers the harmonic oscillator in three dimensions, and motion in arbitrary force laws.

In Book 3 Newton also made clear his heliocentric view of the solar system, modified in a somewhat modern way, since already in the mid-1680s he recognised the "deviation of the Sun" from the centre of gravity of the solar system.[37] For Newton, "the common centre of gravity of the Earth, the Sun and all the Planets is to be esteem'd the Centre of the World",[38] and that this centre "either is at rest, or moves uniformly forward in a right line".[39] Newton rejected the second alternative after adopting the position that "the centre of the system of the world is immoveable", which "is acknowledg'd by all, while some contend that the Earth, others, that the Sun is fix'd in that centre".[39] Newton estimated the mass ratios Sun:Jupiter and Sun:Saturn,[40] and pointed out that these put the centre of the Sun usually a little way off the common center of gravity, but only a little, the distance at most "would scarcely amount to one diameter of the Sun".[41]

10.1.5 Commentary on the Principia

The sequence of definitions used in setting up dynamics in the *Principia* is recognisable in many textbooks today. Newton first set out the definition of mass[6]

> The quantity of matter is that which arises conjointly from its density and magnitude. A body twice as dense in double the space is quadruple in quantity. This quantity I designate by the name of body or of mass.

This was then used to define the "quantity of motion" (today called momentum), and the principle of inertia in which mass replaces the previous Cartesian notion of *intrinsic force*. This then set the stage for the introduction of forces through the change in momentum of a body. Curiously, for today's readers, the exposition looks dimensionally incorrect, since Newton does not introduce the dimension of time in rates of changes of quantities.

He defined space and time "not as they are well known to all". Instead, he defined "true" time and space as "absolute"[42] and explained:

> Only I must observe, that the vulgar conceive those quantities under no other notions but from

the relation they bear to perceptible objects. And it will be convenient to distinguish them into absolute and relative, true and apparent, mathematical and common. [...] instead of absolute places and motions, we use relative ones; and that without any inconvenience in common affairs; but in philosophical discussions, we ought to step back from our senses, and consider things themselves, distinct from what are only perceptible measures of them.

To some modern readers it can appear that some dynamical quantities recognised today were used in the *Principia* but not named. The mathematical aspects of the first two books were so clearly consistent that they were easily accepted; for example, Locke asked Huygens whether he could trust the mathematical proofs, and was assured about their correctness.

However, the concept of an attractive force acting at a distance received a cooler response. In his notes, Newton wrote that the inverse square law arose naturally due to the structure of matter. However, he retracted this sentence in the published version, where he stated that the motion of planets is consistent with an inverse square law, but refused to speculate on the origin of the law. Huygens and Leibniz noted that the law was incompatible with the notion of the aether. From a Cartesian point of view, therefore, this was a faulty theory. Newton's defence has been adopted since by many famous physicists—he pointed out that the mathematical form of the theory had to be correct since it explained the data, and he refused to speculate further on the basic nature of gravity. The sheer number of phenomena that could be organised by the theory was so impressive that younger "philosophers" soon adopted the methods and language of the *Principia*.

10.1.6   Rules of Reasoning in Philosophy

Perhaps to reduce the risk of public misunderstanding, Newton included at the beginning of Book 3 (in the second (1713) and third (1726) editions) a section entitled "Rules of Reasoning in Philosophy." In the four rules, as they came finally to stand in the 1726 edition, Newton effectively offers a methodology for handling unknown phenomena in nature and reaching towards explanations for them. The four Rules of the 1726 edition run as follows (omitting some explanatory comments that follow each):

Rule 1: *We are to admit no more causes of natural things than such as are both true and sufficient to explain their appearances.*

Rule 2: *Therefore to the same natural effects we must, as far*

as possible, assign the same causes.

Rule 3: *The qualities of bodies, which admit neither intensification nor remission of degrees, and which are found to belong to all bodies within the reach of our experiments, are to be esteemed the universal qualities of all bodies whatsoever.*

Rule 4: *In experimental philosophy we are to look upon propositions inferred by general induction from phenomena as accurately or very nearly true, not withstanding any contrary hypothesis that may be imagined, till such time as other phenomena occur, by which they may either be made more accurate, or liable to exceptions.*

This section of Rules for philosophy is followed by a listing of 'Phenomena', in which are listed a number of mainly astronomical observations, that Newton used as the basis for inferences later on, as if adopting a consensus set of facts from the astronomers of his time.

Both the 'Rules' and the 'Phenomena' evolved from one edition of the *Principia* to the next. Rule 4 made its appearance in the third (1726) edition; Rules 1–3 were present as 'Rules' in the second (1713) edition, and predecessors of them were also present in the first edition of 1687, but there they had a different heading: they were not given as 'Rules', but rather in the first (1687) edition the predecessors of the three later 'Rules', and of most of the later 'Phenomena', were all lumped together under a single heading 'Hypotheses' (in which the third item was the predecessor of a heavy revision that gave the later Rule 3).

From this textual evolution, it appears that Newton wanted by the later headings 'Rules' and 'Phenomena' to clarify for his readers his view of the roles to be played by these various statements.

In the third (1726) edition of the *Principia*, Newton explains each rule in an alternative way and/or gives an example to back up what the rule is claiming. The first rule is explained as a philosophers' principle of economy. The second rule states that if one cause is assigned to a natural effect, then the same cause so far as possible must be assigned to natural effects of the same kind: for example respiration in humans and in animals, fires in the home and in the Sun, or the reflection of light whether it occurs terrestrially or from the planets. An extensive explanation is given of the third rule, concerning the qualities of bodies, and Newton discusses here the generalisation of observational results, with a caution against making up fancies contrary to experiments, and use of the rules to illustrate the observation of gravity and space.

Isaac Newton's statement of the four rules revolutionised the investigation of phenomena. With these rules, Newton could in principle begin to address all of the world's present unsolved mysteries. He was able to use his new analytical method to replace that of Aristotle, and he was able to use his method to tweak and update Galileo's experimental method. The re-creation of Galileo's method has never been significantly changed and in its substance, scientists use it today.

10.1.7 General Scholium

Main article: General Scholium

The *General Scholium* is a concluding essay added to the second edition, 1713 (and amended in the third edition, 1726).[43] It is not to be confused with the *General Scholium* at the end of Book 2, Section 6, which discusses his pendulum experiments and resistance due to air, water, and other fluids.

Here Newton used what became his famous expression **Hypotheses non fingo**, "I contrive no hypotheses",[10] in response to criticisms of the first edition of the *Principia*. (*'Fingo'* is sometimes nowadays translated 'feign' rather than the traditional 'frame'.) Newton's gravitational attraction, an invisible force able to act over vast distances, had led to criticism that he had introduced "occult agencies" into science.[44] Newton firmly rejected such criticisms and wrote that it was enough that the phenomena implied gravitational attraction, as they did; but the phenomena did not so far indicate the cause of this gravity, and it was both unnecessary and improper to frame hypotheses of things not implied by the phenomena: such hypotheses "have no place in experimental philosophy", in contrast to the proper way in which "particular propositions are inferr'd from the phenomena and afterwards rendered general by induction".[45]

Newton also underlined his criticism of the vortex theory of planetary motions, of Descartes, pointing to its incompatibility with the highly eccentric orbits of comets, which carry them "through all parts of the heavens indifferently".

Newton also gave theological argument. From the system of the world, he inferred the existence of a Lord God, along lines similar to what is sometimes called the argument from intelligent or purposive design. It has been suggested that Newton gave "an oblique argument for a unitarian conception of God and an implicit attack on the doctrine of the Trinity",[46][47] but the General Scholium appears to say nothing specifically about these matters.

10.2 Writing and publication

See also: Writing of Principia Mathematica

10.2.1 Halley and Newton's initial stimulus

In January 1684, Halley, Wren and Hooke had a conversation in which Hooke claimed to not only have derived the inverse-square law, but also all the laws of planetary motion. Wren was unconvinced, Hooke did not produce the claimed derivation although the others gave him time to do it, and Halley, who could derive the inverse-square law for the restricted circular case (by substituting Kepler's relation into Huygens' formula for the centrifugal force) but failed to derive the relation generally, resolved to ask Newton.[48]

Halley's visits to Newton in 1684 thus resulted from Halley's debates about planetary motion with Wren and Hooke, and they seem to have provided Newton with the incentive and spur to develop and write what became *Philosophiae Naturalis Principia Mathematica* (*Mathematical Principles of Natural Philosophy*). Halley was at that time a Fellow and Council member of the Royal Society in London, (positions that in 1686 he resigned to become the Society's paid Clerk).[49] Halley's visit to Newton in Cambridge in 1684 probably occurred in August.[50] When Halley asked Newton's opinion on the problem of planetary motions discussed earlier that year between Halley, Hooke and Wren,[51] Newton surprised Halley by saying that he had already made the derivations some time ago; but that he could not find the papers. (Matching accounts of this meeting come from Halley and Abraham De Moivre to whom Newton confided.) Halley then had to wait for Newton to 'find' the results, but in November 1684 Newton sent Halley an amplified version of whatever previous work Newton had done on the subject. This took the form of a 9-page manuscript, *De motu corporum in gyrum* (*Of the motion of bodies in an orbit*): the title is shown on some surviving copies, although the (lost) original may have been without title.

Newton's tract *De motu corporum in gyrum*, which he sent to Halley in late 1684, derived what are now known as the three laws of Kepler, assuming an inverse square law of force, and generalised the result to conic sections. It also extended the methodology by adding the solution of a problem on the motion of a body through a resisting medium. The contents of *De motu* so excited Halley by their mathematical and physical originality and far-reaching implications for astronomical theory, that he immediately went to visit Newton again, in November 1684, to ask Newton to let the Royal Society have more of such work.[52] The results of their meetings clearly helped to stimulate Newton with the enthusiasm needed to take his investigations of mathematical problems much further in this area of physical science, and he did so in a period of highly concentrated work that lasted at least until mid-1686.*[53]

Newton's single-minded attention to his work generally, and to his project during this time, is shown by later reminiscences from his secretary and copyist of the period, Humphrey Newton. His account tells of Isaac Newton's absorption in his studies, how he sometimes forgot his food, or his sleep, or the state of his clothes, and how when he took a walk in his garden he would sometimes rush back to his room with some new thought, not even waiting to sit before beginning to write it down.*[54] Other evidence also shows Newton's absorption in the *Principia*: Newton for years kept up a regular programme of chemical or alchemical experiments, and he normally kept dated notes of them, but for a period from May 1684 to April 1686, Newton's chemical notebooks have no entries at all.*[55] So it seems that Newton abandoned pursuits to which he was normally dedicated, and did very little else for well over a year and a half, but concentrated on developing and writing what became his great work.

The first of the three constituent books was sent to Halley for the printer in spring 1686, and the other two books somewhat later. The complete work, published by Halley at his own financial risk,*[56] appeared in July 1687. Newton had also communicated *De motu* to Flamsteed, and during the period of composition he exchanged a few letters with Flamsteed about observational data on the planets, eventually acknowledging Flamsteed's contributions in the published version of the *Principia* of 1687.

10.2.2 Preliminary version

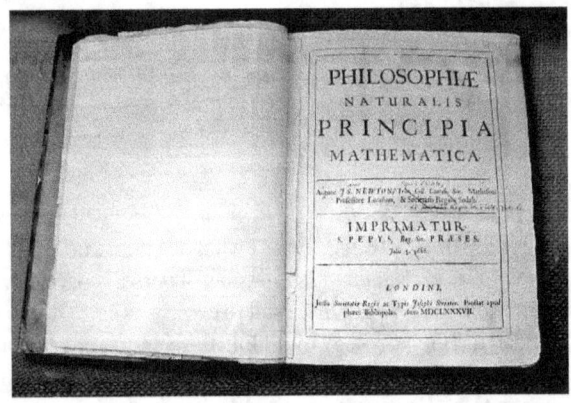

Newton's own first edition copy of his Principia*, with handwritten corrections for the second edition.*

The process of writing that first edition of the *Principia* went through several stages and drafts: some parts of the prelim-

inary materials still survive, others are lost except for fragments and cross-references in other documents.*[57]

Surviving preliminary materials show that Newton (up to some time in 1685) conceived his book as a two-volume work: The first volume was to be *De motu corporum, Liber primus*, with contents that later appeared in extended form as Book 1 of the *Principia*.

A fair-copy draft of Newton's planned second volume *De motu corporum, Liber secundus* still survives, and its completion has been dated to about the summer of 1685. What it covers is the application of the results of *Liber primus* to the Earth, the Moon, the tides, the solar system, and the universe: in this respect it has much the same purpose as the final Book 3 of the *Principia*, but it is written much less formally and is more easily read.

Titlepage and frontispiece of the third edition, London, 1726 (John Rylands Library)

It is not known just why Newton changed his mind so radically about the final form of what had been a readable narrative in *De motu corporum, Liber secundus* of 1685, but he largely started afresh in a new, tighter, and less accessible mathematical style, eventually to produce Book 3 of the *Principia* as we know it. Newton frankly admitted that this change of style was deliberate when he wrote that he had (first) composed this book "in a popular method, that it might be read by many", but to "prevent the disputes" by readers who could not "lay aside the[ir] prejudices", he had "reduced" it "into the form of propositions (in the mathematical way) which should be read by those only, who had first made themselves masters of the principles established in the preceding books".*[58] The final Book 3 also contained in addition some further important quantitative results arrived at by Newton in the meantime, especially about the theory of the motions of comets, and some of the perturbations of the motions of the Moon.

The result was numbered Book 3 of the *Principia* rather than Book 2, because in the meantime, drafts of *Liber primus* had expanded and Newton had divided it into two books.

The new and final Book 2 was concerned largely with the motions of bodies through resisting mediums.

But the *Liber secundus* of 1685 can still be read today. Even after it was superseded by Book 3 of the *Principia*, it survived complete, in more than one manuscript. After Newton's death in 1727, the relatively accessible character of its writing encouraged the publication of an English translation in 1728 (by persons still unknown, not authorised by Newton's heirs). It appeared under the English title *A Treatise of the System of the World*.[59] This had some amendments relative to Newton's manuscript of 1685, mostly to remove cross-references that used obsolete numbering to cite the propositions of an early draft of Book 1 of the *Principia*. Newton's heirs shortly afterwards published the Latin version in their possession, also in 1728, under the (new) title *De Mundi Systemate*, amended to update cross-references, citations and diagrams to those of the later editions of the *Principia*, making it look superficially as if it had been written by Newton after the *Principia*, rather than before.[60] The *System of the World* was sufficiently popular to stimulate two revisions (with similar changes as in the Latin printing), a second edition (1731), and a 'corrected' reprint[61] of the second edition (1740).

10.2.3   Halley's role as publisher

The text of the first of the three books of the *Principia* was presented to the Royal Society at the close of April 1686. Hooke made some priority claims (but failed to substantiate them), causing some delay. When Hooke's claim was made known to Newton, who hated disputes, Newton threatened to withdraw and suppress Book 3 altogether, but Halley, showing considerable diplomatic skills, tactfully persuaded Newton to withdraw his threat and let it go forward to publication. Samuel Pepys, as President, gave his imprimatur on 30 June 1686, licensing the book for publication. The Society had just spent its book budget on a *History of Fishes*,[62] and the cost of publication was borne by Edmund Halley (who was also then acting as publisher of the *Philosophical Transactions of the Royal Society*): the book appeared in summer 1687.[63]

10.3   Historical context

10.3.1   Beginnings of the Scientific Revolution

Nicolaus Copernicus had moved the Earth away from the center of the universe with the heliocentric theory for which he presented evidence in his book *De revolutionibus orbium coelestium* (*On the revolutions of the heavenly spheres*)

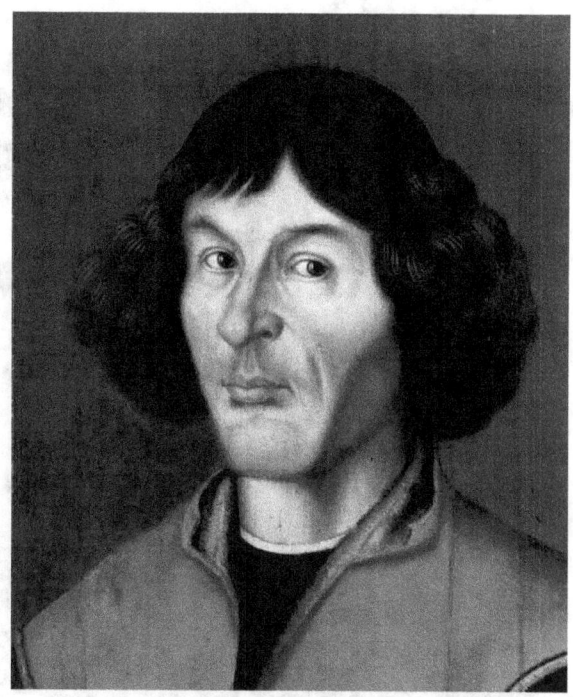

Nicolaus Copernicus (1473–1543) was the first person to formulate a comprehensive heliocentric (or Sun-centered) model of the universe

published in 1543. The structure was completed when Johannes Kepler wrote the book *Astronomia nova* (*A new astronomy*) in 1609, setting out the evidence that planets move in elliptical orbits with the sun at one focus, and that planets do not move with constant speed along this orbit. Rather, their speed varies so that the line joining the centres of the sun and a planet sweeps out equal areas in equal times. To these two laws he added a third a decade later, in his book *Harmonices Mundi* (*Harmonies of the world*). This law sets out a proportionality between the third power of the characteristic distance of a planet from the sun and the square of the length of its year.

The foundation of modern dynamics was set out in Galileo's book *Dialogo sopra i due massimi sistemi del mondo* (*Dialogue on the two main world systems*) where the notion of inertia was implicit and used. In addition, Galileo's experiments with inclined planes had yielded precise mathematical relations between elapsed time and acceleration, velocity or distance for uniform and uniformly accelerated motion of bodies.

Descartes' book of 1644 *Principia philosophiae* (*Principles of philosophy*) stated that bodies can act on each other only through contact: a principle that induced people, among them himself, to hypothesize a universal medium as the carrier of interactions such as light and gravity—the aether. Newton was criticized for apparently introducing forces that acted at distance without any medium.[44] Not until the

Italian physicist Galileo Galilei (1564–1642), a champion of the Copernican model of the universe and an enormously influential figure in the history of kinematics and classical mechanics

development of particle theory was Descartes' notion vindicated when it was possible to describe all interactions, like the strong, weak, and electromagnetic fundamental interactions, using mediating gauge bosons [64] and gravity through hypothesized gravitons. [65] Although he was mistaken in his treatment of circular motion, this effort was more fruitful in the short term when it led others to identify circular motion as a problem raised by the principle of inertia. Christiaan Huygens solved this problem in the 1650s and published it much later in 1673 in his book *Horologium oscillatorium sive de motu pendulorum*.

10.3.2 Newton's role

Newton had studied these books, or, in some cases, secondary sources based on them, and taken notes entitled *Quaestiones quaedam philosophicae* (*Questions about philosophy*) during his days as an undergraduate. During this period (1664–1666) he created the basis of calculus, and performed the first experiments in the optics of colour. At this time, his proof that white light was a combination of primary colours (found via prismatics) replaced the prevailing theory of colours and received an overwhelmingly favourable response, and occasioned bitter disputes with Robert Hooke and others, which forced him to sharpen his

ideas to the point where he already composed sections of his later book *Opticks* by the 1670s in response. Work on calculus is shown in various papers and letters, including two to Leibniz. He became a fellow of the Royal Society and the second Lucasian Professor of Mathematics (succeeding Isaac Barrow) at Trinity College, Cambridge.

10.3.3 Newton's early work on motion

In the 1660s Newton studied the motion of colliding bodies, and deduced that the centre of mass of two colliding bodies remains in uniform motion. Surviving manuscripts of the 1660s also show Newton's interest in planetary motion and that by 1669 he had shown, for a circular case of planetary motion, that the force he called 'endeavour to recede' (now called centrifugal force) had an inverse-square relation with distance from the center. [66] After his 1679–1680 correspondence with Hooke, described below, Newton adopted the language of inward or centripetal force. According to Newton scholar J Bruce Brackenridge, although much has been made of the change in language and difference of point of view, as between centrifugal or centripetal forces, the actual computations and proofs remained the same either way. They also involved the combination of tangential and radial displacements, which Newton was making in the 1660s. The difference between the centrifugal and centripetal points of view, though a significant change of perspective, did not change the analysis. [67] Newton also clearly expressed the concept of linear inertia in the 1660s: for this Newton was indebted to Descartes' work published 1644. [68]

10.3.4 Controversy with Hooke

Hooke published his ideas about gravitation in the 1660s and again in 1674. He argued for an attracting principle of gravitation in *Micrographia* of 1665, in a 1666 Royal Society lecture *On gravity*, and again in 1674, when he published his ideas about the *System of the World* in somewhat developed form, as an addition to *An Attempt to Prove the Motion of the Earth from Observations*. [69] Hooke clearly postulated mutual attractions between the Sun and planets, in a way that increased with nearness to the attracting body, along with a principle of linear inertia. Hooke's statements up to 1674 made no mention, however, that an inverse square law applies or might apply to these attractions. Hooke's gravitation was also not yet universal, though it approached universality more closely than previous hypotheses. [70] Hooke also did not provide accompanying evidence or mathematical demonstration. On these two aspects, Hooke stated in 1674:   "Now what these several degrees [of gravitational attraction] are I have not yet ex-

Imaginary portrait of English polymath Robert Hooke (1635–1703).

perimentally verified" (indicating that he did not yet know what law the gravitation might follow); and as to his whole proposal: "This I only hint at present", "having my self many other things in hand which I would first compleat, and therefore cannot so well attend it" (i.e., "prosecuting this Inquiry").*[69]

In November 1679, Hooke began an exchange of letters with Newton, of which the full text is now published.*[71] Hooke told Newton that Hooke had been appointed to manage the Royal Society's correspondence,*[72] and wished to hear from members about their researches, or their views about the researches of others; and as if to whet Newton's interest, he asked what Newton thought about various matters, giving a whole list, mentioning "compounding the celestial motions of the planets of a direct motion by the tangent and an attractive motion towards the central body", and "my hypothesis of the lawes or causes of springinesse", and then a new hypothesis from Paris about planetary motions (which Hooke described at length), and then efforts to carry out or improve national surveys, the difference of latitude between London and Cambridge, and other items. Newton's reply offered "a fansy of my own" about a terrestrial experiment (not a proposal about celestial motions) which might detect the Earth's motion, by the use of a body first suspended in air and then dropped to let it fall. The main point was to indicate how Newton thought the falling body could experimentally reveal the Earth's motion by its

direction of deviation from the vertical, but he went on hypothetically to consider how its motion could continue if the solid Earth had not been in the way (on a spiral path to the centre). Hooke disagreed with Newton's idea of how the body would continue to move.*[73] A short further correspondence developed, and towards the end of it Hooke, writing on 6 January 1680 to Newton, communicated his "supposition ... that the Attraction always is in a duplicate proportion to the Distance from the Center Reciprocall, and Consequently that the Velocity will be in a subduplicate proportion to the Attraction and Consequently as Kepler Supposes Reciprocall to the Distance." *[74] (Hooke's inference about the velocity was actually incorrect.*[75])

In 1686, when the first book of Newton's *Principia* was presented to the Royal Society, Hooke claimed that Newton had obtained from him the "notion" of "the rule of the decrease of Gravity, being reciprocally as the squares of the distances from the Center". At the same time (according to Edmond Halley's contemporary report) Hooke agreed that "the Demonstration of the Curves generated therby" was wholly Newton's.*[71]

A recent assessment about the early history of the inverse square law is that "by the late 1660s," the assumption of an "inverse proportion between gravity and the square of distance was rather common and had been advanced by a number of different people for different reasons".*[76] Newton himself had shown in the 1660s that for planetary motion under a circular assumption, force in the radial direction had an inverse-square relation with distance from the center.*[66] Newton, faced in May 1686 with Hooke's claim on the inverse square law, denied that Hooke was to be credited as author of the idea, giving reasons including the citation of prior work by others before Hooke.*[71] Newton also firmly claimed that even if it had happened that he had first heard of the inverse square proportion from Hooke, which it had not, he would still have some rights to it in view of his mathematical developments and demonstrations, which enabled observations to be relied on as evidence of its accuracy, while Hooke, without mathematical demonstrations and evidence in favour of the supposition, could only guess (according to Newton) that it was approximately valid "at great distances from the center".*[71]

The background described above shows there was basis for Newton to deny deriving the inverse square law from Hooke. On the other hand, Newton did accept and acknowledge, in all editions of the *Principia*, that Hooke (but not exclusively Hooke) had separately appreciated the inverse square law in the solar system. Newton acknowledged Wren, Hooke and Halley in this connection in the Scholium to Proposition 4 in Book 1.*[77] Newton also acknowledged to Halley that his correspondence with Hooke in 1679–80 had reawakened his dormant interest in astronomical matters, but that did not mean, according to Newton, that Hooke had told Newton anything new or original: "yet am I not beholden to him for any light into that business but only for the diversion he gave me from my other studies to think on these things & for his dogmaticalness in writing as if he had found the motion in the Ellipsis, which inclined me to try it ..." .*[71]) Newton's reawakening interest in astronomy received further stimulus by the appearance of a comet in the winter of 1680/1681, on which he corresponded with John Flamsteed.*[78]

In 1759, decades after the deaths of both Newton and Hooke, Alexis Clairaut, mathematical astronomer eminent in his own right in the field of gravitational studies, made his assessment after reviewing what Hooke had published on gravitation. "One must not think that this idea ... of Hooke diminishes Newton's glory", Clairaut wrote; "The example of Hooke" serves "to show what a distance there is between a truth that is glimpsed and a truth that is demonstrated" .*[79]*[80]

10.4   Location of early-edition copies

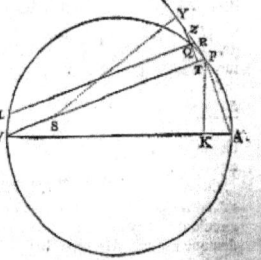

A page from the Principia

Since only between 250 and 400 copies were printed by the Royal Society, the first edition is very rare. Several rare-

book collections contain first edition and other early copies of Newton's *Principia Mathematica*, including:

- Cambridge University Library has Newton's own copy of the first edition, with handwritten notes for the second edition.*[81]

- The Earl Gregg Swem Library at the College of William & Mary has a first edition copy of the Principia.*[82] In it, are notes in Latin throughout by a not yet identified hand.

- The Frederick E. Brasch Collection of Newton and Newtoniana in Stanford University also has a first edition of the Principia.*[83]

- A first edition forms part of the Crawford Collection, housed at the Royal Observatory, Edinburgh.*[84]

- The Uppsala University Library owns a first edition copy, which was stolen in the 1960s and returned to the library in 2009.*[85]

- The Folger Shakespeare Library in Washington, D.C. owns a first edition, as well as a 1713 second edition.

A facsimile edition (based on the 3rd edition of 1726 but with variant readings from earlier editions and important annotations) was published in 1972 by Alexandre Koyré and I. Bernard Cohen.*[5]

10.5   Later editions

Two later editions were published by Newton:

10.5.1   Second edition, 1713

Newton had been urged to make a new edition of the *Principia* since the early 1690s, partly because copies of the first edition had already become very rare and expensive within a few years after 1687.*[86] Newton referred to his plans for a second edition in correspondence with Flamsteed in November 1694:*[87] Newton also maintained annotated copies of the first edition specially bound up with interleaves on which he could note his revisions; two of these copies still survive:*[88] but he had not completed the revisions by 1708, and of two would-be editors, Newton had almost severed connections with one, Fatio de Duillier, and the other, David Gregory seems not to have met with Newton's approval and was also terminally ill, dying later in 1708. Nevertheless, reasons were accumulating not to put off the new edition any longer.*[89] Richard Bentley, master of Trinity College, persuaded Newton to allow him to undertake a second edition, and in June 1708 Bentley wrote to Newton with a specimen print of the first sheet, at the same time expressing the (unfulfilled) hope that Newton had made progress towards finishing the revisions.*[90] It seems that Bentley then realised that the editorship was technically too difficult for him, and with Newton's consent he appointed Roger Cotes, Plumian professor of astronomy at Trinity, to undertake the editorship for him as a kind of deputy (but Bentley still made the publishing arrangements and had the financial responsibility and profit). The correspondence of 1709–1713 shows Cotes reporting to two masters, Bentley and Newton, and managing (and often correcting) a large and important set of revisions to which Newton sometimes could not give his full attention.*[91] Under the weight of Cotes' efforts, but impeded by priority disputes between Newton and Leibniz,*[92] and by troubles at the Mint,*[93] Cotes was able to announce publication to Newton on 30 June 1713.*[94] Bentley sent Newton only six presentation copies; Cotes was unpaid; Newton omitted any acknowledgement to Cotes.

Among those who gave Newton corrections for the Second Edition were: Firmin Abauzit, Roger Cotes and David Gregory. However, Newton omitted acknowledgements to some because of the priority disputes. John Flamsteed, the Astronomer Royal, suffered this especially.

10.5.2 Third edition, 1726

The third edition was published 25 March 1726, under the stewardship of *Henry Pemberton, M.D., a man of the greatest skill in these matters...*; Pemberton later said that this recognition was worth more to him than the two hundred guinea award from Newton.*[95]

10.5.3 Annotated and other editions

In 1739–42, two French priests, Pères Thomas LeSeur and François Jacquier (of the Minim order, but sometimes erroneously identified as Jesuits), produced with the assistance of J.-L. Calandrini an extensively annotated version of the *Principia* in the 3rd edition of 1726. Sometimes this is referred to as the *Jesuit edition*: it was much used, and reprinted more than once in Scotland during the 19th century.*[96]

Émilie du Châtelet also made a translation of Newton's Principia into French. Unlike LeSeur and Jacquier's edition, hers was a complete translation of Newton's three books and their prefaces. She also included a Commentary section where she fused the three books into a much clearer and easier to understand summary. She included an analytical section where she applied the new mathematics of calculus to Newton's most controversial theories. Previously, geometry was the standard mathematics used to analyse theories. Du Chatelet's translation is the only complete one to have been done in French and hers remains the standard French translation to this day. See "Translating Newton's 'Principia': The Marquise du Châtelet's Revisions and Additions for a French Audience." Author(s): Judith P. Zinsser Source: Notes and Records of the Royal Society of London, Vol. 55, No. 2 (May 2001), pp. 227–245.

10.5.4 English translations

Two full English translations of Newton's 'Principia' have appeared, both based on Newton's 3rd edition of 1726.

The first, from 1729, by Andrew Motte,*[3] was described by Newton scholar I. Bernard Cohen (in 1968) as "still of enormous value in conveying to us the sense of Newton's words in their own time, and it is generally faithful to the original: clear, and well written".*[97] The 1729 version was the basis for several republications, often incorporating revisions, among them a widely used modernised English version of 1934, which appeared under the editorial name of Florian Cajori (though completed and published only some years after his death). Cohen pointed out ways in which the 18th-century terminology and punctuation of the 1729 translation might be confusing to modern readers, but he also made severe criticisms of the 1934 modernised English version, and showed that the revisions had been made without regard to the original, also demonstrating gross errors "that provided the final impetus to our decision to produce a wholly new translation".*[98]

The second full English translation, into modern English, is the work that resulted from this decision by collaborating translators I. Bernard Cohen and Anne Whitman; it was published in 1999 with a guide by way of introduction.*[99]

William H. Donahue has published a translation of the work's central argument, published in 1996, along with expansion of included proofs and ample commentary.*[100] The book was developed as a textbook for classes at St. John's College and the aim of this translation is to be faithful to the Latin text.*[101]

10.5.5 Homages

In 2014, British astronaut Tim Peake named his upcoming mission to the International Space Station *Principia* after the book, in "honour of Britain's greatest scientist".*[102] Tim Peake's *Principia* is scheduled to launch on December 15th 2015 aboard Soyuz TMA-19M.*[103]

10.6   See also

- Atomism

- Elements of the Philosophy of Newton

10.7   References

[1] "The Mathematical Principles of Natural Philosophy", *Encyclopædia Britannica*, London

[2] Among versions of the *Principia* online: .

[3] Volume 1 of the 1729 English translation is available as an online scan; limited parts of the 1729 translation (misidentified as based on the 1687 edition) have also been transcribed online.

[4] Newton, Isaac. "Philosophiæ Naturalis Principia Mathematica (Newton's personally annotated 1st edition)".

[5] [In Latin] Isaac Newton's *Philosophiae Naturalis Principia Mathematica: the Third edition (1726) with variant readings*, assembled and ed. by Alexandre Koyré and I Bernard Cohen with the assistance of Anne Whitman (Cambridge, MA, 1972, Harvard UP)

[6] J M Steele, University of Toronto, (review online from Canadian Association of Physicists) of N Guicciardini's "Reading the Principia: The Debate on Newton's Mathematical Methods for Natural Philosophy from 1687 to 1736" (Cambridge UP, 1999), a book which also states (summary before title page) that the "Principia" "is considered one of the masterpieces in the history of science".

[7] (in French) Alexis Clairaut, "Du systeme du monde, dans les principes de la gravitation universelle", in "Histoires (& Memoires) de l'Academie Royale des Sciences" for 1745 (published 1749), at p.329 (according to a note on p.329, Clairaut's paper was read at a session of November 1747).

[8] G E Smith, "Newton's Philosophiae Naturalis Principia Mathematica", The Stanford Encyclopedia of Philosophy (Winter 2008 Edition), E N Zalta (ed.).

[9] The content of infinitesimal calculus in the 'Principia' was recognized both in Newton's lifetime and later, among others by the Marquis de l'Hospital, whose 1696 book "Analyse des infiniment petits" (Infinitesimal analysis) stated in its preface, about the 'Principia', that 'nearly all of it is of this calculus' ('lequel est presque tout de ce calcul'). See also D T Whiteside (1970), "The mathematical principles underlying Newton's *Principia Mathematica*", Journal for the History of Astronomy, vol.1 (1970), 116–138, especially at p.120.

[10] Or "frame" no hypotheses (as traditionally translated at vol.2, p.392, in the 1729 English version).

[11] From Motte's translation of 1729 (at 3rd page of Author's Preface); and see also J. W. Herivel, *The background to Newton's "Principia"*, Oxford University Press, 1965.

[12] The *De motu corporum in gyrum* article indicates the topics that reappear in the *Principia*.

[13] Newton, Sir Isaac (1729). "Definitions". *The Mathematical Principles of Natural Philosophy, Volume I*. p. 1.

[14] Newton, Sir Isaac (1729). "Axioms or Laws of Motion". *The Mathematical Principles of Natural Philosophy, Volume I*. p. 19.

[15] Newton, Sir Isaac (1729). "Section I". *The Mathematical Principles of Natural Philosophy, Volume I*. p. 41.

[16] Newton, Sir Isaac (1729). "Section II". *The Mathematical Principles of Natural Philosophy, Volume I*. p. 57.

[17] This relationship between circular curvature, speed and radial force, now often known as Huygens' formula, was independently found by Newton (in the 1660s) and by Huygens in the 1650s: the conclusion was published (without proof) by Huygens in 1673.This was given by Isaac Newton through his Inverse Square Law.

[18] Newton, Sir Isaac; Machin, John (1729). *The Mathematical Principles of Natural Philosophy, Volume I*. pp. 79–153.

[19] Newton, Sir Isaac (1729). "Section IX". *The Mathematical Principles of Natural Philosophy, Volume I*. p. 177.

[20] Newton, Sir Isaac (1729). "Section XI". *The Mathematical Principles of Natural Philosophy, Volume I*. p. 218.

[21] Newton, Sir Isaac (1729). "Section XI, Proposition LXVI". *The Mathematical Principles of Natural Philosophy, Volume I*. p. 234.

[22] Newton, Sir Isaac; Machin, John (1729). *The Mathematical Principles of Natural Philosophy, Volume I*. pp. 239–256.

[23] Newton, Sir Isaac (1729). "Section XII". *The Mathematical Principles of Natural Philosophy, Volume I*. p. 263.

[24] Eric J Aiton, *The Cartesian vortex theory*, chapter 11 in *Planetary astronomy from the Renaissance to the rise of astrophysics, Part A: Tycho Brahe to Newton*, eds. R Taton & C Wilson, Cambridge (Cambridge University press) 1989; at pp. 207–221.

[25] Newton, Sir Isaac (1729). "Scholium to proposition 53". *The Mathematical Principles of Natural Philosophy, Volume II*. p. 197.

[26] Newton, Sir Isaac (1729). *The Mathematical Principles of Natural Philosophy, Volume II*. p. 252.

[27] Newton, Sir Isaac (1729). *The Mathematical Principles of Natural Philosophy, Volume II*. p. 262.

[28] Newton, Sir Isaac (1729). "The Phaenomena". *The Mathematical Principles of Natural Philosophy, Volume II*. p. 206.

[29] Newton, Sir Isaac (1729). *The Mathematical Principles of Natural Philosophy, Volume II*. p. 213.

[30] Newton, Sir Isaac (1729). *The Mathematical Principles of Natural Philosophy, Volume II*. p. 220.

[31] Newton, Sir Isaac (1729). *The Mathematical Principles of Natural Philosophy, Volume II*. p. 323.

[32] Newton, Sir Isaac (1729). *The Mathematical Principles of Natural Philosophy, Volume II*. p. 332.

[33] Newton, Sir Isaac (1729). *The Mathematical Principles of Natural Philosophy, Volume II*. p. 255.

[34] Newton, Sir Isaac (1729). *The Mathematical Principles of Natural Philosophy, Volume II*. p. 305.

[35] Newton, Sir Isaac (1729). *The Mathematical Principles of Natural Philosophy, Volume II*. p. 306.

[36] Newton, Sir Isaac (1729). *The Mathematical Principles of Natural Philosophy, Volume II*. p. 320.

[37] See Curtis Wilson, "The Newtonian achievement in astronomy", pages 233–274 in R Taton & C Wilson (eds) (1989) *The General History of Astronomy*, Volume, 2A', at page 233).

[38] Newton, Sir Isaac (1729). "Proposition 12, Corollary". *The Mathematical Principles of Natural Philosophy, Volume II*. p. 233.

[39] Newton, Sir Isaac (1729). "Proposition 11 & preceding Hypothesis". *The Mathematical Principles of Natural Philosophy, Volume II*. p. 232.

[40] Newton, Sir Isaac (1729). "Proposition 8, Corollary 2". *The Mathematical Principles of Natural Philosophy, Volume II*. p. 228.

[41] Newton, Sir Isaac (1729). "Proposition 22". *The Mathematical Principles of Natural Philosophy, Volume II*. p. 232. Newton's position is seen to go beyond literal Copernican heliocentrism practically to the modern position in regard to the solar system barycenter.

[42] Mughal, Muhammad Aurang Zeb. 2009. Time, absolute. Birx, H. James (ed.), *Encyclopedia of Time: Science, Philosophy, Theology, and Culture*, Vol. 3. Thousand Oaks, CA: Sage, pp. 1254–1255.

[43] See online *Principia* (1729 translation) vol.2, Books 2 & 3, starting at page 387 of volume 2 (1729).

[44] Edelglass et al., *Matter and Mind*, ISBN 0-940262-45-2, p. 54.

[45] See online *Principia* (1729 translation) vol.2, Books 2 & 3, at page 392 of volume 2 (1729).

[46] Snobelen, Stephen. "The General Scholium to Isaac Newton's *Principia mathematica*". Retrieved 31 May 2008.

[47] Ducheyne, Steffen. "The General Scholium: Some notes on Newton's published and unpublished endeavours" (PDF). *Lias: Sources and Documents Relating to the Early Modern History of Ideas* **33** (2): 223–274. Retrieved 19 November 2008.

[48] Paraphrase of 1686 report by Halley, in H. W. Turnbull (ed.), 'Correspondence of Isaac Newton', Vol.2, cited above, pp. 431–448.

[49] 'Cook, 1998': A. Cook, *Edmond Halley, Charting the Heavens and the Seas*, Oxford University Press 1998, at pp.147 and 152.

[50] As dated e.g. by D. T. Whiteside, in *The Prehistory of the Principia from 1664 to 1686*, Notes and Records of the Royal Society of London, 45 (1991) 11–61.

[51] Cook, 1998; at p. 147.

[52] 'Westfall, 1980': R S Westfall, *Never at Rest: A Biography of Isaac Newton*, Cambridge University Press 1980, at p.404.

[53] Cook, 1998; at p. 151.

[54] Westfall, 1980; at p. 406, also pp. 191–2.

[55] Westfall, 1980; at p.406, n.15.

[56] Westfall, 1980; at pp. 153–156.

[57] The fundamental study of Newton's progress in writing the *Principia* is in I. Bernard Cohen's *Introduction to Newton's 'Principia'*, (Cambridge, Cambridge University Press, 1971), at part 2: "The writing and first publication of the 'Principia' ", pp.47–142.

[58] Newton, Sir Isaac (1729). "Introduction to Book 3". *The Mathematical Principles of Natural Philosophy, Volume II*. p. 200.

[59] Newton, Isaac (1728). *A Treatise of the System of the World*.

[60] I. Bernard Cohen, *Introduction* to Newton's *A Treatise of the System of the World* (facsimile of second English edition of 1731), London (Dawsons of Pall Mall) 1969.

[61] Newton, Sir Isaac (1740). *The System of the World: Demonstrated in an Easy and Popular Manner. Being a Proper Introduction to the Most Sublime Philosophy. By the Illustrious Sir Isaac Newton. Translated into English.* A 'corrected' reprint of the second edition.

[62] Richard Westfall (1980), *Never at Rest*, p. 453, ISBN 0-521-27435-4

[63] "Museum of London exhibit including facsimile of title page from John Flamsteed's copy of 1687 edition of Newton's "Principia"". Museumoflondon.org.uk. Retrieved 16 March 2012.

[64] The Henryk Niewodniczanski Institute of Nuclear Physics. "Particle Physics and Astrophysics Research".

[65] Rovelli, Carlo. "Notes for a brief history of quantum gravity".

[66] D T Whiteside, "The pre-history of the 'Principia' from 1664 to 1686", Notes and Records of the Royal Society of London, 45 (1991), pages 11–61; especially at 13–20.

[67] See J. Bruce Brackenridge, "The key to Newton's dynamics: the Kepler problem and the Principia", (University of California Press, 1995), especially at pages 20–21.

[68] See page 10 in D T Whiteside, "Before the Principia: the maturing of Newton's thoughts on dynamical astronomy, 1664–1684", Journal for the History of Astronomy, i (1970), pages 5–19.

[69] Hooke's 1674 statement in "An Attempt to Prove the Motion of the Earth from Observations", is available in online facsimile here.

[70] See page 239 in Curtis Wilson (1989), "The Newtonian achievement in astronomy", ch.13 (pages 233–274) in "Planetary astronomy from the Renaissance to the rise of astrophysics: 2A: Tycho Brahe to Newton", CUP 1989.

[71] H W Turnbull (ed.), Correspondence of Isaac Newton, Vol 2 (1676–1687), (Cambridge University Press, 1960), giving the Hooke-Newton correspondence (of November 1679 to January 1679/80) at pp.297–314, and the 1686 correspondence over Hooke's priority claim at pp.431–448.

[72] 'Correspondence' vol.2 already cited, at p.297.

[73] Several commentators have followed Hooke in calling Newton's spiral path mistaken, or even a 'blunder', but there are also the following facts: (a) that Hooke left out of account Newton's specific statement that the motion resulted from dropping "a heavy body suspended in the Air" (i.e. a resisting medium), see Newton to Hooke, 28 November 1679, document #236 at page 301, 'Correspondence' vol.2 cited above, and compare Hooke's report to the Royal Society on 11 December 1679 where Hooke reported the matter "supposing no resistance", see D Gjertsen, 'Newton Handbook' (1986), at page 259); and (b) that Hooke's reply of 9 December 1679 to Newton considered the cases of motion both with and without air resistance: The resistance-free path was what Hooke called an 'elliptueid'; but a line in Hooke's diagram showing the path for his case of air resistance was, though elongated, also another inward-spiralling path ending at the Earth's centre: Hooke wrote "where the Medium ... has a power of impeding and destroying its motion the curve in wch it would move would be some what like the Line AIKLMNOP &c and ... would terminate in the center C". Hooke's path including air resistance was therefore to this extent like Newton's (see 'Correspondence' vol.2, cited above, at pages 304–306, document #237, with accompanying figure). The diagrams are also available online: see Curtis Wilson, chapter 13 in "Planetary Astronomy from the Renaissance to the Rise of Astrophysics, Part A, Tycho Brahe to Newton", (Cambridge UP 1989), at page 241 showing Newton's 1679 diagram with spiral, and extract of his letter; also at page 242 showing Hooke's 1679

diagram including two paths, closed curve and spiral. Newton pointed out in his later correspondence over the priority claim that the descent in a spiral "is true in a resisting medium such as our air is", see 'Correspondence', vol.2 cited above, at page 433, document #286.

[74] See page 309 in 'Correspondence of Isaac Newton', Vol 2 cited above, at document #239.

[75] See Curtis Wilson (1989) at page 244.

[76] See "Meanest foundations and nobler superstructures: Hooke, Newton and the 'Compounding of the Celestiall Motions of the Planetts'", Ofer Gal, 2003 at page 9.

[77] See for example the 1729 English translation of the 'Principia', at page 66.

[78] R S Westfall, 'Never at Rest', 1980, at pages 391–2.

[79] The second extract is quoted and translated in W.W. Rouse Ball, "An Essay on Newton's 'Principia'" (London and New York: Macmillan, 1893), at page 69.

[80] The original statements by Clairaut (in French) are found (with orthography here as in the original) in "Explication abregée du système du monde, et explication des principaux phénomenes astronomiques tirée des Principes de M. Newton" (1759), at Introduction (section IX), page 6: "Il ne faut pas croire que cette idée ... de Hook diminue la gloire de M. Newton", [and] "L'exemple de Hook" [serves] "à faire voir quelle distance il y a entre une vérité entrevue & une vérité démontrée".

[81] Newton, Isaac. "Philosophiæ naturalis principia mathematica". Cambridge Digital Library. Retrieved 3 July 2013.

[82] Newton, Isaac (1687). "Philosophiae naturalis principia mathematica" (in Latin). Swem Library: Jussu Societatis Regiae ac Typis Josephi Streater.

[83] "Special Collections & University Archives". stanford.edu.

[84] "The Crawford collection at the Royal Observatory Edinburgh". The Royal Observatory, Edinburgh. Retrieved 3 July 2013.

[85] "Newton's book back in Uppsala University Library". Uppsala University. Retrieved 10 May 2014.

[86] The Correspondence of Isaac Newton, vol.4, Cambridge University Press 1967, at pp.519, n.2.

[87] The Correspondence of Isaac Newton, vol.4, Cambridge University press 1967, at p.42.

[88] I Bernard Cohen, Introduction to the Principia, Cambridge 1971.

[89] Richard S. Westfall. Never at Rest: A Biography of Isaac Newton. Cambridge U. Press. 1980 ISBN 0-521-23143-4, at p.699.

[90] The Correspondence of Isaac Newton, vol.4, Cambridge University press 1967, at pp.518–20.

[91] The Correspondence of Isaac Newton, vol.5, Cambridge University press 1975. Bentley's letter to Newton of October 1709 (at pp. 7–8) describes Cotes' perhaps unenviable position in relation to his master Bentley: "You need not be so shy of giving Mr. Cotes too much trouble: he has more esteem for you, and obligations to you, than to think that trouble too grievous: but however he does it at my Orders, to whom he owes more than that."

[92] Westfall, pp.712–716.

[93] Westfall, pp.751–760.

[94] Westfall, p.750.

[95] Westfall, p.802

[96] [In Latin] Isaac Newton, *Philosophiae naturalis principia mathematica* volume 1 of a facsimile of a reprint (1833) of the 3rd (1726) edition, as annotated in 1740–42 by Thomas LeSeur & François Jacquier, with the assistance of J-L Calandrini

[97] I Bernard Cohen (1968), "Introduction" (at page i) to (facsimile) reprint of 1729 English translation of Newton's "Principia" (London (1968), Dawsons of Pall Mall).

[98] See pages 29–37 in I. Bernard Cohen (1999), "A Guide to Newton's Principia", published as an introduction to "Isaac Newton: The Principia, Mathematical principles of natural philosophy, a new translation" by I Bernard Cohen and Anne Whitman, University of California Press, 1999.

[99] "Isaac Newton: The Principia, Mathematical principles of natural philosophy, a new translation" by I Bernard Cohen and Anne Whitman, preceded by "A Guide to Newton's Principia" by I Bernard Cohen, University of California Press, 1999, ISBN 978-0-520-08816-0, ISBN 978-0-520-08817-7.

[100] Dana Densmore and William H. Donahue, *Newton's Principia: The Central Argument: Translation, Notes, and Expanded Proofs* (Green Lion Press; 3rd edition, 2003) ISBN 978-1-888009-23-1, ISBN 978-1-888009-23-1

[101] Densmore and Donahue, pp. xv–xvi.

[102] "Tim Peake mission name pays tribute to Isaac Newton". *BBC News.*

[103] "Roscosmos Announces New Soyuz/Progress Launch Dates". NASA. 9 June 2015.

10.8 Further reading

- Alexandre Koyré, *Newtonian studies* (London: Chapman and Hall, 1965).

- I. Bernard Cohen, *Introduction to Newton's* Principia (Harvard University Press, 1971).

- Richard S. Westfall, *Force in Newton's physics; the science of dynamics in the seventeenth century* (New York: American Elsevier, 1971).

- S. Chandrasekhar, *Newton's Principia for the common reader* (New York: Oxford University Press, 1995).

- Guicciardini, N., 2005, "Philosophia Naturalis..." in Grattan-Guinness, I., ed., *Landmark Writings in Western Mathematics*. Elsevier: 59–87.

- Andrew Janiak, *Newton as Philosopher* (Cambridge University Press, 2008).

- François De Gandt, *Force and geometry in Newton's Principia* trans. Curtis Wilson (Princeton, NJ: Princeton University Press, c1995).

- Steffen Ducheyne, *The main Business of Natural Philosophy: Isaac Newton's Natural-Philosophical Methodology* (Dordrecht e.a.: Springer, 2012).

- John Herivel, *The background to Newton's Principia; a study of Newton's dynamical researches in the years 1664–84* (Oxford, Clarendon Press, 1965).

- Brian Ellis, "The Origin and Nature of Newton's Laws of Motion" in *Beyond the Edge of Certainty*, ed. R. G. Colodny. (Pittsburgh: University Pittsburgh Press, 1965), 29–68.

- E.A. Burtt, *Metaphysical Foundations of Modern Science* (Garden City, NY: Doubleday and Company, 1954).

- Colin Pask, *Magnificent Principia: Exploring Isaac Newton's Masterpiece* (New York: Prometheus Books, 2013).

10.9 External links

10.9.1 Latin versions

First edition (1687)

- Trinity College Library, Cambridge High resolution digitised version of Newton's own copy of the first edition, with annotations.

- Cambridge University, Cambridge Digital Library High resolution digitised version of Newton's own copy of the first edition, interleaved with blank pages for his annotations and corrections.

- 1687: Newton's 'Principia', first edition (1687, in Latin). High-resolution presentation of the Gunnerus Library's copy.

- 1687: Newton's 'Principia', first edition (1687, in Latin).

- Project Gutenberg.

- ETH-Bibliothek Zürich.

- Philosophiæ Naturalis Principia Mathematica From the Rare Book and Special Collection Division at the Library of Congress

Second edition (1713)

- ETH-Bibliothek Zürich.

- ETH-Bibliothek Zürich (pirated Amsterdam reprint of 1723).

Third edition (1726)

- ETH-Bibliothek Zürich.

Later Latin editions

- Principia (in Latin, annotated). 1833 Glasgow reprint (volume 1) with Books 1 & 2 of the Latin edition annotated by Leseur, Jacquier and Calandrini 1739–42 (described above).

- Archive.org (1871 reprint of the 1726 edition)

10.9.2   English translations

- Andrew Motte, 1729, first English translation of third edition (1726)

  - WikiSource, Partial

  - Google books, vol.1 with Book 1.

  - Google books, vol.2 with Books 2 and 3. (Book 3 starts at p.200.) (Google's metadata wrongly labels this vol.1).

  - Partial HTML

- Robert Thorpe 1802 translation

- N. W. Chittenden, ed., 1846 "American Edition" a partly modernised English version, largely the Motte translation of 1729.

  - Wikisource

- Archive.org #1

- Archive.org #2

- Percival Frost 1863 translation with interpolations Archive.org

- Florian Cajori 1934 modernisation of 1729 Motte and 1802 Thorpe translations

- , Ian Bruce has made a complete translation of the third edition, with notes, on his website.

10.9.3   Other links

- In Search of *Principia*, regarding online editions

- David R. Wilkins of the School of Mathematics at Trinity College, Dublin has transcribed a few sections into TeX and METAPOST and made the source, as well as a formatted .pdf available at Extracts from the Works of Isaac Newton

Chapter 11

Classical mechanics

For the textbooks, see Classical Mechanics (Goldstein book) and Classical Mechanics (Kibble and Berkshire book).

In physics, **classical mechanics** and quantum mechanics

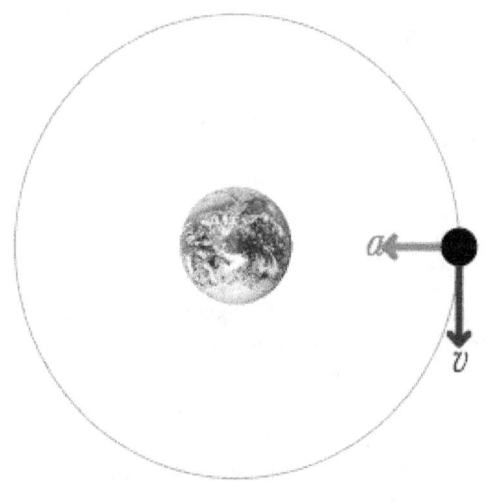

Diagram of orbital motion of a satellite around the earth, showing perpendicular velocity and acceleration (force) vectors.

are the two major sub-fields of mechanics. Classical mechanics is concerned with the set of physical laws describing the motion of bodies under the action of a system of forces. The study of the motion of bodies is an ancient one, making classical mechanics one of the oldest and largest subjects in science, engineering and technology. It is also widely known as **Newtonian mechanics**.

Classical mechanics describes the motion of macroscopic objects, from projectiles to parts of machinery, as well as astronomical objects, such as spacecraft, planets, stars, and galaxies. Besides this, many specializations within the subject deal with solids, liquids and gases and other specific sub-topics. Classical mechanics also provides extremely accurate results as long as the domain of study is restricted

to large objects and the speeds involved do not approach the speed of light. When the objects being dealt with become sufficiently small, it becomes necessary to introduce the other major sub-field of mechanics, quantum mechanics, which reconciles the macroscopic laws of physics with the atomic nature of matter and handles the wave–particle duality of atoms and molecules. When both quantum mechanics and classical mechanics cannot apply, such as at the quantum level with high speeds, quantum field theory (QFT) becomes applicable.

The term *classical mechanics* was coined in the early 20th century to describe the system of physics begun by Isaac Newton and many contemporary 17th century natural philosophers, building upon the earlier astronomical theories of Johannes Kepler, which in turn were based on the precise observations of Tycho Brahe and the studies of terrestrial projectile motion of Galileo. Since these aspects of physics were developed long before the emergence of quantum physics and relativity, some sources exclude Einstein's theory of relativity from this category. However, a number of modern sources *do* include relativistic mechanics, which in their view represents *classical mechanics* in its most developed and most accurate form.[*][note 1]

The initial stage in the development of classical mechanics is often referred to as Newtonian mechanics, and is associated with the physical concepts employed by and the mathematical methods invented by Newton himself, in parallel with Leibniz, and others. This is further described in the following sections. Later, more abstract and general methods were developed, leading to reformulations of classical mechanics known as Lagrangian mechanics and Hamiltonian mechanics. These advances were largely made in the 18th and 19th centuries, and they extend substantially beyond Newton's work, particularly through their use of analytical mechanics.

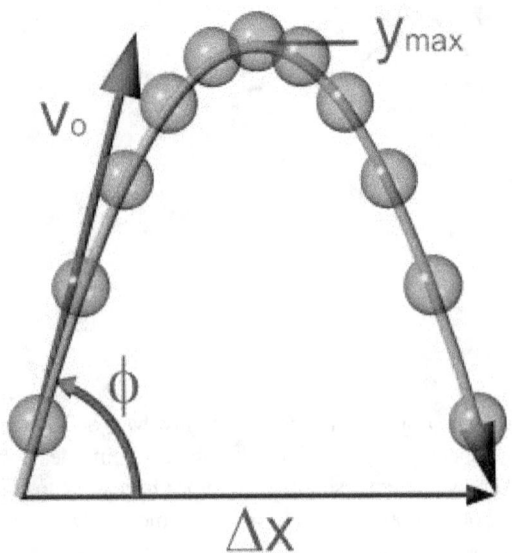

The analysis of projectile motion is a part of classical mechanics.

11.1   Description of the theory

The following introduces the basic concepts of classical mechanics. For simplicity, it often models real-world objects as point particles, objects with negligible size. The motion of a point particle is characterized by a small number of parameters: its position, mass, and the forces applied to it. Each of these parameters is discussed in turn.

In reality, the kind of objects that classical mechanics can describe always have a non-zero size. (The physics of *very* small particles, such as the electron, is more accurately described by quantum mechanics.) Objects with non-zero size have more complicated behavior than hypothetical point particles, because of the additional degrees of freedom: a baseball can spin while it is moving, for example. However, the results for point particles can be used to study such objects by treating them as composite objects, made up of a large number of interacting point particles. The center of mass of a composite object behaves like a point particle.

Classical mechanics uses common-sense notions of how matter and forces exist and interact. It assumes that matter and energy have definite, knowable attributes such as where an object is in space and its speed. It also assumes that objects may be directly influenced only by their immediate surroundings, known as the principle of locality. In quantum mechanics, an object may have either its position or velocity undetermined.

11.1.1   Position and its derivatives

Main article: Kinematics

The *position* of a point particle is defined with respect to an arbitrary fixed reference point, **O**, in space, usually accompanied by a coordinate system, with the reference point located at the *origin* of the coordinate system. It is defined as the vector **r** from **O** to the particle. In general, the point particle need not be stationary relative to **O**, so **r** is a function of *t*, the time elapsed since an arbitrary initial time. In pre-Einstein relativity (known as Galilean relativity), time is considered an absolute, i.e., the time interval between any given pair of events is the same for all observers.[1] In addition to relying on absolute time, classical mechanics assumes Euclidean geometry for the structure of space.[2]

Velocity and speed

Main articles: Velocity and speed

The *velocity*, or the rate of change of position with time, is defined as the derivative of the position with respect to time:

$$\mathbf{v} = \frac{d\mathbf{r}}{dt}$$

In classical mechanics, velocities are directly additive and subtractive. For example, if one car traveling east at 60 km/h passes another car traveling east at 50 km/h, then from the perspective of the slower car, the faster car is traveling east at $60 - 50 = 10$ km/h. Whereas, from the perspective of the faster car, the slower car is moving 10 km/h to the west. Velocities are directly additive as vector quantities; they must be dealt with using vector analysis.

Mathematically, if the velocity of the first object in the previous discussion is denoted by the vector $\mathbf{u} = u\mathbf{d}$ and the velocity of the second object by the vector $\mathbf{v} = v\mathbf{e}$, where u is the speed of the first object, v is the speed of the second object, and **d** and **e** are unit vectors in the directions of motion of each particle respectively, then the velocity of the first object as seen by the second object is

$$\mathbf{u}' = \mathbf{u} - \mathbf{v}.$$

Similarly,

$$\mathbf{v}' = \mathbf{v} - \mathbf{u}.$$

When both objects are moving in the same direction, this equation can be simplified to

$$\mathbf{u}' = (u - v)\mathbf{d}\,.$$

Or, by ignoring direction, the difference can be given in terms of speed only:

$$u' = u - v\,.$$

Acceleration

Main article: Acceleration

The *acceleration*, or rate of change of velocity, is the derivative of the velocity with respect to time (the second derivative of the position with respect to time):

$$\mathbf{a} = \frac{\mathrm{d}\mathbf{v}}{\mathrm{d}t} = \frac{\mathrm{d}^2\mathbf{r}}{\mathrm{d}t^2}.$$

Acceleration represents the velocity's change over time: either of the velocity's magnitude or direction, or both. If only the magnitude v of the velocity decreases, this is sometimes referred to as *deceleration*, but generally any change in the velocity with time, including deceleration, is simply referred to as acceleration.

Frames of reference

Main articles: Inertial frame of reference and Galilean transformation

While the position, velocity and acceleration of a particle can be referred to any observer in any state of motion, classical mechanics assumes the existence of a special family of reference frames in terms of which the mechanical laws of nature take a comparatively simple form. These special reference frames are called inertial frames. An inertial frame is such that when an object without any force interactions (an idealized situation) is viewed from it, it appears either to be at rest or in a state of uniform motion in a straight line. This is the fundamental definition of an inertial frame. They are characterized by the requirement that all forces entering the observer's physical laws originate in identifiable sources (charges, gravitational bodies, and so forth). A non-inertial reference frame is one accelerating with respect to an inertial one, and in such a non-inertial frame a particle is subject to acceleration by fictitious forces that enter the equations of motion solely as a result of its accelerated motion, and do not originate in identifiable sources. These fictitious forces are in addition to the real forces recognized in an inertial

frame. A key concept of inertial frames is the method for identifying them. For practical purposes, reference frames that are unaccelerated with respect to the distant stars (an extremely distant point) are regarded as good approximations to inertial frames.

Consider two reference frames S and S'. For observers in each of the reference frames an event has space-time coordinates of (x,y,z,t) in frame S and (x',y',z',t') in frame S'. Assuming time is measured the same in all reference frames, and if we require x = x' when t = 0, then the relation between the space-time coordinates of the same event observed from the reference frames S' and S, which are moving at a relative velocity of u in the x direction is:

$$\begin{aligned}
x' &= x - u \cdot t \\
y' &= y \\
z' &= z \\
t' &= t.
\end{aligned}$$

This set of formulas defines a group transformation known as the Galilean transformation (informally, the *Galilean transform*). This group is a limiting case of the Poincaré group used in special relativity. The limiting case applies when the velocity u is very small compared to c, the speed of light.

The transformations have the following consequences:

- $\mathbf{v}' = \mathbf{v} - \mathbf{u}$ (the velocity $\mathbf{v}'$ of a particle from the perspective of S' is slower by $\mathbf{u}$ than its velocity $\mathbf{v}$ from the perspective of S)

- $\mathbf{a}' = \mathbf{a}$ (the acceleration of a particle is the same in any inertial reference frame)

- $\mathbf{F}' = \mathbf{F}$ (the force on a particle is the same in any inertial reference frame)

- the speed of light is not a constant in classical mechanics, nor does the special position given to the speed of light in relativistic mechanics have a counterpart in classical mechanics.

For some problems, it is convenient to use rotating coordinates (reference frames). Thereby one can either keep a mapping to a convenient inertial frame, or introduce additionally a fictitious centrifugal force and Coriolis force.

11.1.2   Forces; Newton's second law

Main articles: Force and Newton's laws of motion

Newton was the first to mathematically express the relationship between force and momentum. Some physicists interpret Newton's second law of motion as a definition of force and mass, while others consider it a fundamental postulate, a law of nature. Either interpretation has the same mathematical consequences, historically known as "Newton's Second Law":

$$F = \frac{dp}{dt} = \frac{d(mv)}{dt}.$$

The quantity mv is called the (canonical) momentum. The net force on a particle is thus equal to the rate of change of the momentum of the particle with time. Since the definition of acceleration is $a = dv/dt$, the second law can be written in the simplified and more familiar form:

$$F = ma.$$

So long as the force acting on a particle is known, Newton's second law is sufficient to describe the motion of a particle. Once independent relations for each force acting on a particle are available, they can be substituted into Newton's second law to obtain an ordinary differential equation, which is called the *equation of motion*.

As an example, assume that friction is the only force acting on the particle, and that it may be modeled as a function of the velocity of the particle, for example:

$$F_R = -\lambda v,$$

where λ is a positive constant. Then the equation of motion is

$$-\lambda v = ma = m\frac{dv}{dt}.$$

This can be integrated to obtain

$$v = v_0 e^{-\lambda t/m}$$

where v_0 is the initial velocity. This means that the velocity of this particle decays exponentially to zero as time progresses. In this case, an equivalent viewpoint is that the kinetic energy of the particle is absorbed by friction (which converts it to heat energy in accordance with the conservation of energy), and the particle is slowing down. This expression can be further integrated to obtain the position r of the particle as a function of time.

Important forces include the gravitational force and the Lorentz force for electromagnetism. In addition, Newton's third law can sometimes be used to deduce the forces acting on a particle: if it is known that particle A exerts a force F on another particle B, it follows that B must exert an equal and opposite *reaction force*, $-F$, on A. The strong form of Newton's third law requires that F and $-F$ act along the line connecting A and B, while the weak form does not. Illustrations of the weak form of Newton's third law are often found for magnetic forces.

11.1.3 Work and energy

Main articles: Work (physics), kinetic energy and potential energy

If a constant force F is applied to a particle that achieves a displacement Δr,*[note 2] the *work done* by the force is defined as the scalar product of the force and displacement vectors:

$$W = F \cdot \Delta r.$$

More generally, if the force varies as a function of position as the particle moves from r_1 to r_2 along a path C, the work done on the particle is given by the line integral

$$W = \int_C F(r) \cdot dr.$$

If the work done in moving the particle from r_1 to r_2 is the same no matter what path is taken, the force is said to be conservative. Gravity is a conservative force, as is the force due to an idealized spring, as given by Hooke's law. The force due to friction is non-conservative.

The kinetic energy E_k of a particle of mass m travelling at speed v is given by

$$E_k = \tfrac{1}{2}mv^2.$$

For extended objects composed of many particles, the kinetic energy of the composite body is the sum of the kinetic energies of the particles.

The work–energy theorem states that for a particle of constant mass m the total work W done on the particle from position r_1 to r_2 is equal to the change in kinetic energy E_k of the particle:

$$W = \Delta E_k = E_{k,2} - E_{k,1} = \tfrac{1}{2}m\left(v_2^2 - v_1^2\right).$$

Conservative forces can be expressed as the gradient of a scalar function, known as the potential energy and denoted E_p:

$$\mathbf{F} = -\nabla E_p.$$

If all the forces acting on a particle are conservative, and E_p is the total potential energy (which is defined as a work of involved forces to rearrange mutual positions of bodies), obtained by summing the potential energies corresponding to each force

$$\mathbf{F} \cdot \Delta \mathbf{r} = -\nabla E_p \cdot \Delta \mathbf{r} = -\Delta E_p \Rightarrow -\Delta E_p$$

$$= \Delta E_k \Rightarrow \Delta(E_k + E_p) = 0.$$

This result is known as *conservation of energy* and states that the total energy,

$$\sum E = E_k + E_p,$$

is constant in time. It is often useful, because many commonly encountered forces are conservative.

11.1.4   Beyond Newton's laws

Classical mechanics also includes descriptions of the complex motions of extended non-pointlike objects. Euler's laws provide extensions to Newton's laws in this area. The concepts of angular momentum rely on the same calculus used to describe one-dimensional motion. The rocket equation extends the notion of rate of change of an object's momentum to include the effects of an object "losing mass".

There are two important alternative formulations of classical mechanics: Lagrangian mechanics and Hamiltonian mechanics. These, and other modern formulations, usually bypass the concept of "force", instead referring to other physical quantities, such as energy, speed and momentum, for describing mechanical systems in generalized coordinates.

The expressions given above for momentum and kinetic energy are only valid when there is no significant electromagnetic contribution. In electromagnetism, Newton's second law for current-carrying wires breaks down unless one includes the electromagnetic field contribution to the momentum of the system as expressed by the Poynting vector divided by c^2, where c is the speed of light in free space.

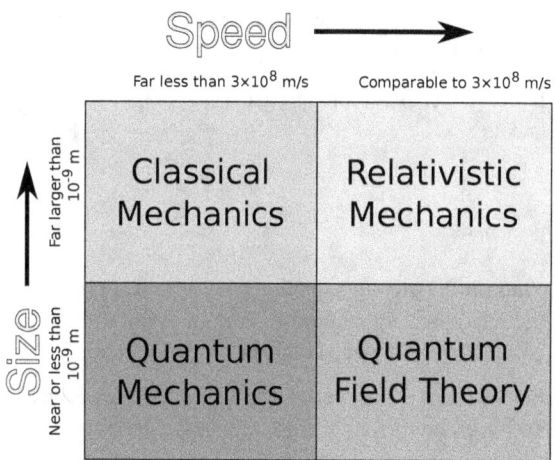

Domain of validity for Classical Mechanics

11.2   Limits of validity

Many branches of classical mechanics are simplifications or approximations of more accurate forms; two of the most accurate being general relativity and relativistic statistical mechanics. Geometric optics is an approximation to the quantum theory of light, and does not have a superior "classical" form.

When both quantum mechanics and classical mechanics cannot apply, such as at the quantum level with many degrees of freedom, quantum field theory (QFT) becomes applicable. QFT deals with small distances and large speeds with many degrees of freedom as well as the possibility of any change in the number of particles throughout the interaction. To deal with large degrees of freedom at the macroscopic level, statistical mechanics becomes valid. Statistical mechanics explores the large number of particles and their interactions as a whole in everyday life. Statistical mechanics is mainly used in thermodynamics. In the case of high velocity objects approaching the speed of light, classical mechanics is enhanced by special relativity. General relativity unifies special relativity with Newton's law of universal gravitation, allowing physicists to handle gravitation at a deeper level.

11.2.1   The Newtonian approximation to special relativity

In special relativity, the momentum of a particle is given by

$$\mathbf{p} = \frac{m\mathbf{v}}{\sqrt{1 - (v^2/c^2)}},$$

where m is the particle's rest mass, $\mathbf{v}$ its velocity, and c is the speed of light.

If v is very small compared to c, v^2/c^2 is approximately zero, and so

$$\mathbf{p} \approx m\mathbf{v}.$$

Thus the Newtonian equation $\mathbf{p} = m\mathbf{v}$ is an approximation of the relativistic equation for bodies moving with low speeds compared to the speed of light.

For example, the relativistic cyclotron frequency of a cyclotron, gyrotron, or high voltage magnetron is given by

$$f = f_c \frac{m_0}{m_0 + T/c^2},$$

where f_c is the classical frequency of an electron (or other charged particle) with kinetic energy T and (rest) mass m_0 circling in a magnetic field. The (rest) mass of an electron is 511 keV. So the frequency correction is 1% for a magnetic vacuum tube with a 5.11 kV direct current accelerating voltage.

11.2.2 The classical approximation to quantum mechanics

The ray approximation of classical mechanics breaks down when the de Broglie wavelength is not much smaller than other dimensions of the system. For non-relativistic particles, this wavelength is

$$\lambda = \frac{h}{p}$$

where h is Planck's constant and p is the momentum.

Again, this happens with electrons before it happens with heavier particles. For example, the electrons used by Clinton Davisson and Lester Germer in 1927, accelerated by 54 volts, had a wavelength of 0.167 nm, which was long enough to exhibit a single diffraction side lobe when reflecting from the face of a nickel crystal with atomic spacing of 0.215 nm. With a larger vacuum chamber, it would seem relatively easy to increase the angular resolution from around a radian to a milliradian and see quantum diffraction from the periodic patterns of integrated circuit computer memory.

More practical examples of the failure of classical mechanics on an engineering scale are conduction by quantum tunneling in tunnel diodes and very narrow transistor gates in integrated circuits.

Classical mechanics is the same extreme high frequency approximation as geometric optics. It is more often accurate because it describes particles and bodies with rest mass. These have more momentum and therefore shorter De Broglie wavelengths than massless particles, such as light, with the same kinetic energies.

11.3 History

Main article: History of classical mechanics
See also: Timeline of classical mechanics

Some Greek philosophers of antiquity, among them Aristotle, founder of Aristotelian physics, may have been the first to maintain the idea that "everything happens for a reason" and that theoretical principles can assist in the understanding of nature. While to a modern reader, many of these preserved ideas come forth as eminently reasonable, there is a conspicuous lack of both mathematical theory and controlled experiment, as we know it. These both turned out to be decisive factors in forming modern science, and they started out with classical mechanics.

In his *Elementa super demonstrationem ponderum*, medieval mathematician Jordanus de Nemore concept of "positional gravity" and the use of component forces.

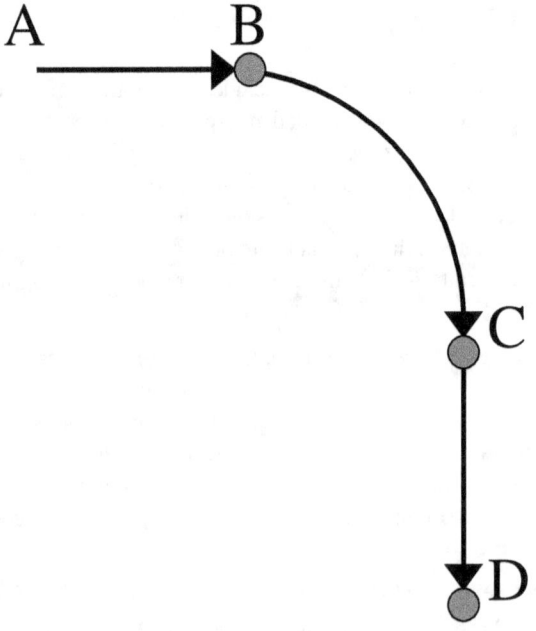

Three stage Theory of impetus according to Albert of Saxony.

The first published causal explanation of the motions of planets was Johannes Kepler's *Astronomia nova* published in 1609. He concluded, based on Tycho Brahe's observations of the orbit of Mars, that the orbits were ellipses. This break with ancient thought was happening around the same time that Galileo was proposing abstract mathematical laws for the motion of objects. He may (or may not) have performed the famous experiment of dropping two cannonballs of different weights from the tower of Pisa, showing that they both hit the ground at the same time. The reality of this experiment is disputed, but, more importantly, he did carry out quantitative experiments by rolling balls on an inclined plane. His theory of accelerated motion derived from the results of such experiments, and forms a cornerstone of classical mechanics.

Sir Isaac Newton (1643–1727), an influential figure in the history of physics and whose three laws of motion form the basis of classical mechanics

As foundation for his principles of natural philosophy, Isaac Newton proposed three laws of motion: the law of inertia, his second law of acceleration (mentioned above), and the law of action and reaction; and hence laid the foundations for classical mechanics. Both Newton's second and third laws were given the proper scientific and mathematical treatment in Newton's *Philosophiæ Naturalis Principia Mathematica*, which distinguishes them from earlier attempts at explaining similar phenomena, which were either incomplete, incorrect, or given little accurate mathematical expression. Newton also enunciated the principles of conservation of momentum and angular momentum. In mechanics, Newton was also the first to provide the first cor-

rect scientific and mathematical formulation of gravity in Newton's law of universal gravitation. The combination of Newton's laws of motion and gravitation provide the fullest and most accurate description of classical mechanics. He demonstrated that these laws apply to everyday objects as well as to celestial objects. In particular, he obtained a theoretical explanation of Kepler's laws of motion of the planets.

Newton previously invented the calculus, of mathematics, and used it to perform the mathematical calculations. For acceptability, his book, the *Principia*, was formulated entirely in terms of the long-established geometric methods, which were soon eclipsed by his calculus. However, it was Leibniz who developed the notation of the derivative and integral preferred*[3] today.

Hamilton's greatest contribution is perhaps the reformulation of Newtonian mechanics, now called Hamiltonian mechanics.

Newton, and most of his contemporaries, with the notable exception of Huygens, worked on the assumption that classical mechanics would be able to explain all phenomena, including light, in the form of geometric optics. Even when discovering the so-called Newton's rings (a wave interference phenomenon) his explanation remained with his own corpuscular theory of light.

After Newton, classical mechanics became a principal field of study in mathematics as well as physics. Several reformulations progressively allowed finding solutions to a far greater number of problems. The first notable reformulation was in 1788 by Joseph Louis Lagrange. Lagrangian mechanics was in turn re-formulated in 1833 by William Rowan Hamilton.

Some difficulties were discovered in the late 19th century that could only be resolved by more modern physics. Some of these difficulties related to compatibility with electromagnetic theory, and the famous Michelson–Morley experiment. The resolution of these problems led to the special theory of relativity, often included in the term classical mechanics.

A second set of difficulties were related to thermodynamics. When combined with thermodynamics, classical mechanics leads to the Gibbs paradox of classical statistical mechanics, in which entropy is not a well-defined quantity. Black-body radiation was not explained without the introduction of quanta. As experiments reached the atomic level, classical mechanics failed to explain, even approximately, such basic things as the energy levels and sizes of atoms and the photo-electric effect. The effort at resolving these problems led to the development of quantum mechanics.

Since the end of the 20th century, the place of classical mechanics in physics has been no longer that of an independent theory. Instead, classical mechanics is now considered an approximate theory to the more general quantum mechanics. Emphasis has shifted to understanding the fundamental forces of nature as in the Standard model and its more modern extensions into a unified theory of everything.[*][4] Classical mechanics is a theory for the study of the motion of non-quantum mechanical, low-energy particles in weak gravitational fields. In the 21st century classical mechanics has been extended into the complex domain and complex classical mechanics exhibits behaviors very similar to quantum mechanics.[*][5]

11.4 Branches

Classical mechanics was traditionally divided into three main branches:

- Statics, the study of equilibrium and its relation to forces

- Dynamics, the study of motion and its relation to forces

- Kinematics, dealing with the implications of observed motions without regard for circumstances causing them

Another division is based on the choice of mathematical formalism:

- Newtonian mechanics

- Lagrangian mechanics

- Hamiltonian mechanics

Alternatively, a division can be made by region of application:

- Celestial mechanics, relating to stars, planets and other celestial bodies

- Continuum mechanics, for materials modelled as a continuum, e.g., solids and fluids (i.e., liquids and gases).

- Relativistic mechanics (i.e. including the special and general theories of relativity), for bodies whose speed is close to the speed of light.

- Statistical mechanics, which provides a framework for relating the microscopic properties of individual atoms and molecules to the macroscopic or bulk thermodynamic properties of materials.

11.5 See also

- Dynamical systems

- History of classical mechanics

- List of equations in classical mechanics

- List of publications in classical mechanics

- Molecular dynamics

- Newton's laws of motion

- Special theory of relativity

- Quantum Mechanics

- Quantum Field Theory

11.6 Notes

[1] . The notion of "classical" may be somewhat confusing, insofar as this term usually refers to the era of classical antiquity in European history. While many discoveries within the mathematics of that period remain in full force today, and of the greatest use, much of the science that emerged then has since been superseded by more accurate models. This in no way detracts from the science of that time, though as most of modern physics is built directly upon the important developments, especially within technology, which took place

in antiquity and during the Middle Ages in Europe and elsewhere. However, the emergence of classical mechanics was a decisive stage in the development of science, in the modern sense of the term. What characterizes it, above all, is its insistence on mathematics (rather than speculation), and its reliance on experiment (rather than observation). With classical mechanics it was established how to formulate quantitative predictions in theory, and how to test them by carefully designed measurement. The emerging globally cooperative endeavor increasingly provided for much closer scrutiny and testing, both of theory and experiment. This was, and remains, a key factor in establishing certain knowledge, and in bringing it to the service of society. History shows how closely the health and wealth of a society depends on nurturing this investigative and critical approach.

[2] The displacement $\Delta\mathbf{r}$ is the difference of the particle's initial and final positions: $\Delta\mathbf{r} = \mathbf{r}_{final} - \mathbf{r}_{initial}$.

11.7   References

[1] Mughal, Muhammad Aurang Zeb. 2009. Time, absolute. Birx, H. James (ed.), *Encyclopedia of Time: Science, Philosophy, Theology, and Culture*, Vol. 3. Thousand Oaks, CA: Sage, pp. 1254-1255.

[2] MIT physics 8.01 lecture notes (page 12) (PDF)

[3] Jesseph, Douglas M. (1998). "Leibniz on the Foundations of the Calculus: The Question of the Reality of Infinitesimal Magnitudes". Perspectives on Science. 6.1&2: 6–40. Retrieved 31 December 2011.

[4] Page 2-10 of the *Feynman Lectures on Physics* says "For already in classical mechanics there was indeterminability from a practical point of view." The past tense here implies that classical physics is no longer fundamental.

[5] Complex Elliptic Pendulum, Carl M. Bender, Daniel W. Hook, Karta Kooner in Asymptotics in Dynamics, Geometry and PDEs; Generalized Borel Summation vol. I

11.8   Further reading

- Alonso, M.; Finn, J. (1992). *Fundamental University Physics*. Addison-Wesley.

- Feynman, Richard (1999). *The Feynman Lectures on Physics*. Perseus Publishing. ISBN 0-7382-0092-1.

- Feynman, Richard; Phillips, Richard (1998). *Six Easy Pieces*. Perseus Publishing. ISBN 0-201-32841-0.

- Goldstein, Herbert; Charles P. Poole; John L. Safko (2002). *Classical Mechanics* (3rd ed.). Addison Wesley. ISBN 0-201-65702-3.

- Kibble, Tom W.B.; Berkshire, Frank H. (2004). *Classical Mechanics (5th ed.)*. Imperial College Press. ISBN 978-1-86094-424-6.

- Kleppner, D.; Kolenkow, R. J. (1973). *An Introduction to Mechanics*. McGraw-Hill. ISBN 0-07-035048-5.

- Landau, L.D.; Lifshitz, E.M. (1972). *Course of Theoretical Physics, Vol. 1—Mechanics*. Franklin Book Company. ISBN 0-08-016739-X.

- Morin, David (2008). *Introduction to Classical Mechanics: With Problems and Solutions* (1st ed.). Cambridge, UK: Cambridge University Press. ISBN 978-0-521-87622-3.*Gerald Jay Sussman; Jack Wisdom (2001). *Structure and Interpretation of Classical Mechanics*. MIT Press. ISBN 0-262-19455-4.

- O'Donnell, Peter J. (2015). *Essential Dynamics and Relativity*. CRC Press. ISBN 978-1-466-58839-4.

- Thornton, Stephen T.; Marion, Jerry B. (2003). *Classical Dynamics of Particles and Systems (5th ed.)*. Brooks Cole. ISBN 0-534-40896-6.

11.9   External links

- Crowell, Benjamin. Newtonian Physics (an introductory text, uses algebra with optional sections involving calculus)

- Fitzpatrick, Richard. Classical Mechanics (uses calculus)

- Hoiland, Paul (2004). Preferred Frames of Reference & Relativity

- Horbatsch, Marko, "*Classical Mechanics Course Notes*".

- Rosu, Haret C., "*Classical Mechanics*". Physics Education. 1999. [arxiv.org : physics/9909035]

- Shapiro, Joel A. (2003). Classical Mechanics

- Sussman, Gerald Jay & Wisdom, Jack & Mayer,Meinhard E. (2001). Structure and Interpretation of Classical Mechanics

- Tong, David. Classical Dynamics (Cambridge lecture notes on Lagrangian and Hamiltonian formalism)

- Kinematic Models for Design Digital Library (KMODDL)
  Movies and photos of hundreds of working mechanical-systems models at Cornell University. Also includes an e-book library of classic texts on mechanical design and engineering.

- MIT OpenCourseWare 8.01: Classical Mechanics Free videos of actual course lectures with links to lecture notes, assignments and exams.

- Alejandro A. Torassa On Classical Mechanics

Chapter 12

History of thermodynamics

*The 1698 **Savery Engine** – the world's first commercially-useful steam engine: built by Thomas Savery*

The **history of thermodynamics** is a fundamental strand in the history of physics, the history of chemistry, and the history of science in general. Owing to the relevance of thermodynamics in much of science and technology, its history is finely woven with the developments of classical mechanics, quantum mechanics, magnetism, and chemical kinetics, to more distant applied fields such as meteorology, information theory, and biology (physiology), and to technological developments such as the steam engine, internal combustion engine, cryogenics and electricity generation. The development of thermodynamics both drove and was driven by atomic theory. It also, albeit in a sub-

tle manner, motivated new directions in probability and statistics; see, for example, the timeline of thermodynamics.

12.1 History

See also: Timeline of thermodynamics

12.1.1 Contributions from ancient and medieval times

See also: History of heat and Vacuum

The ancients viewed heat as that related to fire. In 3000 BC, the ancient Egyptians viewed heat as related to origin mythologies.[1] In the Western philosophical tradition, after much debate about the primal element among earlier pre-Socratic philosophers, Empedocles proposed a four-element theory, in which all substances derive from earth, water, air, and fire. The Empedoclean element of fire is perhaps the principal ancestor of later concepts such as phlogiston and caloric. Around 500 BC, the Greek philosopher Heraclitus became famous as the "flux and fire" philosopher for his proverbial utterance: "All things are flowing." Heraclitus argued that the three principal elements in nature were fire, earth, and water.

Atomism is a central part of today's relationship between thermodynamics and statistical mechanics. Ancient thinkers such as Leucippus and Democritus, and later the Epicureans, by advancing atomism, laid the foundations for the later atomic theory. Until experimental proof of atoms was later provided in the 20th century, the atomic theory was driven largely by philosophical considerations and scientific intuition.

The 5th century BC, Greek philosopher Parmenides, in his only known work, a poem conventionally titled *On Nature*,

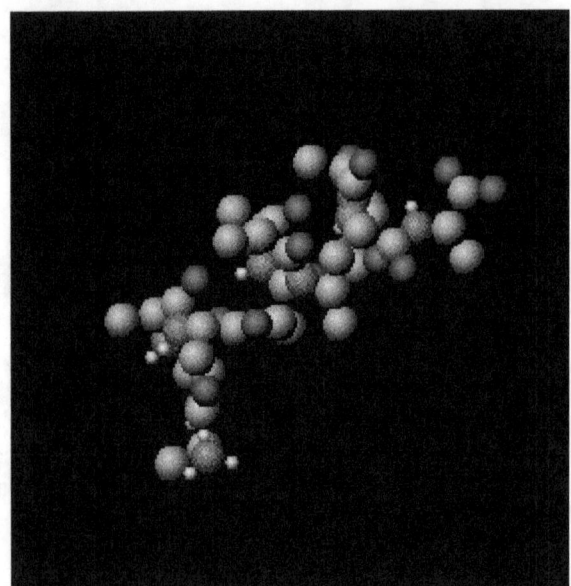

***Heating** a body, such as a segment of protein alpha helix (above), tends to cause its atoms to vibrate more, and to expand or change phase, if heating is continued; an axiom of nature noted by Herman Boerhaave in the in 1700s.*

uses verbal reasoning to postulate that a void, essentially what is now known as a vacuum, in nature could not occur. This view was supported by the arguments of Aristotle, but was criticized by Leucippus and Hero of Alexandria. From antiquity to the Middle Ages various arguments were put forward to prove or disapprove the existence of a vacuum and several attempts were made to construct a vacuum but all proved unsuccessful.

The European scientists Cornelius Drebbel, Robert Fludd, Galileo Galilei and Santorio Santorio in the 16th and 17th centuries were able to gauge the relative "coldness" or "hotness" of air, using a rudimentary air thermometer (or thermoscope). This may have been influenced by an earlier device which could expand and contract the air constructed by Philo of Byzantium and Hero of Alexandria.

Around 1600, the English philosopher and scientist Francis Bacon surmised: "Heat itself, its essence and quiddity is motion and nothing else." In 1643, Galileo Galilei, while generally accepting the 'sucking' explanation of *horror vacui* proposed by Aristotle, believed that nature's vacuum-abhorrence is limited. Pumps operating in mines had already proven that nature would only fill a vacuum with water up to a height of ~30 feet. Knowing this curious fact, Galileo encouraged his former pupil Evangelista Torricelli to investigate these supposed limitations. Torricelli did not believe that vacuum-abhorrence (*Horror vacui*) in the sense of Aristotle's 'sucking' perspective, was responsible for raising the water. Rather, he reasoned, it was the result of the pressure exerted on the liquid by the surrounding air.

To prove this theory, he filled a long glass tube (sealed at one end) with mercury and upended it into a dish also containing mercury. Only a portion of the tube emptied (as shown adjacent); ~30 inches of the liquid remained. As the mercury emptied, and a partial vacuum was created at the top of the tube. The gravitational force on the heavy element Mercury prevented it from filling the vacuum.

12.1.2   Transition from chemistry to thermo-chemistry

See also: History of chemistry

The theory of phlogiston arose in the 17th century, late

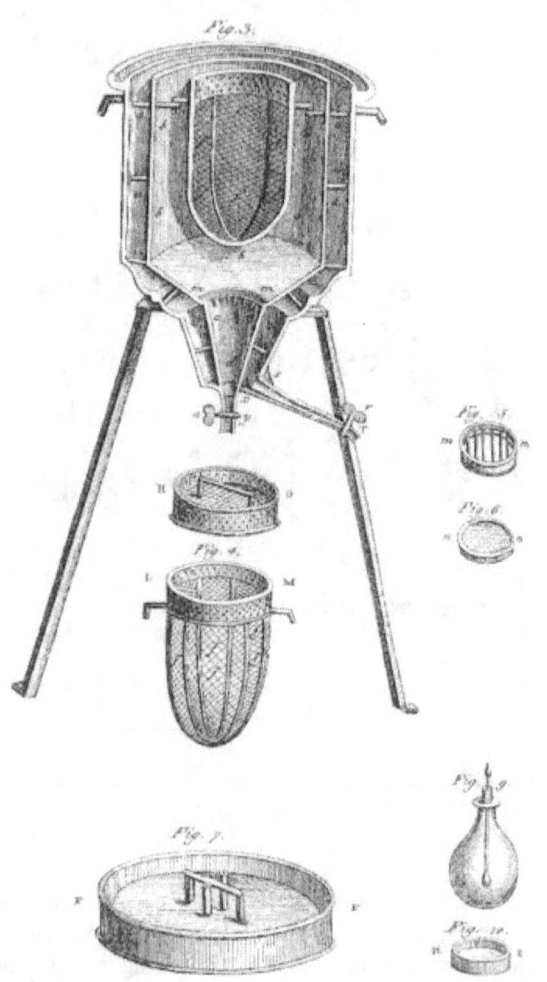

*The world's first **ice-calorimeter**, used in the winter of 1782-83, by Antoine Lavoisier and Pierre-Simon Laplace, to determine the heat evolved in various chemical changes; calculations which were based on Joseph Black's prior discovery of latent heat. These experiments mark the foundation of thermochemistry.*

in the period of alchemy. Its replacement by caloric theory in the 18th century is one of the historical markers of the transition from alchemy to chemistry. Phlogiston was a hypothetical substance that was presumed to be liberated from combustible substances during burning, and from metals during the process of rusting. Caloric, like phlogiston, was also presumed to be the "substance" of heat that would flow from a hotter body to a cooler body, thus warming it.

The first substantial experimental challenges to caloric theory arose in Rumford's 1798 work, when he showed that boring cast iron cannons produced great amounts of heat which he ascribed to friction, and his work was among the first to undermine the caloric theory. The development of the steam engine also focused attention on calorimetry and the amount of heat produced from different types of coal. The first quantitative research on the heat changes during chemical reactions was initiated by Lavoisier using an ice calorimeter following research by Joseph Black on the latent heat of water.

More quantitative studies by James Prescott Joule in 1843 onwards provided soundly reproducible phenomena, and helped to place the subject of thermodynamics on a solid footing. William Thomson, for example, was still trying to explain Joule's observations within a caloric framework as late as 1850. The utility and explanatory power of kinetic theory, however, soon started to displace caloric and it was largely obsolete by the end of the 19th century. Joseph Black and Lavoisier made important contributions in the precise measurement of heat changes using the calorimeter, a subject which became known as thermochemistry.

12.1.3 Phenomenological thermodynamics

- Boyle's law (1662)

- Charles's law was first published by Joseph Louis Gay-Lussac in 1802, but he referenced unpublished work by Jacques Charles from around 1787. The relationship had been anticipated by the work of Guillaume Amontons in 1702.

- Gay-Lussac's law (1802)

12.1.4 Birth of thermodynamics as science

At its origins, thermodynamics was the study of engines. A precursor of the engine was designed by the German scientist Otto von Guericke who, in 1650, designed and built the world's first vacuum pump and created the world's first ever vacuum known as the Magdeburg hemispheres. He was driven to make a vacuum in order to disprove Aristotle's long-held supposition that 'Nature abhors a vacuum'.

Shortly thereafter, Irish physicist and chemist Robert Boyle had learned of Guericke's designs and in 1656, in coordina-

Robert Boyle. 1627-1691

tion with English scientist Robert Hooke, built an air pump. Using this pump, Boyle and Hooke noticed the pressure-volume correlation: P.V=constant. In that time, air was assumed to be a system of motionless particles, and not interpreted as a system of moving molecules. The concept of thermal motion came two centuries later. Therefore Boyle's publication in 1660 speaks about a mechanical concept: the air spring.[2] Later, after the invention of the thermometer, the property temperature could be quantified. This tool gave Gay-Lussac the opportunity to derive his law, which led shortly later to the ideal gas law. But, already before the establishment of the ideal gas law, an associate of Boyle's named Denis Papin built in 1679 a bone digester, which is a closed vessel with a tightly fitting lid that confines steam until a high pressure is generated.

Later designs implemented a steam release valve to keep the machine from exploding. By watching the valve rhythmically move up and down, Papin conceived of the idea of a piston and cylinder engine. He did not however follow through with his design. Nevertheless, in 1697, based on Papin's designs, engineer Thomas Savery built the first engine. Although these early engines were crude and inefficient, they attracted the attention of the leading scientists of the time. One such scientist was Sadi Carnot, the "father of thermodynamics", who in 1824 published *Reflections on the Motive Power of Fire*, a discourse on heat, power, and engine efficiency. This marks the start of thermodynamics as a modern science.

A Watt steam engine, the steam engine that propelled the Industrial Revolution in Britain and the world

Hence, prior to 1698 and the invention of the Savery Engine, horses were used to power pulleys, attached to buckets, which lifted water out of flooded salt mines in England. In the years to follow, more variations of steam engines were built, such as the Newcomen Engine, and later the Watt Engine. In time, these early engines would eventually be utilized in place of horses. Thus, each engine began to be associated with a certain amount of "horse power" depending upon how many horses it had replaced. The main problem with these first engines was that they were slow and clumsy, converting less than 2% of the input fuel into useful work. In other words, large quantities of coal (or wood) had to be burned to yield only a small fraction of work output. Hence the need for a new science of engine dynamics was born.

Most cite Sadi Carnot's 1824 book *Reflections on the Motive Power of Fire* as the starting point for thermodynamics as a modern science. Carnot defined "motive power" to be the expression of the *useful effect* that a motor is capable of producing. Herein, Carnot introduced us to the first modern day definition of "work": *weight lifted through a height*. The desire to understand, via formulation, this *useful effect* in relation to "work" is at the core of all modern day thermodynamics.

In 1843, James Joule experimentally found the mechanical equivalent of heat. In 1845, Joule reported his best-known experiment, involving the use of a falling weight to spin a paddle-wheel in a barrel of water, which allowed him to estimate a mechanical equivalent of heat of 819 ft·lbf/Btu (4.41 J/cal). This led to the theory of conservation of energy and explained why heat can do work.

In 1850, the famed mathematical physicist Rudolf Clausius defined the term entropy *S* to be the heat lost or turned into waste, stemming from the Greek word *entrepein* meaning

Sadi Carnot (1796-1832): the "father" of thermodynamics

to turn.

The name "thermodynamics", however, did not arrive until 1854, when the British mathematician and physicist William Thomson (Lord Kelvin) coined the term *thermo-dynamics* in his paper *On the Dynamical Theory of Heat.*[3]

In association with Clausius, in 1871, the Scottish mathematician and physicist James Clerk Maxwell formulated a new branch of thermodynamics called *Statistical Thermodynamics*, which functions to analyze large numbers of particles at equilibrium, i.e., systems where no changes are occurring, such that only their average properties as temperature *T*, pressure *P*, and volume *V* become important.

Soon thereafter, in 1875, the Austrian physicist Ludwig Boltzmann formulated a precise connection between entropy *S* and molecular motion:

$$S = k \log W$$

being defined in terms of the number of possible states [W] such motion could occupy, where k is the Boltzmann's constant.

The following year, 1876, was a seminal point in the development of human thought. During this essential period, chemical engineer Willard Gibbs, the first person in America to be awarded a PhD in engineering (Yale), published an obscure 300-page paper titled: *On the Equilibrium of Heterogeneous Substances*, wherein he formulated one grand equality, the Gibbs free energy equation, which suggested a measure of the amount of "useful work" attainable in reacting systems. Gibbs also originated the concept we now know as enthalpy *H*, calling it "a heat function for constant pressure".*[4] The modern word *enthalpy* would be coined many years later by Heike Kamerlingh Onnes,*[5] who based it on the Greek word *enthalpein* meaning *to warm*.

Building on these foundations, those as Lars Onsager, Erwin Schrödinger, and Ilya Prigogine, and others, functioned to bring these engine "concepts" into the thoroughfare of almost every modern-day branch of science.

12.1.5 Kinetic theory

Main article: Kinetic theory

The idea that heat is a form of motion is perhaps an ancient one and is certainly discussed by Francis Bacon in 1620 in his *Novum Organum*. The first written scientific reflection on the microscopic nature of heat is probably to be found in a work by Mikhail Lomonosov, in which he wrote:

> "(..) movement should not be denied based on the fact it is not seen. Who would deny that the leaves of trees move when rustled by a wind, despite it being unobservable from large distances? Just as in this case motion remains hidden due to perspective, it remains hidden in warm bodies due to the extremely small sizes of the moving particles. In both cases, the viewing angle is so small that neither the object nor their movement can be seen."

During the same years, Daniel Bernoulli published his book *Hydrodynamics* (1738), in which he derived an equation for the pressure of a gas considering the collisions of its atoms with the walls of a container. He proves that this pressure is two thirds the average kinetic energy of the gas in a unit volume. Bernoulli's ideas, however, made little impact on the dominant caloric culture. Bernoulli made a connection with Gottfried Leibniz's *vis viva* principle, an early formulation of the principle of conservation of energy, and the two theories became intimately entwined throughout their history. Though Benjamin Thompson suggested that heat was a form of motion as a result of his experiments in 1798, no attempt was made to reconcile theoretical and experimental approaches, and it is unlikely that he was thinking of the *vis viva* principle.

John Herapath later independently formulated a kinetic theory in 1820, but mistakenly associated temperature with momentum rather than *vis viva* or kinetic energy. His work ultimately failed peer review and was neglected. John James Waterston in 1843 provided a largely accurate account, again independently, but his work received the same reception, failing peer review even from someone as well-disposed to the kinetic principle as Davy.

Further progress in kinetic theory started only in the middle of the 19th century, with the works of Rudolf Clausius, James Clerk Maxwell, and Ludwig Boltzmann. In his 1857 work *On the nature of the motion called heat*, Clausius for the first time clearly states that heat is the average kinetic energy of molecules. This interested Maxwell, who in 1859 derived the momentum distribution later named after him. Boltzmann subsequently generalized his distribution for the case of gases in external fields.

Boltzmann is perhaps the most significant contributor to kinetic theory, as he introduced many of the fundamental concepts in the theory. Besides the Maxwell–Boltzmann distribution mentioned above, he also associated the kinetic energy of particles with their degrees of freedom. The Boltzmann equation for the distribution function of a gas in non-equilibrium states is still the most effective equation for studying transport phenomena in gases and metals. By introducing the concept of thermodynamic probability as the number of microstates corresponding to the current macrostate, he showed that its logarithm is proportional to entropy.

12.2 Branches of

The following list gives a rough outline as to when the major branches of thermodynamics came into inception:

- Thermochemistry - 1780s

- Classical thermodynamics - 1824

- Chemical thermodynamics - 1876

- Statistical mechanics - c. 1880s

- Equilibrium thermodynamics

- Engineering thermodynamics

- Chemical engineering thermodynamics - c. 1940s

- Non-equilibrium thermodynamics - 1941

- Small systems thermodynamics - 1960s

- Biological thermodynamics - 1957

- Ecosystem thermodynamics - 1959

- Relativistic thermodynamics - 1965

- Quantum thermodynamics - 1968

- Black hole thermodynamics - c. 1970s

- Geological thermodynamics - c. 1970s

- Biological evolution thermodynamics - 1978

- Geochemical thermodynamics - c. 1980s

- Atmospheric thermodynamics - c. 1980s

- Natural systems thermodynamics - 1990s

- Supramolecular thermodynamics - 1990s

- Earthquake thermodynamics - 2000

- Drug-receptor thermodynamics - 2001

- Pharmaceutical systems thermodynamics – 2002

Ideas from thermodynamics have also been applied in other fields, for example:

- Thermoeconomics - c. 1970s

12.3   Entropy and the second law

Main article: History of entropy

Even though he was working with the caloric theory, Sadi Carnot in 1824 suggested that some of the caloric available for generating useful work is lost in any real process. In March 1851, while grappling to come to terms with the work of James Prescott Joule, Lord Kelvin started to speculate that there was an inevitable loss of useful heat in all processes. The idea was framed even more dramatically by Hermann von Helmholtz in 1854, giving birth to the spectre of the heat death of the universe.

In 1854, William John Macquorn Rankine started to make use in calculation of what he called his *thermodynamic function*. This has subsequently been shown to be identical to the concept of entropy formulated by Rudolf Clausius in 1865. Clausius used the concept to develop his classic statement of the second law of thermodynamics the same year.

12.4   Heat transfer

Main article: Heat transfer

The phenomenon of heat conduction is immediately grasped in everyday life. In 1701, Sir Isaac Newton published his law of cooling. However, in the 17th century, it came to be believed that all materials had an identical conductivity and that differences in sensation arose from their different heat capacities.

Suggestions that this might not be the case came from the new science of electricity in which it was easily apparent that some materials were good electrical conductors while others were effective insulators. Jan Ingen-Housz in 1785-9 made some of the earliest measurements, as did Benjamin Thompson during the same period.

The fact that warm air rises and the importance of the phenomenon to meteorology was first realised by Edmund Halley in 1686. Sir John Leslie observed that the cooling effect of a stream of air increased with its speed, in 1804.

Carl Wilhelm Scheele distinguished heat transfer by thermal radiation (radiant heat) from that by convection and conduction in 1777. In 1791, Pierre Prévost showed that all bodies radiate heat, no matter how hot or cold they are. In 1804, Leslie observed that a matte black surface radiates heat more effectively than a polished surface, suggesting the importance of black body radiation. Though it had become to be suspected even from Scheele's work, in 1831 Macedonio Melloni demonstrated that black body radiation could be reflected, refracted and polarised in the same way as light.

James Clerk Maxwell's 1862 insight that both light and radiant heat were forms of electromagnetic wave led to the start of the quantitative analysis of thermal radiation. In 1879, Jožef Stefan observed that the total radiant flux from a blackbody is proportional to the fourth power of its temperature and stated the Stefan–Boltzmann law. The law was derived theoretically by Ludwig Boltzmann in 1884.

12.5   Cryogenics

In 1702 Guillaume Amontons introduced the concept of absolute zero based on observations of gases. In 1810, Sir John Leslie froze water to ice artificially. The idea of absolute zero was generalised in 1848 by Lord Kelvin. In 1906, Walther Nernst stated the third law of thermodynamics.

12.6   See also

- Conservation of energy: Historical development

- History of Chemistry

- History of Physics

- Maxwell's thermodynamic surface

- Timeline of thermodynamics, statistical mechanics, and random processes

- Thermodynamics

- Timeline of heat engine technology

- Timeline of low-temperature technology

12.7   References

[1] J. Gwyn Griffiths (1955). "The Orders of Gods in Greece and Egypt (According to Herodotus)". *The Journal of Hellenic Studies* **75**: 21–23. doi:10.2307/629164. JSTOR 629164.

[2] New Experiments physico-mechanicall, Touching the Spring of the Air and its Effects (1660).

[3] Thomson, W. (1854). "On the Dynamical Theory of Heat". *Transactions of the Royal Society of Edinburgh* **21** (part I): 123. doi:10.1017/s0080456800032014. |chapter= ignored (help) reprinted in Sir William Thomson, LL.D. D.C.L., F.R.S. (1882). *Mathematical and Physical Papers* **1**. London, Cambridge: C.J. Clay, M.A. & Son, Cambridge University Press. p. 232. Hence Thermo-dynamics falls naturally into two Divisions, of which the subjects are respectively, *the relation of heat to the forces acting between contiguous parts of bodies, and the relation of heat to electrical agency.*

[4] Laidler, Keith (1995). *The World of Physical Chemistry*. Oxford University Press. p. 110.

[5] Howard, Irmgard (2002). "H Is for Enthalpy, Thanks to Heike Kamerlingh Onnes and Alfred W. Porter". *Journal of Chemical Education* (ACS Publications) **79** (6): 697. Bibcode:2002JChEd..79..697H. doi:10.1021/ed079p697.

12.8   Further reading

- Cardwell, D.S.L. (1971). *From Watt to Clausius: The Rise of Thermodynamics in the Early Industrial Age*. London: Heinemann. ISBN 0-435-54150-1.

- Leff, H.S. & Rex, A.F. (eds) (1990). *Maxwell's Demon: Entropy, Information and Computing*. Bristol: Adam Hilger. ISBN 0-7503-0057-4.

12.9   External links

- History of Statistical Mechanics and Thermodynamics - Timeline (1575 to 1980) @ Hyperjeff.net

- History of Thermodynamics - University of Waterloo

- Thermodynamic History Notes - Wolfram-Science.com

- Brief History of Thermodynamics - Berkeley [PDF]

- History of Thermodynamics - Thermodynamic-Study.net

- Historical Background of Thermodynamics - Carnegie-Mellon University

- History of Thermodynamics - In Pictures

Chapter 13

History of special relativity

"History of relativity" redirects here. For the history of general relativity, see history of general relativity.

The **history of special relativity** consists of many theoretical results and empirical findings obtained by Albert A. Michelson, Hendrik Lorentz, Henri Poincaré and others. It culminated in the theory of special relativity proposed by Albert Einstein and subsequent work of Max Planck, Hermann Minkowski and others.

13.1 Introduction

Although Isaac Newton based his physics on absolute time and space, he also adhered to the principle of relativity of Galileo Galilei. This can be stated as: as far as the laws of mechanics are concerned, all observers in inertial motion are equally privileged, and no preferred state of motion can be attributed to any particular inertial observer. However, as to electromagnetic theory and electrodynamics, during the 19th century the wave theory of light as a disturbance of a "light medium" or Luminiferous ether was widely accepted, the theory reaching its most developed form in the work of James Clerk Maxwell. According to Maxwell's theory, all optical and electrical phenomena propagate through that medium, which suggested that it should be possible to experimentally determine motion relative to the aether.

The failure of any known experiment to detect motion through the aether led Hendrik Lorentz, starting in 1892, to develop a theory of electrodynamics based on an immobile luminiferous aether (about whose material constitution Lorentz didn't speculate), physical length contraction, and a "local time" in which Maxwell's equations retain their form in all inertial frames of reference. Working with Lorentz's aether theory, Henri Poincaré, having earlier proposed the "relativity principle" as a general law of nature (including electrodynamics and gravitation), used this principle in 1905 to correct Lorentz's preliminary transformation formulas, resulting in an exact set of equations that are now called the Lorentz transformations. A little later in the same year Albert Einstein published his original paper on special relativity in which, again based on the relativity principle, he independently derived and radically reinterpreted the Lorentz transformations by changing the fundamental definitions of space and time intervals, while abandoning the absolute simultaneity of Galilean kinematics, thus avoiding the need for any reference to a luminiferous aether in classical electrodynamics.[*][1] Subsequent work of Hermann Minkowski, in which he introduced a 4-dimensional geometric "spacetime" model for Einstein's version of special relativity, paved the way for Einstein's later development of his general theory of relativity and laid the foundations of relativistic field theories.

13.2 Aether and electrodynamics of moving bodies

13.2.1 Aether models and Maxwell's equations

Following the work of Thomas Young (1804) and Augustin-Jean Fresnel (1816), it was believed that light propagates as a transverse wave within an elastic medium called luminiferous aether. However, a distinction was made between optical and electrodynamical phenomena so it was necessary to create specific aether models for all phenomena. Attempts to unify those models or to create a complete mechanical description of them did not succeed,[*][2] but after considerable work by many scientists, including Michael Faraday and Lord Kelvin, James Clerk Maxwell (1864) developed an accurate theory of electromagnetism by deriving a set of equations in electricity, magnetism and inductance, named Maxwell's equations. He first proposed that light was in fact undulations (electromagnetic radiation) in the *same* aetherial medium that is the cause of electric and magnetic phenomena. However, Maxwell's theory was unsatisfactory regarding the optics of moving bodies, and while he was able to present a complete mathematical model, he was not able to provide a coherent mechanical description of the aether.[*][3]

After Heinrich Hertz in 1887 demonstrated the existence of electromagnetic waves, Maxwell's theory was widely accepted. In addition, Oliver Heaviside and Hertz further developed the theory and introduced modernized versions of Maxwell's equations. The "Maxwell-Hertz" or "Heaviside-Hertz" equations subsequently formed an important basis for the further development of electrodynamics, and Heaviside's notation is still used today. Other important contributions to Maxwell's theory were made by George FitzGerald, Joseph John Thomson, John Henry Poynting, Hendrik Lorentz, and Joseph Larmor.[*][4][*][5]

13.2.2   Search for the aether

Regarding the relative motion and the mutual influence of matter and aether, there were two controversial theories.
One of them was developed by Fresnel (and subsequently Lorentz). This model (Stationary Aether Theory) supposed that light propagates as a transverse wave and aether is partially dragged with a certain coefficient by matter. Based on this assumption, Fresnel was able to explain the aberration of light and many optical phenomena.[*][6]
The other hypothesis was proposed by George Gabriel Stokes, who stated in 1845 that the aether was *fully* dragged by matter (later this view was also shared by Hertz). In this model the aether might be (by analogy with pine pitch) rigid for fast objects and fluid for slower objects. Thus the Earth could move through it fairly freely, but it would be rigid enough to transport light.[*][7] Fresnel's theory was preferred because his dragging coefficient was confirmed by the Fizeau experiment in 1851, who measured the speed of light in moving liquids.[*][8]

Albert A. Michelson (1881) tried to measure the relative motion of the Earth and aether (Aether-Wind), as it was expected in Fresnel's theory, by using an interferometer. He could not determine any relative motion, so he interpreted the result as a confirmation of the thesis of Stokes.[*][9] However, Lorentz (1886) showed Michelson's calculations were wrong and that he had overestimated the accuracy of the measurement. This, together with the large margin of error, made the result of Michelson's experiment inconclusive. In addition, Lorentz showed that Stokes' completely dragged aether led to contradictory consequences, and therefore he supported an aether theory similar to Fresnel's.[*][10] To check Fresnel's theory again, Michelson and Edward W. Morley (1886) performed a repetition of the Fizeau experiment. Fresnel's dragging coefficient was confirmed very exactly on that occasion, and Michelson was now of the opinion that Fresnel's stationary aether theory was correct.[*][11]

A. A. Michelson

To clarify the situation, Michelson and Morley (1887) repeated Michelson's 1881-experiment, and they substantially increased the accuracy of the measurement. However, this now famous Michelson–Morley experiment again yielded a negative result, i.e., no motion of the apparatus through the aether was detected (although the Earth's velocity is 60 km/s different in winter than summer). So the physicists were confronted with two seemingly contradictory experiments: the 1886-experiment as an apparent confirmation of Fresnel's stationary aether, and the 1887-experiment as an apparent confirmation of Stokes' completely dragged aether.[*][12]

A possible solution to the problem was shown by Woldemar Voigt (1887), who investigated the Doppler effect for waves propagating in an incompressible elastic medium and deduced transformation relations that left the wave equation in free space unchanged, and explained the negative result of the Michelson–Morley experiment. The Voigt transformations include the Lorentz factor $1/\sqrt{1-v^2/c^2}$ for the y- and z-coordinates, and a new time variable $t' = t - vx/c^2$ which later was called "local time". However, Voigt's work was completely ignored by his contemporaries.[*][13][*][14]

FitzGerald (1889) offered another explanation of the negative result of the Michelson–Morley experiment. Contrary to Voigt, he speculated that the intermolecular forces are possibly of electrical origin so that material bodies would contract in the line of motion (length contraction). This was in connection with the work of Heaviside (1887), who determined that the electrostatic fields in motion were deformed (Heaviside Ellipsoid), which leads to physically undetermined conditions at the speed of light.[15] However, FitzGerald's idea remained widely unknown and was not discussed before Oliver Lodge published a summary of the idea in 1892.[16] Also Lorentz (1892b) proposed length contraction independently from FitzGerald in order to explain the Michelson–Morley experiment. For plausibility reasons, Lorentz referred to the analogy of the contraction of electrostatic fields. However, even Lorentz admitted that that was not a necessary reason and length-contraction consequently remained an ad hoc hypothesis.[17][18]

13.2.3   Lorentz's theory of electrons

Hendrik Antoon Lorentz

Lorentz (1892a) set the foundations of Lorentz aether theory, by assuming the existence of electrons which he separated from the aether, and by replacing the "Maxwell-Hertz" Equations by the "Maxwell-Lorentz" Equations. In his model, the aether is completely motionless and, contrary to Fresnel's theory, also is not partially dragged by matter. An important consequence of this notion was that the velocity of light is totally independent of the velocity

of the source. Lorentz gave no statements about the mechanical nature of the aether and the electromagnetic processes, but, vice versa, tried to explain the mechanical processes by electromagnetic ones and therefore created an abstract electromagnetic æther. In the framework of his theory, Lorentz calculated, like Heaviside, the contraction of the electrostatic fields.[19] Lorentz (1895) also introduced what he called the "Theorem of Corresponding States" for terms of first order in v/c. This theorem states that a moving observer (relative to the aether) in his "fictitious" field makes the same observations as a resting observer in his "real" field. An important part of it was local time $t'=t-vx/c^2$, which paved the way to the Lorentz transformation and which he introduced independently of Voigt. With the help of this concept, Lorentz could explain the aberration of light, the Doppler effect and the Fizeau experiment as well. However, Lorentz's local time was only an auxiliary mathematical tool to simplify the transformation from one system into another – it was Poincaré in 1900 who recognized that "local time" is actually indicated by moving clocks.[20][21][22] Lorentz also recognized that his theory violated the principle of action and reaction, since the aether acts on matter, but matter cannot act on the immobile aether.[23]

A very similar model was created by Joseph Larmor (1897, 1900). Larmor was the first to put Lorentz's 1895-transformation into a form algebraically equivalent to the modern Lorentz transformations, however, he stated that his transformations preserved the form of Maxwell's equations only to second order of v/c. Lorentz later noted that these transformations did in fact preserve the form of Maxwell's equations to all orders of v/c. Larmor noticed on that occasion, that not only can length-contraction be derived from it, but he also calculated some sort of time dilation for electron orbits. Larmor specified his considerations in 1900 and 1904.[14][24] Independently of Larmor, also Lorentz (1899) extended his transformation for second order terms and noted a (mathematical) Time Dilation effect as well.

Other physicists besides Lorentz and Larmor also tried to develop a consistent model of electrodynamics. For example, Emil Cohn (1900, 1901) created an alternative Electrodynamics in which he, as one of the first, discarded the existence of the aether (at least in the previous form) and would use, like Ernst Mach, the fixed stars as a reference frame instead. Due to inconsistencies within his theory, like different light speeds in different directions, it was superseded by Lorentz's and Einstein's.[25]

13.2.4   Electromagnetic mass

Main article: Electromagnetic mass

During his development of Maxwell's Theory, J. J. Thomson (1881) recognized that charged bodies are harder to set in motion than uncharged bodies. Electrostatic fields behave as if they add an "electromagnetic mass" to the mechanical mass of the bodies. I.e., according to Thomson, electromagnetic energy corresponds to a certain mass. This was interpreted as some form of self-inductance of the electromagnetic field.[26][27] He also noticed that the mass of a body *in motion* is increased by a constant quantity. Thomson's work was continued and perfected by FitzGerald, Heaviside (1888), and George Frederick Charles Searle (1896, 1897). For the electromagnetic mass they gave — in modern notation —the formula $m=(4/3)E/c^2$, where m is the electromagnetic mass and E is the electromagnetic energy. Heaviside and Searle also recognized that the increase of the mass of a body is not constant and varies with its velocity. Consequently, Searle noted the impossibility of superluminal velocities, because infinite energy would be needed to exceed the speed of light. Also for Lorentz (1899), the integration of the speed-dependence of masses recognized by Thomson was especially important. He noticed that the mass not only varied due to speed, but is also dependent on the direction, and he introduced what Abraham later called "longitudinal" and "transverse" mass. (The transversal mass corresponds to what later was called relativistic mass.[28])

Wilhelm Wien (1900) assumed (following the works of Thomson, Heaviside, and Searle) that the *entire* mass is of electromagnetic origin, which was formulated in the context that all forces of nature are electromagnetic ones (the "Electromagnetic World View"). Wien stated that, if it is assumed that gravitation is an electromagnetic effect too, then there has to be a proportionality between electromagnetic energy, inertial mass and gravitational mass.[29] In the same paper Henri Poincaré (1900b) found another way of combining the concepts of mass and energy. He recognized that electromagnetic energy behaves like a fictitious fluid with mass density of $m=E/c^2$ (or $E=mc^2$) and defined a fictitious electromagnetic momentum as well. However, he arrived at a radiation paradox which was fully explained by Einstein in 1905.[30]

Walter Kaufmann (1901–1903) was the first to confirm the velocity dependence of electromagnetic mass by analyzing the ratio e/m (where e is the charge and m the mass) of cathode rays. He found that the value of e/m decreased with the speed, showing that, assuming the charge constant, the mass of the electron increased with the speed. He also believed that those experiments confirmed the assumption of Wien, that there is no "real" mechanical mass, but only

the "apparent" electromagnetic mass, or in other words, the mass of all bodies is of electromagnetic origin.[31]

Max Abraham (1902–1904), who was a supporter of the electromagnetic world view, quickly offered an explanation for Kaufmann's experiments by deriving expressions for the electromagnetic mass. Together with this concept, Abraham introduced (like Poincaré in 1900) the notion of "Electromagnetic Momentum" which is proportional to E/c^2 . But unlike the fictitious quantities introduced by Poincaré, he considered it as a *real* physical entity. Abraham also noted (like Lorentz in 1899) that this mass also depends on the direction and coined the names "Longitudinal" and "Transverse" Mass. In contrast to Lorentz, he didn't incorporate the Contraction Hypothesis into his theory, and therefore his mass terms differed from those of Lorentz.[32]

Based on the preceding work on electromagnetic mass, Friedrich Hasenöhrl suggested that part of the mass of a body (which he called apparent mass) can be thought of as radiation bouncing around a cavity. The "apparent mass" of radiation depends on the temperature (because every heated body emits radiation) and is proportional to its energy. Hasenöhrl stated that this energy-apparent-mass relation only holds as long as the body radiates, i.e., if the temperature of a body is greater than 0 K. At first he gave the expression $m=(8/3)E/c^2$ for the apparent mass; however, Abraham and Hasenöhrl himself in 1905 changed the result to $m=(4/3)E/c^2$, the same value as for the electromagnetic mass for a body at rest.[33]

13.2.5   Absolute space and time

Some scientists and philosophers of science were critical of Newton's definitions of absolute space and time.[34][35][36] Ernst Mach (1883) argued that absolute time and space are essentially metaphysical concepts and thus scientifically meaningless, and suggested that only relative motion between material bodies is a useful concept in physics. Mach argued that even effects that according to Newton depend on accelerated motion with respect to absolute space, such as rotation, could be described purely with reference to material bodies, and that the inertial effects cited by Newton in support of absolute space might instead be related purely to acceleration with respect to the fixed stars. Carl Neumann (1870) introduced a "Body alpha" , which represents some sort of rigid and fixed body for defining inertial motion. Based on the definition of Neumann, Heinrich Streintz (1883) argued that in a coordinate system where gyroscopes don't measure any signs of rotation, then one can speak of inertial motion which is related to a "Fundamental body" and a "Fundamental Coordinate System" . Eventually, Ludwig Lange (1885) was the first to coin the expression inertial frame of reference and "inertial time scale" as operational replacements for absolute space and time; he defined "inertial frame" as *a reference frame in which a mass point thrown from the same point in three different (non-co-planar) directions follows rectilinear paths each time it is thrown*". In 1902, Henri Poincaré published a collection of essays titled *Science and Hypothesis*, which included: detailed philosophical discussions on the relativity of space, time, and on the conventionality of distant simultaneity; the conjecture that a violation of the relativity principle can never be detected; the possible non-existence of the aether, together with some arguments supporting the aether; and many remarks on non-Euclidean vs. Euclidean geometry.

There were also some attempts to use time as a fourth dimension.[37][38] This was done as early as 1754 by Jean le Rond d'Alembert in the Encyclopédie, and by some authors in the 19th century like H. G. Wells in his novel The Time Machine (1895). In 1901 a philosophical model was developed by Menyhért Palágyi, in which space and time were only two sides of some sort of "spacetime".[39] He used time as an imaginary fourth dimension, which he gave the form it (where $i=\sqrt{-1}$, i.e. imaginary number). However, Palagyi's time coordinate is not connected to the speed of light. He also rejected any connection with the existing constructions of n-dimensional spaces and non-Euclidean geometry, so his philosophical model bears only little resemblance with spacetime physics, as it was later developed by Minkowski.[40]

13.2.6 Light constancy and the principle of relative motion

In the second half of the 19th century there were many attempts to develop a worldwide clock network synchronized by electrical signals. For that endeavor, the finite propagation speed of light had to be considered, because synchronization signals could travel no faster than the speed of light. So Henri Poincaré (1898) in his paper The Measure of Time drew some important consequences of this process and explained that astronomers, in determining the speed of light, simply assume that light has a constant speed and that this speed is the same in all directions. Without this postulate it would be impossible to infer the speed of light from astronomical observations, as Ole Rømer did based on observations of the moons of Jupiter. Poincaré also noted that the propagation speed of light can be (and in practice often is) used to define simultaneity between spatially separate events. He concluded by saying that "*The simultaneity of two events, or the order of their succession, the equality of two durations, are to be so defined that the enunciation of the natural laws may be as simple as possible. In other words, all these rules, all these definitions are only the fruit*

Henri Poincaré

of an unconscious opportunism."[41]

In some other papers, Poincaré (1895, 1900b) argued that experiments like that of Michelson–Morley show the impossibility of detecting the absolute motion of matter, i.e., the relative motion of matter in relation to the aether. He called this the "principle of relative motion".[42] In the same year he interpreted Lorentz's local time as the result of a synchronization procedure based on light signals. He assumed that 2 observers A and B, which are moving in the aether, synchronize their clocks by optical signals. Since they believe themselves to be at rest, they must consider only the transmission time of the signals and then cross-reference their observations to examine whether their clocks are synchronous. However, from the point of view of an observer at rest in the aether, the clocks are not synchronous and indicate the local time $t'=t-vx/c^2$. But because the moving observers do not know anything about their movement, they do not recognize this. So, contrary to Lorentz, Poincaré-defined local time can be measured and indicated by clocks.[43] Therefore, in his recommendation of Lorentz for the Nobel Prize in 1902, Poincaré argued that Lorentz has convincingly explained the negative outcome of the aether drift experiments by inventing the "diminished" or "local" time, i.e. a time coordinate in which two events at different places could appear as simultaneous, although they are not simultaneous in reality.[*][44]

Like Poincaré, Alfred Bucherer (1903) believed in the validity of the relativity principle within the domain of electrodynamics, but contrary to Poincaré, Bucherer even assumed that this implies the nonexistence of the aether. However, the theory that was created by him later in 1906 was incorrect and not self-consistent, and the Lorentz transformation was absent within his theory as well.[*][45]

13.2.7 Lorentz's 1904 model

In his paper Electromagnetic phenomena in a system moving with any velocity smaller than that of light, Lorentz (1904) was following the suggestion of Poincaré and attempted to create a formulation of Electrodynamics, which explains the failure of all known aether drift experiments, i.e. the validity of the relativity principle. He tried to prove the applicability of the Lorentz transformation for all orders, although he didn't succeed completely. Like Wien and Abraham, he argued that there exists only electromagnetic mass, not mechanical mass, and derived the correct expression for longitudinal and transverse mass, which were in agreement with Kaufmann's experiments (even though those experiments were not precise enough to distinguish between the theories of Lorentz and Abraham). And using the electromagnetic momentum, he could explain the negative result of the Trouton–Noble experiment, in which a charged parallel-plate capacitor moving through the aether should orient itself perpendicular to the motion. Also the experiments of Rayleigh and Brace could be explained. Another important step was the postulate that the Lorentz transformation has to be valid for non-electrical forces as well.[*][46]

At the same time, when Lorentz worked out his theory, Wien (1903) recognized an important consequence of the velocity dependence of mass. He argued that superluminal velocities were impossible, because that would require an infinite amount of energy —the same was already noted by Thomson (1893) and Searle (1897). And in June 1904, after he had read Lorentz's 1904 paper, he noticed the same in relation to length contraction, because at superluminal velocities the factor $\sqrt{1-v^2/c^2}$ becomes imaginary.[*][47]

Lorentz's theory was criticized by Abraham, who demonstrated that on one side the theory obeys the relativity principle, and on the other side the electromagnetic origin of all forces is assumed. Abraham showed, that both assumptions were incompatible, because in Lorentz's theory of the contracted electrons, non-electric forces were needed in order to guarantee the stability of matter. However, in Abraham's theory of the rigid electron, no such forces were needed. Thus the question arose whether the Electromagnetic conception of the world (compatible with Abraham's theory) or

the Relativity Principle (compatible with Lorentz's Theory) was correct.[*][48]

In a September 1904 lecture in St. Louis named The Principles of Mathematical Physics, Poincaré drew some consequences from Lorentz's theory and defined (in modification of Galileo's Relativity Principle and Lorentz's Theorem of Corresponding States) the following principle: "*The Principle of Relativity, according to which the laws of physical phenomena must be the same for a stationary observer as for one carried along in a uniform motion of translation, so that we have no means, and can have none, of determining whether or not we are being carried along in such a motion.*" He also specified his clock synchronization method and explained the possibility of a "new method" or "new mechanics", in which no velocity can surpass that of light for *all* observers. However, he critically noted that the Relativity Principle, Newton's action and reaction, the conservation of mass, and the conservation of energy are not fully established and are even threatened by some experiments.[*][49]

Also Emil Cohn (1904) continued to develop his alternative model (as described above), and while comparing his theory with that of Lorentz, he discovered some important physical interpretations of the Lorentz transformations. He illustrated (like Joseph Larmor in the same year) this transformation by using rods and clocks: If they are at rest in the aether, they indicate the true length and time, and if they are moving, they indicate contracted and dilated values. Like Poincaré, Cohn defined local time as the time, which is based on the assumption of isotropic propagation of light. Contrary to Lorentz and Poincaré it was noticed by Cohn, that within Lorentz's theory the separation of "real" and "apparent" coordinates is artificial, because no experiment can distinguish between them. Yet according to Cohn's own theory, the Lorentz transformed quantities would only be valid for optical phenomena, while mechanical clocks would indicate the "real" time.[*][25]

13.2.8 Poincaré's dynamics of the electron

On June 5, 1905, Henri Poincaré submitted the summary of a work which closed the existing gaps of Lorentz's work. (This short paper contained the results of a more complete work which would be published later, in January 1906.) He showed that Lorentz's equations of electrodynamics were not fully Lorentz-covariant. So he pointed out the group characteristics of the transformation, and he corrected Lorentz's formulas for the transformations of charge density and current density (which implicitly contained the relativistic velocity-addition formula, which he elaborated in May in a letter to Lorentz). Poincaré used for the first time the term "Lorentz transformation", and he gave the transformations their symmetrical form used to this day. He introduced a non-electrical binding force (the so-called "Poincaré stresses") to ensure the stability of the electrons and to explain length contraction. He also sketched a Lorentz-invariant model of gravitation (including gravitational waves) by extending the validity of Lorentz-invariance to non-electrical forces.[*][50][*][51]

Eventually Poincaré (independently of Einstein) finished a substantially extended work of his June paper (the so-called "Palermo paper", received July 23, printed December 14, published January 1906). He spoke literally of "the postulate of relativity". He showed that the transformations are a consequence of the principle of least action and developed the properties of the Poincaré stresses. He demonstrated in more detail the group characteristics of the transformation, which he called the Lorentz group, and he showed that the combination $x^2+y^2+z^2-c^2t^2$ is invariant. While elaborating his gravitational theory, he said the Lorentz transformation is merely a rotation in four-dimensional space about the origin, by introducing $ct\sqrt{-1}$ as a fourth imaginary coordinate (contrary to Palagyi, he included the speed of light), and he already used four-vectors. He wrote that the discovery of magneto-cathode rays by Paul Ulrich Villard (1904) seemed to threaten the entire theory of Lorentz, but this problem was quickly solved.[*][52] However, although in his philosophical writings Poincaré rejected the ideas of absolute space and time, in his physical papers he continued to refer to an (undetectable) aether. He also continued (1900b, 1904, 1906, 1908b) to describe coordinates and phenomena as local/apparent (for moving observers) and true/real (for observers at rest in the aether).[*][22][*][53] So, with a few exceptions,[*][54][*][55][*][56][*][57] most historians of science argue that Poincaré did not invent what is now called special relativity, although it is admitted that Poincaré anticipated much of Einstein's methods and terminology.[*][58][*][59][*][60][*][61][*][62][*][63]

13.3 Special relativity

13.3.1 Einstein 1905

Electrodynamics of moving bodies

On September 26, 1905 (received June 30), Albert Einstein published his annus mirabilis paper on what is now called *special relativity*. Einstein's paper includes a fundamental new definition of space and time (all time and space coordinates in all reference frames are on an equal footing, so there is no physical basis for distinguishing "true" from "apparent" time) and makes the aether an unnecessary concept, at least in regard to inertial motion. Einstein identified two fundamental principles, the principle of relativity and the *principle of the constancy of light* (*light principle*),

Albert Einstein, 1921

which ostensibly served as the axiomatic basis of his theory. To better understand Einstein's step, a summary of the situation before 1905, as it was described above, shall be given[*][64] (it must be remarked that Einstein was familiar with the 1895 theory of Lorentz, and *Science and Hypothesis* by Poincaré, but not their papers of 1904-1905):

a) Maxwell's electrodynamics, as presented by Lorentz in 1895, was the most successful theory at this time. Here, the speed of light is constant in all directions in the stationary aether and completely independent of the velocity of the source;

b) The inability to find an absolute state of motion, *i.e.* the validity of the relativity principle as the consequence of the negative results of all aether drift experiments and effects like the moving magnet and conductor problem which only depend on relative motion;

c) The Fizeau experiment;

d) The aberration of light;

with the following consequences for the speed of light and the theories known at that time:

1. The speed of light is not composed of the speed of light in vacuum and the velocity of a preferred frame

of reference, by *b*. This contradicts the theory of the (nearly) stationary aether.

2. The speed of light is not composed of the speed of light in vacuum and the velocity of the light source, by *a* and *c*. This contradicts the emission theory.

3. The speed of light is not composed of the speed of light in vacuum and the velocity of an aether that would be dragged within or in the vicinity of matter, by *a*, *c*, and *d*. This contradicts the hypothesis of the complete aether drag.

4. The speed of light in moving media is not composed of the speed of light when the medium is at rest and the velocity of the medium, but is determined by Fresnel's dragging coefficient, by *c*.*[W 1]

In order to make the principle of relativity as required by Poincaré an exact law of nature in the immobile aether theory of Lorentz, the introduction of a variety ad hoc hypotheses was required, such as the contraction hypothesis, local time, the Poincaré stresses, etc.. This method was criticized by many scholars, since the assumption of a conspiracy of effects which completely prevent the discovery of the aether drift is considered to be very improbable, and it would violate Occam's razor as well.*[20]*[65]*[66]*[67] Einstein is considered the first who completely dispensed with such auxiliary hypotheses and drew the direct conclusions from the facts stated above:*[20]*[65]*[66]*[67] that the relativity principle is correct and the directly observed speed of light is the same in all inertial reference frames. Based on his axiomatic approach, Einstein was able to derive *all results* obtained by his predecessors – and in addition the formulas for the relativistic Doppler effect and relativistic aberration – in a few pages, while prior to 1905 his competitors had devoted years of long, complicated work to arrive at the same mathematical formalism. Before 1905 Lorentz and Poincaré had adopted these same principles, as necessary to achieve their final results, but didn't recognize that they were also sufficient in the sense that there was no immediate logical need to assume the existence of a stationary aether in order to arrive at the Lorentz transformations.*[62]*[68] Another reason for Einstein's early rejection of the aether in any form (which he later partially retracted) may have been related to his work on quantum physics. Einstein discovered that light can also be described (at least heuristically) as a kind of particle, so the aether as the medium for electromagnetic "waves" (which was highly important for Lorentz and Poincaré) no longer fitted into his conceptual scheme.*[69]

It's notable that Einstein's paper contains no direct references to other papers. However, many historians of science like Holton,*[65] Miller,*[59] Stachel,*[70] have tried to find out possible influences on Einstein. He stated that his thinking was influenced by the empiricist philosophers David Hume and Ernst Mach. Regarding the Relativity Principle, the moving magnet and conductor problem (possibly after reading a book of August Föppl) and the various negative aether drift experiments were important for him to accept that principle —but he denied any significant influence of the *most important* experiment: the Michelson–Morley experiment.*[70] Other likely influences include Poincaré's *Science and Hypothesis*, where Poincaré presented the Principle of Relativity (which, as has been reported by Einstein's friend Maurice Solovine, was closely studied and discussed by Einstein and his friends over a period of years before the publication of Einstein's 1905 paper),*[71] and the writings of Max Abraham, from whom he borrowed the terms "Maxwell-Hertz equations" and "longitudinal and transverse mass" .*[72]

Regarding his views on Electrodynamics and the Principle of the Constancy of Light, Einstein stated that Lorentz's theory of 1895 (or the Maxwell-Lorentz electrodynamics) and also the Fizeau experiment had considerable influence on his thinking. He said in 1909 and 1912 that he borrowed that principle from Lorentz's stationary aether (which implies validity of Maxwell's equations and the constancy of light in the aether frame), but he recognized that this principle together with the principle of relativity makes any reference to an aether unnecessary (at least as to the description of electrodynamics in inertial frames).*[73] As he wrote in 1907 and in later papers, the apparent contradiction between those principles can be resolved if it is admitted that Lorentz's local time is not an auxiliary quantity, but can simply be defined as *time* and is connected with signal velocity. Before Einstein, Poincaré also developed a similar physical interpretation of local time and noticed the connection with signal velocity, but contrary to Einstein he continued to argue that clocks at rest in the stationary aether show the true time, while clocks in inertial motion relative to the aether show only the apparent time. Eventually, near the end of his life in 1953 Einstein described the advantages of his theory over that of Lorentz as follows (although Poincaré had already stated in 1905 that Lorentz invariance is an exact condition for any physical theory):*[73]

Mass-energy equivalence

Main article: Mass–energy equivalence § History

Already in §10 of his paper on electrodynamics, Einstein used the formula

$$E_{kin} = mc^2 \left(\frac{1}{\sqrt{1 - \frac{v^2}{c^2}}} - 1 \right)$$

for the kinetic energy of an electron. In elaboration of this he published a paper (received September 27, November 1905), in which Einstein showed that when a material body lost energy (either radiation or heat) of amount E, its mass decreased by the amount E/c^2. This led to the famous mass–energy equivalence formula: $E = mc^2$. Einstein considered the equivalency equation to be of paramount importance because it showed that a massive particle possesses an energy, the "rest energy", distinct from its classical kinetic and potential energies.[*][30] As it was shown above, many authors before Einstein arrived at similar formulas (including a 4/3-factor) for the relation of mass to energy. However, their work was focused on electromagnetic energy which (as we know today) only represents a small part of the entire energy within matter. So it was Einstein who was the first to: (a) ascribe this relation to all forms of energy, and (b) understand the connection of Mass-energy equivalence with the relativity principle.

13.3.2 Early reception

First assessments

Walter Kaufmann (1905, 1906) was probably the first who referred to Einstein's work. He compared the theories of Lorentz and Einstein and, although he said Einstein's method is to be preferred, he argued that both theories are observationally equivalent. Therefore, he spoke of the relativity principle as the "Lorentz-Einsteinian" basic assumption.[*][74] Shortly afterwards, Max Planck (1906a) was the first who publicly defended the theory and interested his students, Max von Laue and Kurd von Mosengeil, in this formulation. He described Einstein's theory as a "generalization" of Lorentz's theory and, to this "Lorentz-Einstein-Theory", he gave the name "relative theory"; while Alfred Bucherer changed Planck's nomenclature into the now common "theory of relativity". On the other hand, Einstein himself and many others continued to refer simply to the new method as the "relativity principle". And in an important overview article on the relativity principle (1908a), Einstein described SR as a "union of Lorentz's theory and the relativity principle", including the fundamental assumption that Lorentz's local time can be described as real time. (Yet, Poincaré's contributions were rarely mentioned in the first years after 1905.) All of those expressions, (Lorentz-Einstein theory, relativity principle, relativity theory) were used by different physicists alternately in the next years.[*][75]

Kaufmann-Bucherer experiments

Kaufmann (1905, 1906) announced the results of his new experiments on the charge to mass ratio, i.e. the veloc-

ity dependence of mass. They represented, in his opinion, a clear refutation of the relativity principle and the Lorentz-Einstein-Theory, and a confirmation of Abraham's theory. For some years Kaufmann's experiments represented a weighty objection against the relativity principle, although it was criticized by Planck and Adolf Bestelmeyer (1906). Following Kaufmann other physicists, like Alfred Bucherer (1908) and Günther Neumann (1914), also examined the velocity-dependence of mass and this time it was thought that the "Lorentz-Einstein theory" and the relativity principle were confirmed, and Abraham's theory disproved. However, it was later pointed out that the Kaufmann–Bucherer–Neumann experiments only showed a qualitative mass increase of moving electrons, but they were not precise enough to distinguish between the models of Lorentz-Einstein and Abraham. So it continued until 1940, when experiments of this kind were repeated with sufficient accuracy for confirming the Lorentz-Einstein formula.[*][74] However, this problem occurred only with this kind of experiment. The investigations of the fine structure of the hydrogen lines already in 1917 provided a clear confirmation of the Lorentz-Einstein formula and the refutation of Abraham's theory.[*][76]

Relativistic momentum and mass

Planck (1906a) defined the relativistic momentum and gave the correct values for the longitudinal and transverse mass by correcting a slight mistake of the expression given by Einstein in 1905. Planck's expressions were in principle equivalent to those used by Lorentz in 1899.[*][77] Based on the work of Planck, the concept of relativistic mass was developed by Gilbert Newton Lewis and Richard C. Tolman (1908, 1909) by defining mass as the ratio of momentum to velocity. So the older definition of longitudinal and transverse mass, in which mass was defined as the ratio of force to acceleration, became superfluous. Finally, Tolman (1912) interpreted relativistic mass simply as *the* mass of the body.[*][78] However, many modern textbooks on relativity don't use the concept of relativistic mass anymore, and mass in special relativity is considered as an invariant quantity.

Mass and energy

Einstein (1906) showed that the inertia of energy (mass-energy-equivalence) is a necessary and sufficient condition for the conservation of the center of mass theorem. On that occasion, he noted that the formal mathematical content of Poincaré paper on the center of mass (1900b) and his own paper were mainly the same, although the physical interpretation was different in light of relativity.[*][30]

Max Planck

Kurd von Mosengeil (1906) by extending Hasenöhrl's calculation of black-body-radiation in a cavity, derived the same expression for the additional mass of a body due to electromagnetic radiation as Hasenöhrl. Hasenöhrl's idea was that the mass of bodies included a contribution from the electromagnetic field, he imagined a body as a cavity containing light. His relationship between mass and energy, like all other pre-Einstein ones, contained incorrect numerical prefactors (see Electromagnetic mass). Eventually Planck (1907) derived the mass-energy-equivalence in general within the framework of special relativity, including the binding forces within matter. He acknowledged the priority of Einstein's 1905 work on $E = mc^2$, but Planck judged his own approach as more general than Einstein's.[*79]

Experiments by Fizeau and Sagnac

As was explained above, already in 1895 Lorentz succeeded in deriving Fresnel's dragging coefficient (to first order of v/c) and the Fizeau experiment by using the electromagnetic theory and the concept of local time. After first attempts by Jakob Laub (1907) to create a relativistic "optics of

moving bodies", it was Max von Laue (1907) who derived the coefficient for terms of all orders by using the colinear case of the relativistic velocity addition law. In addition, Laue's calculation was much simpler than the complicated methods used by Lorentz.[*23]

In 1911 Laue also discussed a situation where on a platform a beam of light is split and the two beams are made to follow a trajectory in opposite directions. On return to the point of entry the light is allowed to exit the platform in such a way that an interference pattern is obtained. Laue calculated a displacement of the interference pattern if the platform is in rotation – because the speed of light is independent of the velocity of the source, so one beam has covered less distance than the other beam. An experiment of this kind was performed by Georges Sagnac in 1913, who actually measured a displacement of the interference pattern (Sagnac effect). While Sagnac himself concluded that his theory confirmed the theory of an aether at rest, Laue's earlier calculation showed that it is compatible with special relativity as well because in *both* theories the speed of light is independent of the velocity of the source. This effect can be understood as the electromagnetic counterpart of the mechanics of rotation, for example in analogy to a Foucault pendulum.[*80] Already in 1909–11, Franz Harress (1912) performed an experiment which can be considered as a synthesis of the experiments of Fizeau and Sagnac. He tried to measure the dragging coefficient within glass. Contrary to Fizeau he used a rotating device so he found the same effect as Sagnac. While Harress himself misunderstood the meaning of the result, it was shown by Laue that the theoretical explanation of Harress' experiment is in accordance with the Sagnac effect.[*81] Eventually, the Michelson–Gale–Pearson experiment (1925, a variation of the Sagnac experiment) indicated the angular velocity of the Earth itself in accordance with special relativity and a resting aether.

Relativity of simultaneity

The first derivations of relativity of simultaneity by synchronization with light signals were also simplified.[*82] Daniel Frost Comstock (1910) placed an observer in the middle between two clocks A and B. From this observer a signal is sent to both clocks, and in the frame in which A and B are at rest, they synchronously start to run. But from the perspective of a system in which A and B are moving, clock B is first set in motion, and then comes clock A – so the clocks are not synchronized. Also Einstein (1917) created a model with an observer in the middle between A and B. However, in his description two signals are sent *from* A and B to the observer. From the perspective of the frame in which A and B are at rest, the signals are sent at the same time and the observer "*is hastening towards the beam of light com-*

ing from B, whilst he is riding on ahead of the beam of light coming from A. Hence the observer will see the beam of light emitted from B earlier than he will see that emitted from A. Observers who take the railway train as their reference-body must therefore come to the conclusion that the lightning flash B took place earlier than the lightning flash A."

13.3.3   Spacetime physics

Minkowski's spacetime

Hermann Minkowski

Poincaré's attempt of a four-dimensional reformulation of the new mechanics was not continued by himself,[52] so it was Hermann Minkowski (1907), who worked out the consequences of that notion (other contributions were made by Roberto Marcolongo (1906) and Richard Hargreaves (1908)[83]). This was based on the work of many mathematicians of the 19th century like Arthur Cayley, Felix Klein, or William Kingdon Clifford, who contributed to group theory, invariant theory and projective geometry.[84] Using similar methods, Minkowski succeeded in formulating a geometrical interpretation of the Lorentz transformation. He completed, for example, the concept of four vectors; he created the Minkowski diagram for the depiction of space-time; he was the first to use expressions like world line, proper time, Lorentz invariance/covariance,

etc.; and most notably he presented a four-dimensional formulation of electrodynamics. Similar to Poincaré he tried to formulate a Lorentz-invariant law of gravity, but that work was subsequently superseded by Einstein's elaborations on gravitation.

In 1907 Minkowski named four predecessors who contributed to the formulation of the relativity principle: Lorentz, Einstein, Poincaré and Planck. And in his famous lecture Space and Time (1908) he mentioned Voigt, Lorentz and Einstein. Minkowski himself considered Einstein's theory as a generalization of Lorentz's and credited Einstein for completely stating the relativity of time, but he criticized his predecessors for not fully developing the relativity of space. However, modern historians of science argue that Minkowski's claim for priority was unjustified, because Minkowski (like Wien or Abraham) adhered to the electromagnetic world-picture and apparently didn't fully understand the difference between Lorentz's electron theory and Einstein's kinematics.[85][86] In 1908, Einstein and Laub rejected the four-dimensional electrodynamics of Minkowski as overly complicated "learned superfluousness" and published a "more elementary", non-four-dimensional derivation of the basic-equations for moving bodies. But it was Minkowski's geometric model that (a) showed that the special relativity is a complete and internally self-consistent theory, (b) added the Lorentz invariant proper time interval (which accounts for the actual readings shown by moving clocks), and (c) served as a basis for further development of relativity.[83] Eventually, Einstein (1912) recognized the importance of Minkowski's geometric spacetime model and used it as the basis for his work on the foundations of general relativity.

Today special relativity is seen as an application of linear algebra, but at the time special relativity was being developed the field of linear algebra was still in its infancy. There were no textbooks on linear algebra as modern vector space and transformation theory, and the matrix notation of Arthur Cayley (that unifies the subject) had not yet come into widespread use. Cayleys matrix calculus notation was used by Minkowski (1908) in formulating relativistic electrodynamics, even though it was later replaced by Sommerfeld using vector notation.[87] In retrospect, we can see that the Lorentz transformations are equivalent to hyperbolic rotations.[88]

Vector notation and closed systems

Minkowski's space-time formalism was quickly accepted and further developed.[86] For example, Arnold Sommerfeld (1910) replaced Minkowski's matrix notation by an elegant vector notation and coined the terms "four vector" and "six vector". He also introduced a trigonometric formulation of the relativistic velocity addition rule, which according to Sommerfeld, removes much of the strangeness of that concept. Other important contributions were made by Laue (1911, 1913), who used the spacetime formalism to create a relativistic theory of deformable bodies and an elementary particle theory.*[89]*[90] He extended Minkowski's expressions for electromagnetic processes to all possible forces and thereby clarified the concept of mass-energy-equivalence. Laue also showed that non-electrical forces are needed to ensure the proper Lorentz transformation properties, and for the stability of matter – he could show that the "Poincaré stresses" (as mentioned above) are a natural consequence of relativity theory so that the electron can be a closed system.

Lorentz transformation without second postulate

There were some attempts to derive the Lorentz transformation without the postulate of the constancy of the speed of light. Vladimir Ignatowski (1910) for example used for this purpose (a) the principle of relativity, (b) homogeneity and isotropy of space, and (c) the requirement of reciprocity. Philipp Frank and Hermann Rothe (1911) argued that this derivation is incomplete and needs additional assumptions. Their own calculation was based on the assumptions that: (a) the Lorentz transformation forms a homogeneous linear group, (b) when changing frames, only the sign of the relative speed changes, (c) length contraction solely depends on the relative speed. However, according to Pauli and Miller such models were insufficient to identify the invariant speed in their transformation with the speed of light —for example, Ignatowski was forced to seek recourse in electrodynamics to include the speed of light. So Pauli and others argued that both postulates are needed to derive the Lorentz transformation.*[91]*[92] However, until today, others continued the attempts to derive special relativity without the light postulate.

Non-euclidean formulations without imaginary time coordinate

It was noted by Minkowski (1907) that his space-time formalism represents a four-dimensional non-euclidean manifold, but in order to emphasize the formal similarity to the more familiar Euclidean geometry, Minkowski noted that the time coordinate could be treated as imaginary. This was just a way of representing a non-Euclidean metric while emphasizing the formal similarity to a Euclidean metric. However, subsequent writers*[93] have dispensed with the imaginary time coordinate, and simply written the metric in explicitly non-Euclidean form (i.e., with a negative signature): Sommerfeld (1910) gave a trigonometric formulation of velocities; Vladimir Varićak (1912) emphasized

the similarity of this formulation to (Bolyai-Lobachevskian) hyperbolic geometry and tried to reformulate relativity using that non-euclidean geometry; Alfred Robb (1911) introduced the concept of rapidity as a hyperbolic angle to characterize frame velocity; Edwin Bidwell Wilson and Gilbert N. Lewis (1912) introduced a vector notation for spacetime; Émile Borel (1913) derived the kinematic basis of Thomas precession; Felix Klein (1910) and Ludwik Silberstein (1914) employed such methods as well. One historian argues that the non-Euclidean style had little to show "in the way of creative power of discovery", but it offered notational advantages in some cases, particularly in the law of velocity addition.*[94] So in the years before World War I, the acceptance of the non-Euclidean style was approximately equal to that of the initial spacetime formalism, and it continued to be employed in relativity textbooks of the 20th century.*[94]

Time dilation and twin paradox

Einstein (1907a) proposed a method for detecting the transverse Doppler effect as a direct consequence of time dilation. And in fact, that effect was measured in 1938 by Herbert E. Ives and G. R. Stilwell (Ives–Stilwell experiment).*[95] And Lewis and Tolman (1909) described the reciprocity of time dilation by using two light clocks A and B, traveling with a certain relative velocity to each other. The clocks consist of two plane mirrors parallel to one another and to the line of motion. Between the mirrors a light signal is bouncing, and for the observer resting in the same reference frame as A, the period of clock A is the distance between the mirrors divided by the speed of light. But if the observer looks at clock B, he sees that within that clock the signal traces out a longer, angled path, thus clock B is slower than A. However, for the observer moving alongside with B the situation is completely in reverse: Clock B is faster and A is slower. Also Lorentz (1910–1912) discussed the reciprocity of time dilation and analyzed a clock "paradox", which apparently occurs as a consequence of the reciprocity of time dilation. Lorentz showed that there is no paradox if one considers that in one system only one clock is used, while in the other system two clocks are necessary, and the relativity of simultaneity is fully taken into account.

A similar situation was created by Paul Langevin in 1911 with what was later called the "twin paradox", where he replaced the clocks by persons (Langevin never used the word "twins" but his description contained all other features of the paradox). Langevin solved the paradox by alluding to the fact that one twin accelerates and changes direction, so Langevin could show that the symmetry is broken and the accelerated twin is younger. However, Langevin himself interpreted this as a hint to the existence of an aether. Although Langevin's explanation is used in principle un-

Max von Laue

til today, his deductions regarding the aether were not accepted. Laue (1913) pointed out that the acceleration can be made arbitrarily small in relation to the inertial motion of the twin. So it is much more important that one twin travels within two inertial frames during his journey, while the other twin remains in one frame. Laue was also the first to visualize the situation using Minkowski spacetime-formalism – he demonstrated how the world lines of inertially moving bodies maximize the proper time elapsed between two events.[*][96]

Acceleration

Einstein (1908) tried – as a preliminary in the framework of special relativity – also to include accelerated frames within the relativity principle. In the course of this attempt he recognized that for any single moment of acceleration of a body one can define an inertial reference frame in which the accelerated body is temporarily at rest. It follows that in accelerated frames defined in this way, the application of the constancy of the speed of light to define simultaneity is restricted to small localities. However, the equivalence principle that was used by Einstein in the course of that investigation, which expresses the equality of inertial and gravita-

tional mass and the equivalence of accelerated frames and homogeneous gravitational fields, transcended the limits of special relativity and resulted in the formulation of general relativity.[*][97]

Nearly simultaneously with Einstein, also Minkowski (1908) considered the special case of uniform accelerations within the framework of his space-time formalism. He recognized that the world-line of such an accelerated body corresponds to a hyperbola. This notion was further developed by Born (1909) and Sommerfeld (1910), with Born introducing the expression "hyperbolic motion". He noted that uniform acceleration can be used as an approximation for any form of acceleration within special relativity.[*][98] In addition, Harry Bateman and Ebenezer Cunningham (1910) showed that Maxwell's equations are invariant under a much wider group of transformation than the Lorentz-group, i.e., the spherical wave transformations, being a form of conformal transformations. Under those transformations the equations preserve their form for some types of accelerated motions.[*][99] A general covariant formulation of electrodynamics in Minkowski space was eventually given by Friedrich Kottler (1912), whereby his formulation is also valid for general relativity.[*][100] Concerning the further development of the description of accelerated motion in special relativity, the works by Langevin and others for rotating frames (Born coordinates), and by Wolfgang Rindler and others for uniform accelerated frames (Rindler coordinates) must be mentioned.[*][101]

Rigid bodies and Ehrenfest paradox

Einstein (1907b) discussed the question of whether, in rigid bodies, as well as in all other cases, the velocity of information can exceed the speed of light, and explained that information could be transmitted under these circumstances into the past, thus causality would be violated. Since this contravenes radically against every experience, superluminal velocities are thought impossible. He added that a dynamics of the rigid body must be created in the framework of SR. Eventually, Max Born (1909) in the course of his above-mentioned work concerning accelerated motion, tried to include the concept of rigid bodies into SR. However, Paul Ehrenfest (1909) showed that Born's concept lead the so-called Ehrenfest paradox, in which, due to length contraction, the circumference of a rotating disk is shortened while the radius stays the same. This question was also considered by Gustav Herglotz (1910), Fritz Noether (1910), and von Laue (1911). It was recognized by Laue that the classic concept is not applicable in SR since a "rigid" body possesses infinitely many degrees of freedom. Yet, while Born's definition was not applicable on rigid bodies, it was very useful in describing rigid *motions* of bodies.[*][102] In connection to the Ehrenfest paradox, it was also discussed (by Vladimir

Varićak and others) whether length contraction is "real" or "apparent", and whether there is a difference between the dynamic contraction of Lorentz and the kinematic contraction of Einstein. However, it was rather a dispute over words because, as Einstein said, the kinematic length contraction is "apparent" for a co-moving observer, but for an observer at rest it is "real" and the consequences are measurable.*[103]

Acceptance of special relativity

Planck, in 1909, compared the implications of the modern relativity principle —he particularly referred to the relativity of time – with the revolution by the Copernican system.*[104] An important factor in the adoption of special relativity by physicists was its development by Minkowski into a spacetime theory.*[86] Consequently, by about 1911, most theoretical physicists accepted special relativity.*[105]*[106] In 1912 Wilhelm Wien recommended both Lorentz (for the mathematical framework) and Einstein (for reducing it to a simple principle) for the Nobel Prize in Physics – although it was decided by the Nobel committee not to award the prize for special relativity.*[107] Only a minority of theoretical physicists such as Abraham, Lorentz, Poincaré, or Langevin still believed in the existence of an aether.*[105] (Einstein later (1918–1920) qualified his position by arguing that one can speak about a relativistic aether, but the "idea of motion" cannot be applied to it.*[108] Lorentz and Poincaré had always argued that motion through the aether was undetectable.) Einstein used the expression "special theory of relativity" in 1915, to distinguish it from general relativity.

13.3.4 Relativistic theories

Gravitation

The first attempt to formulate a relativistic theory of gravitation was undertaken by Poincaré (1905). He tried to modify Newton's law of gravitation so that it assumes a Lorentz-covariant form. He noted that there were many possibilities for a relativistic law, and he discussed two of them. It was shown by Poincaré that the argument of Pierre-Simon Laplace, who argued that the speed of gravity is many times faster than the speed of light, is not valid within a relativistic theory. That is, in a relativistic theory of gravitation, planetary orbits are stable even when the speed of gravity is equal to that of light. Similar models as that of Poincaré were discussed by Minkowski (1907b) and Sommerfeld (1910). However, it was shown by Abraham (1912) that those models belong to the class of "vector theories" of gravitation. The fundamental defect of those theories is that they implicitly contain a negative value for the gravitational energy in the vicinity of matter, which would violate the energy principle. As an alternative, Abraham (1912) and Gustav Mie (1913) proposed different "scalar theories" of gravitation. While Mie never formulated his theory in a consistent way, Abraham completely gave up the concept of Lorentz-covariance (even locally), and therefore it was irreconcilable with relativity.

In addition, all of those models violated the equivalence principle, and Einstein argued that it is impossible to formulate a theory which is both Lorentz-covariant and satisfies the equivalence principle. However, Gunnar Nordström (1912, 1913) was able to create a model which fulfilled both conditions. This was achieved by making both the gravitational and the inertial mass dependent on the gravitational potential. Nordström's theory of gravitation was remarkable because it was shown by Einstein and Adriaan Fokker (1914), that in this model gravitation can be completely described in terms of space-time curvature. Although Nordström's theory is without contradiction, from Einstein's point of view a fundamental problem persisted: It doesn't fulfill the important condition of general covariance, as in this theory preferred frames of reference can still be formulated. So contrary to those "scalar theories", Einstein (1911–1915) developed a "tensor theory" (i.e. general relativity), which fulfills both the equivalence principle and general covariance. As a consequence, the notion of a complete "special relativistic" theory of gravitation had to be given up, as in general relativity the constancy of light speed (and Lorentz covariance) is only locally valid. The decision between those models was brought about by Einstein, when he was able to exactly derive the perihelion precession of Mercury, while the other theories gave erroneous results. In addition, Einstein's theory was the only theory which gave the correct value for the deflection of light near the sun.*[109]*[110]

Quantum field theory

The need to put together relativity and quantum mechanics was one of the major motivations in the development of quantum field theory. Pascual Jordan and Wolfgang Pauli showed in 1928 that quantum fields could be made to be relativistic, and Paul Dirac produced the Dirac equation for electrons, and in so doing predicted the existence of antimatter.*[111]

Many other domains have since been reformulated with relativistic treatments: relativistic thermodynamics, relativistic statistical mechanics, relativistic hydrodynamics, relativistic quantum chemistry, relativistic heat conduction, etc.

13.3.5   Experimental evidence

Main article: Tests of special relativity

Important early experiments confirming special relativity as mentioned above were the Fizeau experiment, the Michelson–Morley experiment, the Kaufmann–Bucherer–Neumann experiments, the Trouton–Noble experiment, the experiments of Rayleigh and Brace, and the Trouton–Rankine experiment.

In the 1920s, a series of Michelson–Morley type experiments were conducted, confirming relativity to even higher precision than the original experiment. Another type of interferometer experiment was the Kennedy–Thorndike experiment in 1932, by which the independence of the speed of light from the velocity of the apparatus was confirmed. Also time dilation was directly measured in the Ives–Stilwell experiment in 1938 and by measuring the decay rates of moving particles in 1940. All of those experiments have been repeated several times with increased precision. In addition, that the speed of light is unreachable for massive bodies was measured in many tests of relativistic energy and momentum. Therefore, knowledge of those relativistic effects is required in the construction of particle accelerators.

In 1962 J. G. Fox pointed out that all previous experimental tests of the constancy of the speed of light were conducted using light which had passed through stationary material: glass, air, or the incomplete vacuum of deep space. As a result, all were thus subject to the effects of the extinction theorem. This implied that the light being measured would have had a velocity different than that of the original source. He concluded that there was likely as yet no acceptable proof of the second postulate of special relativity. This surprising gap in the experimental record was quickly closed in the ensuing years, by experiments by Fox, and by Alvager et al., which used gamma rays sourced from high energy mesons. The high energy levels of the measured photons, along with very careful accounting for extinction effects, eliminated any significant doubt from their results.

Many other tests of special relativity have been conducted, testing possible violations of Lorentz invariance in some variants of quantum gravity. However, no sign of anisotropy of the speed of light has been found even at the $10^{*}-17$ level, and some experiments even ruled out Lorentz violations at the $10^{*}-40$ level, see Modern searches for Lorentz violation.

13.3.6   Priority

Some claim that Poincaré and Lorentz, not Einstein, are the true founders of special relativity. For more see the article on relativity priority dispute.

13.3.7   Criticisms

Main article: Criticism of relativity theory

Some criticized Special Relativity for various reasons, such as lack of empirical evidence, internal inconsistencies, rejection of mathematical physics *per se*, or philosophical reasons. Although there still are critics of relativity outside the scientific mainstream, the overwhelming majority of scientists agree that Special Relativity has been verified in many different ways and there are no inconsistencies within the theory.

13.4   See also

- History of Lorentz transformations
- Tests of special relativity

13.5   References

13.5.1   Primary sources

- Abraham, Max (1902), "Dynamik des Electrons", *Nachrichten von der Gesellschaft der Wissenschaften zu Göttingen, Mathematisch-Physikalische Klasse*: 20–41

- Abraham, Max (1903), "Prinzipien der Dynamik des Elektrons", *Annalen der Physik* **315** (1): 105–179, Bibcode:1902AnP...315..105A, doi:10.1002/andp.19023150105

- Abraham, Max (1904), "Die Grundhypothesen der Elektronentheorie" [The Fundamental Hypotheses of the Theory of Electrons], *Physikalische Zeitschrift* **5**: 576–579

- Abraham, Max (1914), "Neuere Gravitationstheorien", *Jahrbuch für Radioaktivität und Elektronik* **11** (4): 470–520.

- Alväger, Farley, Kjellmann, Walle (1964), "Test of the second postulate of special relativity in the GeV region", *Phys. Rev. Letters* **12** (3): 260–262, Bibcode:1964PhL....12..260A, doi:10.1016/0031-9163(64)91095-9

- Bartoli, Adolfo (1876/1884), "Il calorico raggiante e il secondo principio di termodinamica" (PDF), *Nuovo Cimento* **15**: 196–202, doi:10.1007/bf02737234 Check date values in: |date= (help)

- Bateman, Harry (1909/10), "The Transformation of the Electrodynamical Equations", *Proceedings of the London Mathematical Society* **8** (1): 223–264, doi:10.1112/plms/s2-8.1.223. Check date values in: |date= (help)

- Borel, Émile (1913), "La théorie de la relativité et la cinématique", *Comptes Rendus des Séances de l'Académie des Sciences* **156**: 215–218

- Borel, Émile (1913), "La cinématique dans la théorie de la relativité", *Comptes Rendus des Séances de l'Académie des Sciences* **157**: 703–705

- Born, Max (1909), "Die Theorie des starren Körpers in der Kinematik des Relativitätsprinzips" [The Theory of the Rigid Electron in the Kinematics of the Principle of Relativity], *Annalen der Physik* **335** (11): 1–56, Bibcode:1909AnP...335....1B, doi:10.1002/andp.19093351102

- Brecher, Kenneth (1977), "Is the Speed of Light Independent of the Velocity of the Source?", *Phys. Rev. Letters* **39** (17): 1051–1054, Bibcode:1977PhRvL..39.1051B, doi:10.1103/PhysRevLett.39.1051

- Bucherer, A. H. (1903), "Über den Einfluß der Erdbewegung auf die Intensität des Lichtes", *Annalen der Physik* **316** (6): 270–283, Bibcode:1903AnP...316..270B, doi:10.1002/andp.19033160604

- Bucherer, A. H. (1908), "Messungen an Becquerelstrahlen. Die experimentelle Bestätigung der Lorentz-Einsteinschen Theorie. (Measurements of Becquerel rays. The Experimental Confirmation of the Lorentz-Einstein Theory)", *Physikalische Zeitschrift* **9** (22): 755–762

- Cohn, Emil (1901), "Über die Gleichungen der Electrodynamik für bewegte Körper", *Archives néerlandaises des sciences exactes et naturelles* **5**: 516–523

- Cohn, Emil (1904), "Zur Elektrodynamik bewegter Systeme I" [On the Electrodynamics of Moving Systems I], *Sitzungsberichte der Königlich Preussischen Akademie der Wissenschaften*, 1904/2 (40): 1294–1303

- Cohn, Emil (1904), "Zur Elektrodynamik bewegter Systeme II" [On the Electrodynamics of Moving Systems II], *Sitzungsberichte der Königlich Preussischen Akademie der Wissenschaften*, 1904/2 (43): 1404–1416

- Comstock, Daniel Frost (1910), "The Principle of Relativity", *Science* **31** (803): 767–772, Bibcode:1910Sci....31..767C, doi:10.1126/science.31.803.767, PMID 17758464

- Cunningham, Ebenezer (1909/10), "The principle of Relativity in Electrodynamics and an Extension Thereof", *Proceedings of the London Mathematical Society* **8** (1): 77–98, doi:10.1112/plms/s2-8.1.77. Check date values in: |date= (help)

- De Sitter, Willem (1913), "A proof of the constancy of the velocity of light", *Proceedings of the Royal Netherlands Academy of Arts and Sciences* **15** (2): 1297–1298

- De Sitter, Willem (1913), "On the constancy of the velocity of light", *Proceedings of the Royal Netherlands Academy of Arts and Sciences* **16** (1): 395–396

- Ehrenfest, Paul (1909), "Gleichförmige Rotation starrer Körper und Relativitätstheorie" [Uniform Rotation of Rigid Bodies and the Theory of Relativity], *Physikalische Zeitschrift* **10**: 918

- Einstein, Albert (1905a), "Zur Elektrodynamik bewegter Körper" (PDF), *Annalen der Physik* **322** (10): 891–921, Bibcode:1905AnP...322..891E, doi:10.1002/andp.19053221004. See also: English translation.

- Einstein, Albert (1905b), "Ist die Trägheit eines Körpers von seinem Energieinhalt abhängig?" (PDF), *Annalen der Physik* **323** (13): 639–641, Bibcode:1905AnP...323..639E, doi:10.1002/andp.19053231314. See also the English translation.

- Einstein, Albert (1906), "Das Prinzip von der Erhaltung der Schwerpunktsbewegung und die Trägheit der Energie" (PDF), *Annalen der Physik* **325** (8): 627–633, Bibcode:1906AnP...325..627E, doi:10.1002/andp.19063250814

- Einstein, Albert (1907), "Über die vom Relativitätsprinzip geforderte Trägheit der Energie" (PDF), *Annalen der Physik* **328** (7): 371–384, Bibcode:1907AnP...328..371E, doi:10.1002/andp.19073280713

- Einstein, Albert (1908a), "Über das Relativität-sprinzip und die aus demselben gezogenen Folgerungen" (PDF), *Jahrbuch der Radioaktivität und Elektronik* **4**: 411–462, Bibcode:1908JRE.....4..411E

- Einstein, Albert & Laub, Jakob (1908b), "Über die elektromagnetischen Grundgleichungen für bewegte Körper" (PDF), *Annalen der Physik* **331** (8): 532–540, Bibcode:1908AnP...331..532E, doi:10.1002/andp.19083310806

- Einstein, Albert (1909), "The Development of Our Views on the Composition and Essence of Radiation", *Physikalische Zeitschrift* **10** (22): 817–825

- Einstein, Albert (1912), "Relativität und Gravitation. Erwiderung auf eine Bemerkung von M. Abraham" (PDF), *Annalen der Physik* **38** (10): 1059–1064, Bibcode:1912AnP...343.1059E, doi:10.1002/andp.19123431014

- Einstein A. (1916), *Relativity: The Special and General Theory*, Springery

- Einstein, Albert (1922), *Ether and the Theory of Relativity*, Methuen & Co.

- FitzGerald, George Francis (1889), "The Ether and the Earth's Atmosphere", *Science* **13** (328): 390, Bibcode:1889Sci....13..390F, doi:10.1126/science.ns-13.328.390, PMID 17819387

- Fizeau, H. (1851). "The Hypotheses Relating to the Luminous Aether, and an Experiment which Appears to Demonstrate that the Motion of Bodies Alters the Velocity with which Light Propagates itself in their Interior". *Philosophical Magazine* **2**: 568–573.

- Fox, J.G. (1962), "Experimental Evidence for the Second Postulate of Special Relativity", *American Journal of Physics* **30** (1): 297–300, Bibcode:1965AmJPh..33....1F, doi:10.1119/1.1941992.

- Filippas, T.A.; Fox, J.G. (1964), "Velocity of Gamma Rays from a Moving Source", *Physical Review* **135** (4B): B1071–1075, Bibcode:1964PhRv..135.1071F, doi:10.1103/PhysRev.135.B1071

- Frank, Philipp & Rothe, Hermann (1910), "Über die Transformation der Raum-Zeitkoordinaten von ruhenden auf bewegte Systeme", *Annalen der Physik* **339** (5): 825–855, Bibcode:1911AnP...339..825F, doi:10.1002/andp.19113390502

- Augustin Fresnel (1816), "Sur la diffraction de la lumière", *Annales de chimie et de physique* **1**: 239–281

- Hasenöhrl, Friedrich (1904), "Zur Theorie der Strahlung in bewegten Körpern" [On the Theory of Radiation in Moving Bodies], *Annalen der Physik* **320** (12): 344–370, Bibcode:1904AnP...320..344H, doi:10.1002/andp.19043201206

- Hasenöhrl, Friedrich (1905), "Zur Theorie der Strahlung in bewegten Körpern. Berichtigung" [On the Theory of Radiation in Moving Bodies. Correction], *Annalen der Physik* **321** (3): 589–592, Bibcode:1905AnP...321..589H, doi:10.1002/andp.19053210312

- Heaviside, Oliver (1888/1894), "Electromagnetic waves, the propagation of potential, and the electromagnetic effects of a moving charge", *Electrical papers* **2**, pp. 490–499 Check date values in: |date= (help)

- Heaviside, Oliver (1889), "On the Electromagnetic Effects due to the Motion of Electrification through a Dielectric", *Philosophical Magazine*, 5 **27** (167): 324–339, doi:10.1080/14786448908628362

- Herglotz, Gustav (1909), "Über den vom Standpunkt des Relativitätsprinzips aus als starr zu bezeichnenden Körper" [On bodies that are to be designated as "rigid" from the standpoint of the relativity principle], *Annalen der Physik* **336** (2): 393–415, Bibcode:1910AnP...336..393H, doi:10.1002/andp.19103360208

- Hertz, Heinrich (1890a), "Über die Grundgleichungen der Elektrodynamik für ruhende Körper", *Annalen der Physik* **276** (8): 577, Bibcode:1890AnP...276..577H, doi:10.1002/andp.18902760803

- Hertz, Heinrich (1890b), "Über die Grundgleichungen der Elektrodynamik für bewegte Körper", *Annalen der Physik* **277** (11): 369–399, Bibcode:1890AnP...277..369H, doi:10.1002/andp.18902771102

- Ignatowsky, W. v. (1910). "Einige allgemeine Bemerkungen über das Relativitätsprinzip" [Some General Remarks on the Relativity Principle]. *Physikalische Zeitschrift* **11**: 972–976.

- Ignatowsky, W. v. (1911). "Das Relativitätsprinzip". *Archiv der Mathematik und Physik*. 17, 18: 1–24, 17–40.

- Kaufmann, Walter (1902), "Die elektromagnetische Masse des Elektrons" [The Electromagnetic Mass of the Electron], *Physikalische Zeitschrift* **4** (1b): 54–56

- Kaufmann, Walter (1905), "Über die Konstitution des Elektrons" [On the Constitution of the Electron], *Sitzungsberichte der Königlich Preußische Akademie der Wissenschaften* **45**: 949–956

- Kaufmann, Walter (1906), "Über die Konstitution des Elektrons" [On the Constitution of the Electron], *Annalen der Physik* **324** (3): 487–553, Bibcode:1906AnP...324..487K, doi:10.1002/andp.19063240303

- Lange, Ludwig (1885), "Ueber die wissenschaftliche Fassung des Galileischen Beharrungsgesetzes", *Philosophische Studien* **2**: 266–297

- Langevin, Paul (1904/1908), "The Relations of Physics of Electrons to Other Branches of Science", *International Congress of Arts and Science* **7**: 121–156 Check date values in: |date= (help)

- Langevin, Paul (1905), "Sur l'impossibilité physique de mettre en évidence le mouvement de translation de la Terre", *Comptes Rendus des Séances de l'Académie des Sciences* **140**: 1171–1173

- Langevin, P. (1911), "The evolution of space and time", *Scientia* **X**: 31–54 (translated by J. B. Sykes, 1973).

- Larmor, Joseph (1897), "On a Dynamical Theory of the Electric and Luminiferous Medium, Part 3, Relations with material media", *Philosophical Transactions of the Royal Society* **190**: 205–300, Bibcode:1897RSPTA.190..205L, doi:10.1098/rsta.1897.0020

- Larmor, Joseph (1900), *Aether and Matter*, Cambridge University Press

- Laub, Jakob (1907), "Zur Optik der bewegten Körper", *Annalen der Physik* **328** (9): 738–744, Bibcode:1907AnP...328..738L, doi:10.1002/andp.19073280910

- Laue, Max von (1907), "Die Mitführung des Lichtes durch bewegte Körper nach dem Relativitätsprinzip" [The Entrainment of Light by Moving Bodies in Accordance with the Principle of Relativity], *Annalen der Physik* **328** (10): 989–990, Bibcode:1907AnP...328..989L, doi:10.1002/andp.19073281015

- Laue, Max von (1911a), Das Relativitätsprinzip *on Internet Archive*, Braunschweig: Vieweg External link in |title= (help)

- Laue, Max von (1911b), "Zur Diskussion über den starren Körper in der Relativitätstheorie" [On the Discussion Concerning Rigid Bodies in the Theory of Relativity], *Physikalische Zeitschrift* **12**: 85–87

- Laue, Max von (1911c), "Über einen Versuch zur Optik der bewegten Körper" [On an Experiment on the Optics of Moving Bodie], *Münchener Sitzungsberichte* **1911**: 405–412

- Laue, Max von (1913), *Das Relativitätsprinzip* (2 ed.), Braunschweig: Vieweg

- Lewis, Gilbert N. (1908), "A revision of the Fundamental Laws of Matter and Energy", *Philosophical Magazine* **16**: 705–717, doi:10.1080/14786441108636549

- Lewis, Gilbert N. & Tolman, Richard C. (1909), "The Principle of Relativity, and Non-Newtonian Mechanics", *Proceedings of the American Academy of Arts and Sciences* **44**: 709–726, doi:10.2307/20022495

- Lewis, Gilbert N. & Wilson, Edwin B. (1912), "*The Space-time Manifold of Relativity. The Non-Euclidean Geometry of Mechanics and Electromagnetics* on Internet Archive", *Proceedings of the American Academy of Arts and Sciences* **48**: 387–507, doi:10.2307/20022840 External link in |title= (help)

- Lorentz, Hendrik Antoon (1886), "De l'influence du mouvement de la terre sur les phénomènes lumineux", *Archives néerlandaises des sciences exactes et naturelles* **21**: 103–176

- Lorentz, Hendrik Antoon (1892a), "*La Théorie electromagnétique de Maxwell et son application aux corps mouvants* on Internet Archive", *Archives néerlandaises des sciences exactes et naturelles* **25**: 363–552 External link in |title= (help)

- Lorentz, Hendrik Antoon (1892b), "De relatieve beweging van de aarde en den aether" [The Relative Motion of the Earth and the Aether], *Zittingsverlag Akad. V. Wet.* **1**: 74–79

- Lorentz, Hendrik Antoon (1895), *Versuch einer Theorie der electrischen und optischen Erscheinungen in bewegten Körpern* [Attempt of a Theory of Electrical and Optical Phenomena in Moving Bodies], Leiden: E.J. Brill

- Lorentz, Hendrik Antoon (1899), "Simplified Theory of Electrical and Optical Phenomena in Moving Systems", *Proceedings of the Royal Netherlands Academy of Arts and Sciences* **1**: 427–442

- Lorentz, Hendrik Antoon (1900), "Considerations on Gravitation", *Proceedings of the Royal Netherlands Academy of Arts and Sciences* **2**: 559–574

- Lorentz, Hendrik Antoon (1904a), "Weiterbildung der Maxwellschen Theorie. Elektronentheorie.", *Encyclopädie der mathematischen Wissenschaften* **5** (2): 145–288

- Lorentz, Hendrik Antoon (1904b), "Electromagnetic phenomena in a system moving with any velocity smaller than that of light", *Proceedings of the Royal Netherlands Academy of Arts and Sciences* **6**: 809–831

- Lorentz, Hendrik Antoon (1931) [1910], *Lecture on theoretical physics, Vol.3*, London: MacMillan

- Lorentz, Hendrik Antoon & Einstein, Albert & Minkowski, Hermann (1913), Das Relativitätsprinzip. Eine Sammlung von Abhandlungen. *on Internet Archive*, Leipzig & Berlin: B.G. Teubner External link in |title= (help)

- Lorentz, Hendrik Antoon (1914), *Das Relativitätsprinzip. Drei Vorlesungen gehalten in Teylers Stiftung zu Haarlem*, Leipzig and Berlin: B.G. Teubner

- Lorentz, Hendrik Antoon (1914), "La Gravitation", *Scientia* **16**: 28–59

- Lorentz, Hendrik Antoon (1916), The theory of electrons and its applications to the phenomena of light and radiant heat *on Internet Archive*, Leipzig & Berlin: B.G. Teubner External link in |title= (help)

- Lorentz, Hendrik Antoon (1921), "Deux Mémoires de Henri Poincaré sur la Physique Mathématique" [Two Papers of Henri Poincaré on Mathematical Physics], *Acta Mathematica* **38** (1): 293–308, doi:10.1007/BF02392073;

- Lorentz, Hendrik Antoon; Lorentz, H. A.; Miller, D. C.; Kennedy, R. J.; Hedrick, E. R.; Epstein, P. S. (1928), "Conference on the Michelson-Morley Experiment", *The Astrophysical Journal* **68**: 345–351, Bibcode:1928ApJ....68..341M, doi:10.1086/143148

- Mach, Ernst (1883/1912), *Die Mechanik in ihrer Entwicklung* (PDF), Leipzig: Brockhaus Check date values in: |date= (help)

- Maxwell, James Clerk (1864), "A Dynamical Theory of the Electromagnetic Field", *Philosophical Transactions of the Royal Society* **155**: 459–512, Bibcode:1865RSPT..155..459C, doi:10.1098/rstl.1865.0008

- Maxwell, James Clerk (1873), "§ 792", A Treatise on electricity and magnetism *on Internet Archive* **2**, London: Macmillan & Co., p. 391 External link in |title= (help)

- Michelson, Albert A. (1881), "The Relative Motion of the Earth and the Luminiferous Ether", *American Journal of Science* **22**: 120–129, doi:10.2475/ajs.s3-22.128.120

- Michelson, Albert A. & Morley, Edward W. (1886), "Influence of Motion of the Medium on the Velocity of Light", *American Journal of Science* **31**: 377–386, doi:10.2475/ajs.s3-31.185.377

- Michelson, Albert A. & Morley, Edward W. (1887), "On the Relative Motion of the Earth and the Luminiferous Ether", *American Journal of Science* **34**: 333–345, doi:10.2475/ajs.s3-34.203.333

- Michelson, Albert A. & Gale, Henry G. (1925), "The Effect of the Earth's Rotation on the Velocity of Light", *The Astrophysical Journal* **61**: 140–145, Bibcode:1925ApJ....61..140M, doi:10.1086/142879

- Minkowski, Hermann (1915) [1907], "Das Relativitätsprinzip", *Annalen der Physik* **352** (15): 927–938, Bibcode:1915AnP...352..927M, doi:10.1002/andp.19153521505

- Minkowski, Hermann (1908), "Die Grundgleichungen für die elektromagnetischen Vorgänge in bewegten Körpern" [The Fundamental Equations for Electromagnetic Processes in Moving Bodies], *Nachrichten von der Gesellschaft der Wissenschaften zu Göttingen, Mathematisch-Physikalische Klasse*: 53–111 (English translation in 1920 by Meghnad Saha).

- Minkowski, Hermann (1909), "Raum und Zeit", *Physikalische Zeitschrift* **10**: 75–88

  - Various English translations on Wikisource: Space and Time

- Mosengeil, Kurd von (1907), "Theorie der stationären Strahlung in einem gleichförmig bewegten Hohlraum", *Annalen der Physik* **327** (5): 867–904, Bibcode:1907AnP...327..867V, doi:10.1002/andp.19073270504

- Neumann, Carl (1870), Ueber die Principien der Galilei-Newtonschen Theorie *on Internet Archive*, Leipzig: B.G. Teubner External link in |title= (help)

- Neumann, Günther (1914), "Die träge Masse schnell bewegter Elektronen", *Annalen der Physik* **350** (20): 529–579, Bibcode:1914AnP...350..529N, doi:10.1002/andp.19143502005

- Nordström, Gunnar (1913), "Zur Theorie der Gravitation vom Standpunkt des Relativitätsprinzips", *Annalen der Physik* **347** (13): 533–554, Bibcode:1913AnP...347..533N, doi:10.1002/andp.19133471303.

- Palagyi, Menyhért (1901), *Neue Theorie des Raumes und der Zeit*, Leipzig: Wilhelm Engelmann

- Planck, Max (1906a), "Das Prinzip der Relativität und die Grundgleichungen der Mechanik" [The Principle of Relativity and the Fundamental Equations of Mechanics], *Verhandlungen Deutsche Physikalische Gesellschaft* **8**: 136–141

- Planck, Max (1906b), "Die Kaufmannschen Messungen der Ablenkbarkeit der β-Strahlen in ihrer Bedeutung für die Dynamik der Elektronen" [The Measurements of Kaufmann on the Deflectability of β-Rays in their Importance for the Dynamics of the Electrons], *Physikalische Zeitschrift* **7**: 753–761

- Planck, Max (1907), "Zur Dynamik bewegter Systeme" [On the Dynamics of Moving Systems], *Sitzungsberichte der Königlich-Preussischen Akademie der Wissenschaften, Berlin*, Erster Halbband (29): 542–570

- Planck, Max (1908), "Bemerkungen zum Prinzip der Aktion und Reaktion in der allgemeinen Dynamik" [Notes on the Principle of Action and Reaction in General Dynamics], *Physikalische Zeitschrift* **9** (23): 828–830

- Planck, Max (1915) [1909], "General Dynamics. Principle of Relativity", *Eight lectures on theoretical physics*, New York: Columbia University Press

- Poincaré, Henri (1889), *Théorie mathématique de la lumière* **1**, Paris: G. Carré & C. Naud Preface partly reprinted in "Science and Hypothesis", Ch. 12.

- Poincaré, Henri (1895), "A propos de la Théorie de M. Larmor", *L'Éclairage électrique* **5**: 5–14 Reprinted in Poincaré, Oeuvres, tome IX, pp. 395–413

- Poincaré, Henri (1898/1913), "The Measure of Time", *The Foundations of Science (The Value of Science)*, New York: Science Press, pp. 222–234 Check date values in: |date= (help)

- Poincaré, Henri (1900a), "Les relations entre la physique expérimentale et la physique mathématique", *Revue générale des sciences pures et appliquées* **11**: 1163–1175. Reprinted in "Science and Hypothesis", Ch. 9–10.

- Poincaré, Henri (1900b), "La théorie de Lorentz et le principe de réaction", *Archives néerlandaises des sciences exactes et naturelles* **5**: 252–278. See also the English translation.

- Poincaré, Henri (1901a), "Sur les principes de la mécanique", *Bibliothèque du Congrès international de philosophie*: 457–494. Reprinted in "Science and Hypothesis", Ch. 6–7.

- Poincaré, Henri (1901b), Électricité et optique *on Internet Archive*, Paris: Gauthier-Villars External link in |title= (help)

- Poincaré, Henri (1902), *Science and Hypothesis*, London and Newcastle-on-Cyne (1905): The Walter Scott publishing Co.

- Poincaré, Henri (1904/6), "The Principles of Mathematical Physics", *Congress of arts and science, universal exposition, St. Louis, 1904* **1**, Boston and New York: Houghton, Mifflin and Company, pp. 604–622 Check date values in: |date= (help)

- Poincaré, Henri (1905b), "Sur la dynamique de l'électron" [On the Dynamics of the Electron], *Comptes Rendus* **140**: 1504–1508.

- Poincaré, Henri (1906), "Sur la dynamique de l'électron" [On the Dynamics of the Electron], *Rendiconti del Circolo matematico di Palermo* **21**: 129–176, doi:10.1007/BF03013466

- Poincaré, Henri (1908/13), "The New Mechanics", *The foundations of science (Science and Method)*, New York: Science Press, pp. 486–522 Check date values in: |date= (help)

- Poincaré, Henri (1909), "La Mécanique nouvelle (Lille)", *Revue scientifique* (Paris) **47**: 170–177

- Poincaré, Henri (1909/10), "The New Mechanics (Göttingen)", *Sechs Vorträge über ausgewählte Gegenstände aus der reinen Mathematik und mathematischen Physik*, Leipzig und Berlin: B.G.Teubner, pp. 41–47 Check date values in: |date= (help)

- Poincaré, Henri (1910/1), *Die neue Mechanik (Berlin)*, Leipzig & Berlin: B.G. Teubner Check date values in: |date= (help)

- Poincaré, Henri (1912), "L'hypothèse des quanta", *Revue scientifique* **17**: 225–232 Reprinted in Poincaré 1913, Ch. 6.

- Poincaré, Henri (1913), Last Essays *on Internet Archive*, New York: Dover Publication (1963) External link in |title= (help)

- Ritz, Walter (1908), "Recherches critiques sur l'Électrodynamique Générale", *Annales de Chimie et de Physique* **13**: 145–275, see English translation.

- Robb, Alfred A. (1911), Optical Geometry of Motion: A New View of the Theory of Relativity *on Internet Archive*, Cambridge: W. Heffer External link in |title= (help)

- Sagnac, Georges (1913), "L'éther lumineux démontré par l'effet du vent relatif d'éther dans un interféromètre en rotation uniforme" [The demonstration of the luminiferous aether by an interferometer in uniform rotation], *Comptes Rendus* **157**: 708–710

- Sagnac, Georges (1913), "Sur la preuve de la réalité de l'éther lumineux par l'expérience de l'interférographe tournant" [On the proof of the reality of the luminiferous aether by the experiment with a rotating interferometer], *Comptes Rendus* **157**: 1410–1413

- Searle, George Frederick Charles (1897), "On the Steady Motion of an Electrified Ellipsoid", *Philosophical Magazine*, 5 **44** (269): 329–341, doi:10.1080/14786449708621072

- Sommerfeld, Arnold (1910), "Zur Relativitätstheorie I: Vierdimensionale Vektoralgebra" [On the Theory of Relativity I: Four-dimensional Vector Algebra], *Annalen der Physik* **337** (9): 749–776, Bibcode:1910AnP...337..749S, doi:10.1002/andp.19103370904

- Sommerfeld, Arnold (1910), "Zur Relativitätstheorie II: Vierdimensionale" [On the Theory of Relativity II: Four-dimensional Vector Analysis], *Annalen der Physik* **338** (14): 649–689, Bibcode:1910AnP...338..649S, doi:10.1002/andp.19103381402

- Stokes, George Gabriel (1845), "On the Aberration of Light", *Philosophical Magazine* **27**: 9–15, doi:10.1080/14786444508645215

- Streintz, Heinrich (1883), Die physikalischen Grundlagen der Mechanik *on Internet Archive*, Leipzig: B.G. Teubner External link in |title= (help)

- Thomson, Joseph John (1881), "On the Electric and Magnetic Effects produced by the Motion of Electrified Bodies", *Philosophical Magazine*, 5 **11** (68): 229–249, doi:10.1080/14786448108627008

- Tolman, Richard Chase (1912), "The mass of a moving body", *Philosophical Magazine* **23**: 375–380, doi:10.1080/14786440308637231

- Varičak, Vladimir (1911), "Zum Ehrenfestschen Paradoxon" [On Ehrenfest's Paradox], *Physikalische Zeitschrift* **12**: 169

- Varičak, Vladimir (1912), "Über die nichteuklidische Interpretation der Relativtheorie" [On the Non-Euclidean Interpretation of the Theory of Relativity], *Jahresbericht der Deutschen Mathematiker-Vereinigung* **21**: 103–127

- Voigt, Woldemar (1887), "Ueber das Doppler'sche Princip" [On the Principle of Doppler], *Nachrichten von der Königl. Gesellschaft der Wissenschaften und der Georg-Augusts-Universität zu Göttingen* (2): 41–51

- Wien, Wilhelm (1900), "Über die Möglichkeit einer elektromagnetischen Begründung der Mechanik" [On the Possibility of an Electromagnetic Foundation of Mechanics], *Annalen der Physik* **310** (7): 501–513, Bibcode:1901AnP...310..501W, doi:10.1002/andp.19013100703

- Wien, Wilhelm (1904a), "Über die Differentialgleichungen der Elektrodynamik für bewegte Körper. I", *Annalen der Physik* **318** (4): 641–662, Bibcode:1904AnP...318..641W, doi:10.1002/andp.18943180402

- Wien, Wilhelm (1904a), "Über die Differentialgleichungen der Elektrodynamik für bewegte Körper. II", *Annalen der Physik* **318** (4): 663–668, Bibcode:1904AnP...318..663W, doi:10.1002/andp.18943180403

- Wien, Wilhelm (1904b), "Erwiderung auf die Kritik des Hrn. M. Abraham", *Annalen der Physik* **319** (8): 635–637, Bibcode:1904AnP...319..635W, doi:10.1002/andp.19043190817

13.5.2 Notes and Secondary sources

13.6 External links

- O'Connor, John J.; Robertson, Edmund F., "Special relativity", *MacTutor History of Mathematics archive*, University of St Andrews.

- Mathpages: Corresponding States, The End of My Latin, Who Invented Relativity?, Poincaré Contemplates Copernicus

[1] For many other experiments on light constancy and relativity, see PhysicsFaq: What is the experimental basis of special relativity?

Chapter 14

History of general relativity

See also: history of special relativity

General relativity (GR) is a theory of gravitation that was developed by Albert Einstein between 1907 and 1915, with contributions by many others after 1915. According to general relativity, the observed gravitational attraction between masses results from the warping of space and time by those masses.

Before the advent of general relativity, Newton's law of universal gravitation had been accepted for more than two hundred years as a valid description of the gravitational force between masses, even though Newton himself did not regard the theory as the final word on the nature of gravity. Within a century of Newton's formulation, careful astronomical observation revealed unexplainable variations between the theory and the observations. Under Newton's model, gravity was the result of an attractive force between massive objects. Although even Newton was bothered by the unknown nature of that force, the basic framework was extremely successful at describing motion.

However, experiments and observations show that Einstein's description accounts for several effects that are unexplained by Newton's law, such as minute anomalies in the orbits of Mercury and other planets. General relativity also predicts novel effects of gravity, such as gravitational waves, gravitational lensing and an effect of gravity on time known as gravitational time dilation. Many of these predictions have been confirmed by experiment, while others are the subject of ongoing research. For example, although there is indirect evidence for gravitational waves, direct evidence of their existence is still being sought by several teams of scientists in experiments such as the LIGO and GEO 600 projects.

General relativity has developed into an essential tool in modern astrophysics. It provides the foundation for the current understanding of black holes, regions of space where gravitational attraction is so strong that not even light can escape. Their strong gravity is thought to be responsible for the intense radiation emitted by certain types of astronomical objects (such as active galactic nuclei or microquasars). General relativity is also part of the framework of the standard Big Bang model of cosmology.

14.1   Creation of general relativity

Albert Einstein developed the theories of special and general relativity. Picture from 1921.

14.1.1 Early investigations

As Einstein later said, the reason for the development of general relativity was the preference of inertial motion within special relativity, while a theory which from the outset prefers no state of motion (even accelerated ones) appeared more satisfactory to him.[1] So, while still working at the patent office in 1907, Einstein had what he would call his "happiest thought". He realized that the principle of relativity could be extended to gravitational fields.

Consequently, in 1907 (published 1908) he wrote an article on acceleration under special relativity.[2] In that article, he argued that free fall is really inertial motion, and that for a freefalling observer the rules of special relativity must apply. This argument is called the Equivalence principle. In the same article, Einstein also predicted the phenomenon of gravitational time dilation.

In 1911, Einstein published another article expanding on the 1907 article.[3] There, he thought about the case of a uniformly accelerated box not in a gravitational field, and noted that it would be indistinguishable from a box sitting still in an unchanging gravitational field. He used special relativity to see that the rate of clocks at the top of a box accelerating upward would be faster than the rate of clocks at the bottom. He concludes that the rates of clocks depend on their position in a gravitational field, and that the difference in rate is proportional to the gravitational potential to first approximation.

Also the deflection of light by massive bodies was predicted. Although the approximation was crude, it allowed him to calculate that the deflection is nonzero. German astronomer Erwin Finlay-Freundlich publicized Einstein's challenge to scientists around the world.[4] This urged astronomers to detect the deflection of light during a solar eclipse, and gave Einstein confidence that the scalar theory of gravity proposed by Gunnar Nordström was incorrect. But the actual value for the deflection that he calculated was too small by a factor of two, because the approximation he used doesn't work well for things moving at near the speed of light. When Einstein finished the full theory of general relativity, he would rectify this error and predict the correct amount of light deflection by the sun.

Another of Einstein's notable thought experiments about the nature of the gravitational field is that of the rotating disk (a variant of the Ehrenfest paradox). He imagined an observer performing experiments on a rotating turntable. He noted that such an observer would find a different value for the mathematical constant π than the one predicted by Euclidean geometry. The reason is that the radius of a circle would be measured with an uncontracted ruler, but, according to special relativity, the circumference would seem to be longer because the ruler would be contracted. Since Ein-

stein believed that the laws of physics were local, described by local fields, he concluded from this that spacetime could be locally curved. This led him to study Riemannian geometry, and to formulate general relativity in this language.

14.1.2 Developing general relativity

Eddington's photograph of a solar eclipse, which confirmed Einstein's theory that light "bends".

In 1912, Einstein returned to Switzerland to accept a professorship at his *alma mater*, the ETH. Once back in Zurich, he immediately visited his old ETH classmate Marcel Grossmann, now a professor of mathematics, who introduced him to Riemannian geometry and, more generally, to differential geometry. On the recommendation of Italian mathematician Tullio Levi-Civita, Einstein began exploring the usefulness of general covariance (essentially the use of tensors) for his gravitational theory. For a while Einstein thought that there were problems with the approach, but he later returned to it and, by late 1915, had published his general theory of relativity in the form in which it is used today.[5] This theory explains gravitation as distortion of the structure of spacetime by matter, affecting the inertial motion of other matter.

During World War I, the work of Central Powers scientists was available only to Central Powers academics, for national security reasons. Some of Einstein's work did reach the United Kingdom and the United States through the efforts of the Austrian Paul Ehrenfest and physicists in the Netherlands, especially 1902 Nobel Prize-winner Hendrik Lorentz and Willem de Sitter of Leiden University. After the war ended, Einstein maintained his relationship with Leiden University, accepting a contract as an *Extraordinary Professor*; for ten years, from 1920 to 1930, he travelled to Holland regularly to lecture.[*][6]

In 1917, several astronomers accepted Einstein's 1911 challenge from Prague. The Mount Wilson Observatory in California, U.S., published a solar spectroscopic analysis that showed no gravitational redshift.[*][7] In 1918, the Lick Observatory, also in California, announced that it too had disproved Einstein's prediction, although its findings were not published.[*][8]

However, in May 1919, a team led by the British astronomer Arthur Stanley Eddington claimed to have confirmed Einstein's prediction of gravitational deflection of starlight by the Sun while photographing a solar eclipse with dual expeditions in Sobral, northern Brazil, and Príncipe, a west African island.[*][4] Nobel laureate Max Born praised general relativity as the "greatest feat of human thinking about nature";[*][9] fellow laureate Paul Dirac was quoted saying it was "probably the greatest scientific discovery ever made".[*][10] The international media guaranteed Einstein's global renown.

There have been claims that scrutiny of the specific photographs taken on the Eddington expedition showed the experimental uncertainty to be comparable to the same magnitude as the effect Eddington claimed to have demonstrated, and that a 1962 British expedition concluded that the method was inherently unreliable.[*][11] The deflection of light during a solar eclipse was confirmed by later, more accurate observations.[*][12] Some resented the newcomer's fame, notably among some German physicists, who later started the *Deutsche Physik* (German Physics) movement.[*][13][*][14]

14.1.3 General covariance and the hole argument

By 1912, Einstein was actively seeking a theory in which gravitation was explained as a geometric phenomenon. At the urging of Tullio Levi-Civita, Einstein began by exploring the use of general covariance (which is essentially the use of curvature tensors) to create a gravitational theory. However, in 1913 Einstein abandoned that approach, arguing that it is inconsistent based on the "hole argument". In 1914 and much of 1915, Einstein was trying to create field equations based on another approach. When that approach was proven to be inconsistent, Einstein revisited the concept of general covariance and discovered that the hole

argument was flawed.

14.1.4 The development of the Einstein field equations

Main article: Einstein field equations

When Einstein realized that general covariance was actually tenable, he quickly completed the development of the field equations that are named after him. However, he made a now-famous mistake. The field equations he published in October 1915 were

$$R_{\mu\nu} = T_{\mu\nu}$$

where $R_{\mu\nu}$ is the Ricci tensor, and $T_{\mu\nu}$ the energy–momentum tensor. This predicted the non-Newtonian perihelion precession of Mercury, and so had Einstein very excited. However, it was soon realized that they were inconsistent with the local conservation of energy–momentum unless the universe had a constant density of mass–energy–momentum. In other words, air, rock and even a vacuum should all have the same density. This inconsistency with observation sent Einstein back to the drawing board. However, the solution was all but obvious, and on November 25, 1915 Einstein presented the actual Einstein field equations to the Prussian Academy of Sciences:[*][15]

$$R_{\mu\nu} - \frac{1}{2}Rg_{\mu\nu} = T_{\mu\nu}$$

where R is the Ricci scalar and $g_{\mu\nu}$ the metric tensor. With the publication of the field equations, the issue became one of solving them for various cases and interpreting the solutions. This and experimental verification have dominated general relativity research ever since.

14.1.5 Einstein and Hilbert

See also: Relativity priority dispute

Although Einstein is credited with finding the field equations, the German mathematician David Hilbert published them in an article before Einstein's article. This has resulted in accusations of plagiarism against Einstein, although not from Hilbert, and assertions that the field equations should be called the "Einstein–Hilbert field equations". However, Hilbert did not press his claim for priority and some have asserted that Einstein submitted the correct equations before Hilbert amended his own work to include them. This suggests that Einstein developed the correct field equations first, though Hilbert may have reached them later independently (or even learned of them afterwards through his correspondence with Einstein).[16] However, others have criticized those assertions.[17]

14.1.6 Sir Arthur Eddington

In the early years after Einstein's theory was published, Sir Arthur Eddington lent his considerable prestige in the British scientific establishment in an effort to champion the work of this German scientist. Because the theory was so complex and abstruse (even today it is popularly considered the pinnacle of scientific thinking; in the early years it was even more so), it was rumored that only three people in the world understood it. There was an illuminating, though probably apocryphal, anecdote about this. As related by Ludwik Silberstein,[18] during one of Eddington's lectures he asked "Professor Eddington, you must be one of three persons in the world who understands general relativity." Eddington paused, unable to answer. Silberstein continued "Don't be modest, Eddington!" Finally, Eddington replied "On the contrary, I'm trying to think who the third person is."

14.2 Solutions

14.2.1 The Schwarzschild solution

Since the field equations are non-linear, Einstein assumed that they were unsolvable. However, in 1915 Karl Schwarzschild discovered an exact solution for the case of a spherically symmetric spacetime surrounding a massive object in spherical coordinates. This is now known as the Schwarzschild solution. Since then, many other exact solutions have been found.

14.2.2 The expanding universe and the cosmological constant

Main article: Cosmological constant

In 1922, Alexander Friedmann found a solution in which the universe may expand or contract, and later Georges Lemaître derived a solution for an expanding universe. However, Einstein believed that the universe was apparently static, and since a static cosmology was not supported by the general relativistic field equations, he added a cosmological constant Λ to the field equations, which became

$$R_{\mu\nu} - \frac{1}{2}Rg_{\mu\nu} + \Lambda g_{\mu\nu} = T_{\mu\nu}$$

This permitted the creation of steady-state solutions, but they were unstable: the slightest perturbation of a static state would result in the universe expanding or contracting. In 1929, Edwin Hubble found evidence for the idea that the universe is expanding. This resulted in Einstein dropping the cosmological constant, referring to it as "the biggest blunder in my career". At the time, it was an ad hoc hypothesis to add in the cosmological constant, as it was only intended to justify one result (a static universe).

14.2.3 More exact solutions

Progress in solving the field equations and understanding the solutions has been ongoing. The solution for a spherically symmetric charged object was discovered by Reissner and later rediscovered by Nordström, and is called the Reissner–Nordström solution. The black hole aspect of the Schwarzschild solution was very controversial, and Einstein did not believe that singularities could be real. However, in 1957 (two years after Einstein's death in 1955), Martin Kruskal published a proof that black holes are called for by the Schwarzschild Solution. Additionally, the solution for a rotating massive object was obtained by Kerr in the 1960s and is called the Kerr solution. The Kerr–Newman solution for a rotating, charged massive object was published a few years later.

14.3 Testing the theory

Main article: Tests of general relativity

The perihelion precession of Mercury was the first evidence that general relativity is correct. Sir Arthur Stanley Eddington's 1919 expedition in which he confirmed Einstein's prediction for the deflection of light by the Sun during the total solar eclipse of 29 May 1919 helped to cement the status of general relativity as a likely true theory. Since then many observations have confirmed the correctness of general relativity. These include studies of binary pulsars, observations of radio signals passing the limb of the Sun, and even the GPS system.

14.4 Alternative theories

Main article: Alternatives to general relativity

There have been various attempts to find modifications to general relativity. The most famous of these are the Brans–Dicke theory (also known as scalar-tensor theory), and Rosen's bimetric theory. Both of these theories proposed changes to the field equations of general relativity, and both suffer from these changes permitting the presence of bipolar gravitational radiation. As a result, Rosen's original theory has been refuted by observations of binary pulsars. As for Brans–Dicke (which has a tunable parameter ω such that $\omega = \infty$ is the same as general relativity), the amount by which it can differ from general relativity has been severely constrained by these observations.

In addition, general relativity is inconsistent with quantum mechanics, the physical theory that describes the wave–particle duality of matter, and quantum mechanics does not currently describe gravitational attraction at relevant (microscopic) scales. There is a great deal of speculation in the physics community as to the modifications that might be needed to both general relativity and quantum mechanics in order to unite them consistently. The speculative theory that unites general relativity and quantum mechanics is usually called quantum gravity, prominent examples of which include String Theory and Loop Quantum Gravity.

14.5   More about GR history

Kip Thorne identifies the "golden age of general relativity" as the period roughly from 1960 to 1975 during which the study of general relativity,*[19] which had previously been regarded as something of a curiosity, entered the mainstream of theoretical physics. During this period, many of the concepts and terms which continue to inspire the imagination of gravitation researchers and the general public were introduced, including black holes and 'gravitational singularity'. At the same time, in a closely related development, the study of physical cosmology entered the mainstream and the Big Bang became well established.

14.6   See also

- Contributors to general relativity

- Golden age of physics

- Golden age of cosmology

14.7   Notes

[1] Albert Einstein, Nobel lecture in 1921

[2] Einstein, A., "Relativitätsprinzip und die aus demselben gezogenen Folgerungen (On the Relativity Principle and the Conclusions Drawn from It)", *Jahrbuch der Radioaktivität (Yearbook of Radioactivity)* **4**: 411–462 page 454 (Wir betrachen zwei Bewegung systeme ...)

[3] Einstein, Albert (1911), "Einfluss der Schwerkraft auf die Ausbreitung des Lichtes (On the Influence of Gravity on the Propagation of Light)", *Annalen der Physik* **35**: 898–908, Bibcode:1911AnP...340..898E, doi:10.1002/andp.19113401005 (also in *Collected Papers* Vol. 3, document 23)

[4] Crelinsten, Jeffrey. "Einstein's Jury: The Race to Test Relativity". *Princeton University Press.* 2006. Retrieved on 13 March 2007. ISBN 978-0-691-12310-3

[5] O'Connor, J.J. and E.F. Robertson (1996), "General relativity". *Mathematical Physics index*, School of Mathematics and Statistics, University of St. Andrews, Scotland, May, 1996. Retrieved 2015-02-04.

[6] *Two friends in Leiden*, retrieved 11 June 2007

[7] Crelinsten, Jeffrey (2006), *Einstein's Jury: The Race to Test Relativity*, Princeton University Press, pp. 103–108, ISBN 978-0-691-12310-3, retrieved 13 March 2007

[8] Crelinsten, Jeffrey (2006), *Einstein's Jury: The Race to Test Relativity*, Princeton University Press, pp. 114–119, ISBN 978-0-691-12310-3, retrieved 13 March 2007

[9] Smith, PD (17 September 2005), *The genius of space and time*, London: The Guardian, retrieved 31 March 2007

[10] Jürgen Schmidhuber. "Albert Einstein (1879–1955) and the 'Greatest Scientific Discovery Ever'". 2006. Retrieved on 4 October 2006.

[11] Andrzej, Stasiak (2003), "Myths in science", *EMBO Reports* **4** (3): 236, doi:10.1038/sj.embor.embor779, retrieved 31 March 2007

[12] See the table in MathPages Bending Light

[13] Hentschel, Klaus and Ann M. (1996), *Physics and National Socialism: An Anthology of Primary Sources*, Birkhaeuser Verlag, xxi, ISBN 3-7643-5312-0

[14] For a discussion of astronomers' attitudes and debates about relativity, see Crelinsten, Jeffrey (2006), *Einstein's Jury: The Race to Test Relativity*, Princeton University Press, ISBN 0-691-12310-1, especially chapters 6, 9, 10 and 11.

[15] Pais, Abraham (1982). "14. The Field Equations of Gravitation". *Subtle is the Lord : The Science and the Life of Albert Einstein: The Science and the Life of Albert Einstein.* Oxford University Press. p. 239. ISBN 9780191524028.

[16] Leo Corry, Jürgen Renn, John Stachel: "Belated Decision in the Hilbert-Einstein Priority Dispute", SCIENCE, Vol. 278, 14 November 1997 - article text

[17] Friedwart Winterberg's response to the Cory-Renn-Stachel paper as printed in "Zeitschrift für Naturforschung" 59a, 715-719.

[18] John Waller (2002), *Einstein's Luck*, Oxford University Press, ISBN 0-19-860719-9

[19] Thorne, Kip (2003). "Warping spacetime". *The future of theoretical physics and cosmology: celebrating Stephen Hawking's 60th birthday*. Cambridge University Press. p. 74. ISBN 0-521-82081-2., Extract of page 74

14.8 References

- Pais, Abraham (1982). *Subtle is the lord: the science and life of Albert Einstein*. Oxford: Oxford University Press. ISBN 0-19-853907-X.

- Einstein, A.; Grossmann, M. (1913). "Entwurf einer verallgemeinerten Relativitätstheorie und einer Theorie der Gravitation" [Outline of a Generalized Theory of Relativity and of a Theory of Gravitation]. *Zeitschrift für Mathematik und Physik* **62**: 225–261.

- *Einstein and the Changing Worldviews of Physics* (editors —Lehner C., Renn J., Schemmel M.) 2012 (Birkhäuser).

- Genesis of general relativity series

Chapter 15

History of quantum mechanics

See also: Timeline of quantum mechanics and History of physics

The **history of quantum mechanics** is a fundamental part of the history of modern physics. Quantum mechanics' history, as it interlaces with the history of quantum chemistry, began essentially with a number of different scientific discoveries: the 1838 discovery of cathode rays by Michael Faraday; the 1859–60 winter statement of the black-body radiation problem by Gustav Kirchhoff; the 1877 suggestion by Ludwig Boltzmann that the energy states of a physical system could be *discrete*; the discovery of the photoelectric effect by Heinrich Hertz in 1887; and the 1900 quantum hypothesis by Max Planck that any energy-radiating atomic system can theoretically be divided into a number of discrete "energy elements" ε (epsilon) such that each of these energy elements is proportional to the frequency ν with which each of them individually radiate energy, as defined by the following formula:

$$\epsilon = h\nu$$

where h is a numerical value called Planck's constant.

Then, Albert Einstein in 1905, in order to explain the photoelectric effect previously reported by Heinrich Hertz in 1887, postulated consistently with Max Planck's quantum hypothesis that light itself is made of individual quantum particles, which in 1926 came to be called photons by Gilbert N. Lewis. The photoelectric effect was observed upon shining light of particular wavelengths on certain materials, such as metals, which caused electrons to be ejected from those materials only if the light quantum energy was greater than the work function of the metal's surface.

The phrase "quantum mechanics" was coined (in German, *Quantenmechanik*) by the group of physicists including Max Born, Werner Heisenberg, and Wolfgang Pauli, at the University of Göttingen in the early 1920s, and was first used in Born's 1924 paper "*Zur Quantenmechanik*".[1] In the years to follow, this theoretical basis slowly began to be applied to chemical structure, reactivity, and bonding.

15.1 Overview

Ludwig Eduard Boltzmann suggested in 1877 that the energy levels of a physical system, such as a molecule, could be discrete. He was a founder of the Austrian Mathematical Society, together with the mathematicians Gustav von Escherich and Emil Müller. Boltzmann's rationale for the presence of discrete energy levels in molecules such as those of iodine gas had its origins in his statistical thermodynamics and statistical mechanics theories and was backed up by mathematical arguments, as would also be the case twenty years later with the first quantum theory put forward by Max Planck.

In 1900, the German physicist Max Planck reluctantly introduced the idea that energy is *quantized* in order to derive a formula for the observed frequency dependence of the energy emitted by a black body, called Planck's Law, that included a Boltzmann distribution (applicable in the classical limit). Planck's law[2] can be stated as follows: $I(\nu, T) = \frac{2h\nu^3}{c^2} \frac{1}{e^{\frac{h\nu}{kT}} - 1}$, where:

$I(\nu,T)$ is the energy per unit time (or the power) radiated per unit area of emitting surface in the normal direction per unit solid angle per unit frequency by a black body at temperature T;

h is the Planck constant;

c is the speed of light in a vacuum;

k is the Boltzmann constant;

ν is the frequency of the electromagnetic radiation; and

T is the temperature of the body in kelvins.

The earlier Wien approximation may be derived from Planck's law by assuming $h\nu \gg kT$.

Moreover, the application of Planck's quantum theory to the electron allowed Ştefan Procopiu in 1911–1913, and subsequently Niels Bohr in 1913, to calculate the magnetic moment of the electron, which was later called the "magneton"; similar quantum computations, but with numerically quite different values, were subsequently made possible for both the magnetic moments of the proton and the neutron that are three orders of magnitude smaller than that of the electron.

In 1905, Einstein explained the photoelectric effect by postulating that light, or more generally all electromagnetic radiation, can be divided into a finite number of "energy quanta" that are localized points in space. From the introduction section of his March 1905 quantum paper, "On a heuristic viewpoint concerning the emission and transformation of light", Einstein states:

> "According to the assumption to be contemplated here, when a light ray is spreading from a point, the energy is not distributed continuously over ever-increasing spaces, but consists of a finite number of 'energy quanta' that are localized in points in space, move without dividing, and can be absorbed or generated only as a whole."

This statement has been called the most revolutionary sentence written by a physicist of the twentieth century.[*][3] These *energy quanta* later came to be called "photons", a term introduced by Gilbert N. Lewis in 1926. The idea that each photon had to consist of energy in terms of quanta was a remarkable achievement; it effectively solved the problem of black-body radiation attaining infinite energy, which occurred in theory if light were to be explained only in terms of waves. In 1913, Bohr explained the spectral lines of the hydrogen atom, again by using quantization, in his paper of July 1913 *On the Constitution of Atoms and Molecules*.

These theories, though successful, were strictly phenomenological: during this time, there was no rigorous justification for quantization, aside, perhaps, from Henri Poincaré's discussion of Planck's theory in his 1912 paper *Sur la théorie des quanta*.[*][4][*][5] They are collectively known as the *old quantum theory*.

The phrase "quantum physics" was first used in Johnston's *Planck's Universe in Light of Modern Physics* (1931).

In 1923, the French physicist Louis de Broglie put forward his theory of matter waves by stating that particles can exhibit wave characteristics and vice versa. This theory was for a single particle and derived from special relativity theory. Building on de Broglie's approach, modern quantum mechanics was born in 1925, when the German physicists Werner Heisenberg, Max Born, and Pascual Jordan[*][6][*][7] developed matrix mechanics and the Austrian physicist Erwin Schrödinger invented wave mechanics and the non-relativistic Schrödinger equation as an approximation to the generalised case of de Broglie's theory.[*][8]

Schrödinger subsequently showed that the two approaches were equivalent.

Heisenberg formulated his uncertainty principle in 1927, and the Copenhagen interpretation started to take shape at about the same time. Starting around 1927, Paul Dirac began the process of unifying quantum mechanics with special relativity by proposing the Dirac equation for the electron. The Dirac equation achieves the relativistic description of the wavefunction of an electron that Schrödinger failed to obtain. It predicts electron spin and led Dirac to predict the existence of the positron. He also pioneered the use of operator theory, including the influential bra–ket notation, as described in his famous 1930 textbook. During the same period, Hungarian polymath John von Neumann formulated the rigorous mathematical basis for quantum mechanics as the theory of linear operators on Hilbert spaces, as described in his likewise famous 1932 textbook. These, like many other works from the founding period, still stand, and remain widely used.

The field of quantum chemistry was pioneered by physicists Walter Heitler and Fritz London, who published a study of the covalent bond of the hydrogen molecule in 1927. Quantum chemistry was subsequently developed by a large number of workers, including the American theoretical chemist Linus Pauling at Caltech, and John C. Slater into various theories such as Molecular Orbital Theory or Valence Theory.

Beginning in 1927, researchers made attempts at applying quantum mechanics to fields instead of single particles, resulting in quantum field theories. Early workers in this area include P.A.M. Dirac, W. Pauli, V. Weisskopf, and P. Jordan. This area of research culminated in the formulation of quantum electrodynamics by R.P. Feynman, F. Dyson, J. Schwinger, and S.I. Tomonaga during the 1940s. Quantum electrodynamics describes a quantum theory of electrons, positrons, and the electromagnetic field, and served as a model for subsequent quantum field theories.[*][6][*][7][*][9]

The theory of quantum chromodynamics was formulated beginning in the early 1960s. The theory as we know it today was formulated by Politzer, Gross and Wilczek in 1975.

Building on pioneering work by Schwinger, Higgs and Goldstone, the physicists Glashow, Weinberg and Salam independently showed how the weak nuclear force and quantum electrodynamics could be merged into a single electroweak force, for which they received the 1979 Nobel Prize in Physics.

15.2 Founding experiments

- Thomas Young's double-slit experiment demonstrating the wave nature of light. (c1805)

- Henri Becquerel discovers radioactivity. (1896)

- J. J. Thomson's cathode ray tube experiments (discovers the electron and its negative charge). (1897)

- The study of black-body radiation between 1850 and 1900, which could not be explained without quantum concepts.

- The photoelectric effect: Einstein explained this in 1905 (and later received a Nobel prize for it) using the concept of photons, particles of light with quantized energy.

- Robert Millikan's oil-drop experiment, which showed that electric charge occurs as *quanta* (whole units). (1909)

- Ernest Rutherford's gold foil experiment disproved the plum pudding model of the atom which suggested that the mass and positive charge of the atom are almost uniformly distributed. (1911)

- James Franck and Gustav Hertz's electron collision experiment shows that energy absorption by mercury atoms is quantized. (1914)

- Otto Stern and Walther Gerlach conduct the Stern–Gerlach experiment, which demonstrates the quantized nature of particle spin. (1920)

- Clinton Davisson and Lester Germer demonstrate the wave nature of the electron[*][10] in the Electron diffraction experiment. (1927)

- Clyde L. Cowan and Frederick Reines confirm the existence of the neutrino in the neutrino experiment. (1955)

- Clauss Jönsson's double-slit experiment with electrons. (1961)

- The Quantum Hall effect, discovered in 1980 by Klaus von Klitzing. The quantized version of the Hall effect has allowed for the definition of a new practical standard for electrical resistance and for an extremely precise independent determination of the fine structure constant.

- The experimental verification of quantum entanglement by Alain Aspect. (1982)

- The Mach-Zehnder Interferometer experiment conducted by Paul Kwiat, Harold Wienfurter, Thomas Herzog, Anton Zeilinger, and Mark Kasevich, providing experimental verification of the Elitzur-Vadiman bomb tester, proving Interaction-free measurement is possible. (1994)

15.3   See also

- Golden age of physics

- History of quantum field theory

- History of chemistry

- History of the molecule

- History of thermodynamics

- Timeline of atomic and subatomic physics

15.4   References

[1] Max Born, *My Life: Recollections of a Nobel Laureate*, Taylor & Francis, London, 1978. ("We became more and more convinced that a radical change of the foundations of physics was necessary, i.e., a new kind of mechanics for which we used the term quantum mechanics. This word appears for the first time in physical literature in a paper of mine...")

[2] M. Planck (1914). *The theory of heat radiation*, second edition, translated by M. Masius, Blakiston's Son & Co, Philadelphia, pages 22, 26, 42, 43.

[3] Folsing, Albrecht (1997), *Albert Einstein: A Biography*, trans. Ewald Osers, Viking

[4] McCormmach, Russell (Spring 1967), "Henri Poincaré and the Quantum Theory", *Isis* **58** (1): 37–55, doi:10.1086/350182

[5] Irons, F. E. (August 2001), "Poincaré's 1911–12 proof of quantum discontinuity interpreted as applying to atoms", *American Journal of Physics* **69** (8): 879–884, Bibcode:2001AmJPh..69..879I, doi:10.1119/1.1356056

[6] David Edwards, *The Mathematical Foundations of Quantum Mechanics*, Synthese, Volume 42, Number 1/September, 1979, pp. 1–70.

[7] D. Edwards, *The Mathematical Foundations of Quantum Field Theory: Fermions, Gauge Fields, and Super-symmetry, Part I: Lattice Field Theories*, International J. of Theor. Phys., Vol. 20, No. 7 (1981).

[8] Hanle, P.A. (December 1977), "Erwin Schrodinger's Reaction to Louis de Broglie's Thesis on the Quantum Theory.", *Isis* **68** (4): 606–609, doi:10.1086/351880

[9] S. Auyang, *How is Quantum Field Theory Possible?*, Oxford University Press, 1995.

[10] The Davisson-Germer experiment, which demonstrates the wave nature of the electron

15.5 Further reading

- Bacciagaluppi, Guido; Valentini; Valentini, Antony (2009), *Quantum theory at the crossroads: reconsidering the 1927 Solvay conference*, Cambridge, UK: Cambridge University Press, p. 9184, arXiv:quant-ph/0609184, Bibcode:2006quant.ph..9184B, ISBN 978-0-521-81421-8, OCLC 227191829

- Bernstein, Jeremy (2009), *Quantum Leaps*, Cambridge, Massachusetts: Belknap Press of Harvard University Press, ISBN 978-0-674-03541-6

- Jammer, Max (1966), *The conceptual development of quantum mechanics*, New York: McGraw-Hill, OCLC 534562

- Jammer, Max (1974), *The philosophy of quantum mechanics: The interpretations of quantum mechanics in historical perspective*, New York: Wiley, ISBN 0-471-43958-4, OCLC 969760

- F. Bayen, M. Flato, C. Fronsdal, A. Lichnerowicz and D. Sternheimer, Deformation theory and quantization I,and II, *Ann. Phys. (N.Y.)*, **111** (1978) pp. 61–110, 111-151.

- D. Cohen, *An Introduction to Hilbert Space and Quantum Logic*, Springer-Verlag, 1989. This is a thorough and well-illustrated introduction.

- Finkelstein, D., "Matter, Space and Logic", *Boston Studies in the Philosophy of Science* **V**: 1969, doi:10.1007/978-94-010-3381-7_4.

- A. Gleason. Measures on the Closed Subspaces of a Hilbert Space, *Journal of Mathematics and Mechanics*, 1957.

- R. Kadison. Isometries of Operator Algebras, *Annals of Mathematics*, Vol. 54, pp. 325–338, 1951

- G. Ludwig. *Foundations of Quantum Mechanics*, Springer-Verlag, 1983.

- G. Mackey. *Mathematical Foundations of Quantum Mechanics*, W. A. Benjamin, 1963 (paperback reprint by Dover 2004).

- R. Omnès. *Understanding Quantum Mechanics*, Princeton University Press, 1999. (Discusses logical and philosophical issues of quantum mechanics, with careful attention to the history of the subject).

- N. Papanikolaou. *Reasoning Formally About Quantum Systems: An Overview*, ACM SIGACT News, 36(3), pp. 51–66, 2005.

- C. Piron. *Foundations of Quantum Physics*, W. A. Benjamin, 1976.

- Hermann Weyl. *The Theory of Groups and Quantum Mechanics*, Dover Publications, 1950.

- A. Whitaker. *The New Quantum Age: From Bell's Theorem to Quantum Computation and Teleportation*, Oxford University Press, 2011, ISBN 978-0-19-958913-5

- Stephen Hawking. *The Dreams that Stuff is Made of*, Running Press, 2011, ISBN 978-0-76-243434-3

- A. Douglas Stone. *Einstein and the Quantum, the Quest of the Valiant Swabian*, Princeton University Press, 2013, ISBN 978-0-691-13968-5

15.6 External links

- A History of Quantum Mechanics

- A Brief History of Quantum Mechanics

- Homepage of the Quantum History Project

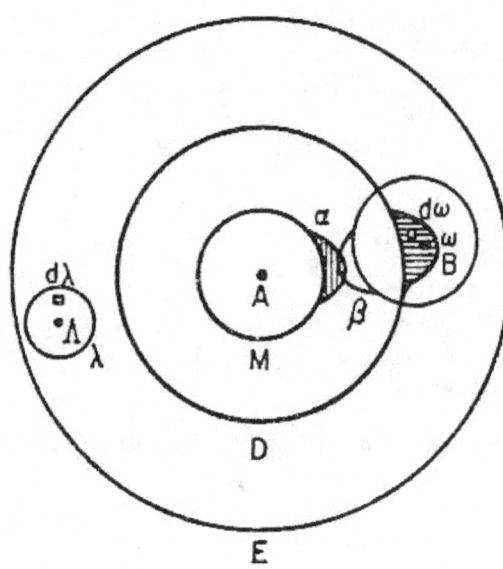

*Ludwig Boltzmann's **diagram of the I₂ molecule** proposed in 1898 showing the atomic "sensitive region" (α, β) of overlap.*

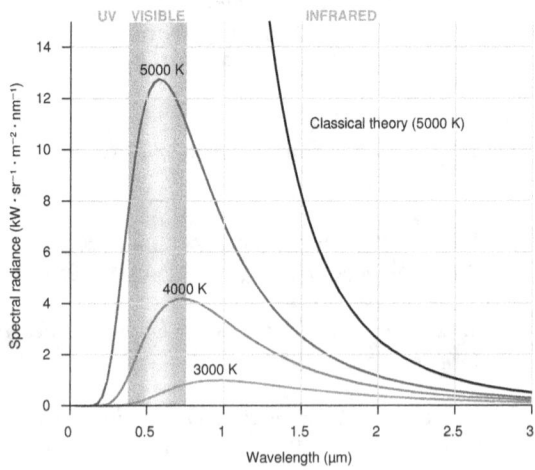

*With decreasing temperature, the peak of the **blackbody radiation** curve shifts to longer wavelengths and also has lower intensities. The blackbody radiation curves (1862) at left are also compared with the early, classical limit model of Rayleigh and Jeans (1900) shown at right. The short wavelength side of the curves was already approximated in 1896 by the Wien distribution law.*

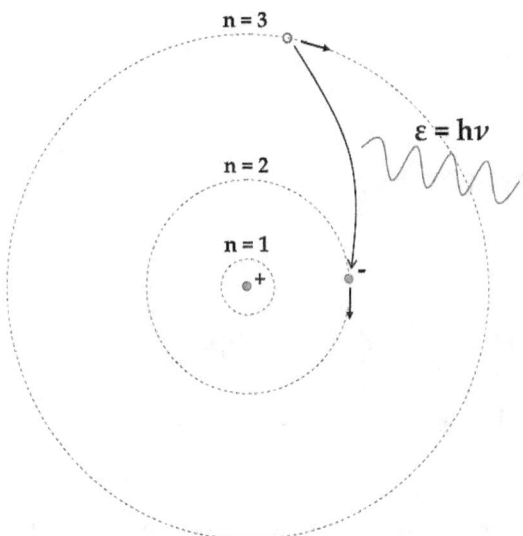

Niels Bohr's 1913 quantum model of the atom, which incorporated an explanation of Johannes Rydberg's 1888 formula, Max Planck's 1900 quantum hypothesis, i.e. that atomic energy radiators have discrete energy values ($\varepsilon = h\nu$), J. J. Thomson's 1904 plum pudding model, Albert Einstein's 1905 light quanta postulate, and Ernest Rutherford's 1907 discovery of the atomic nucleus. Note that the electron does not travel along the black line when emitting a photon. It jumps, disappearing from the outer orbit and appearing in the inner one and cannot exist in the space between orbits 2 and 3.

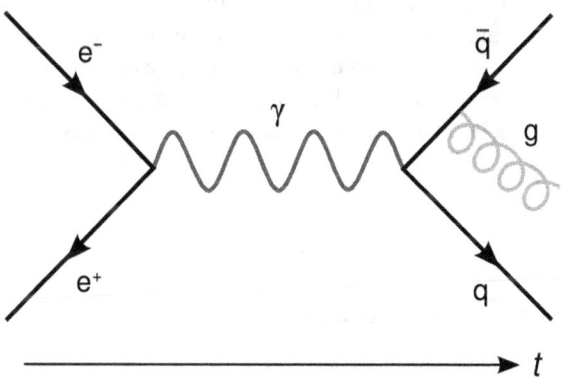

*Feynman diagram of **gluon radiation** in quantum chromodynamics*

Chapter 16

History of subatomic physics

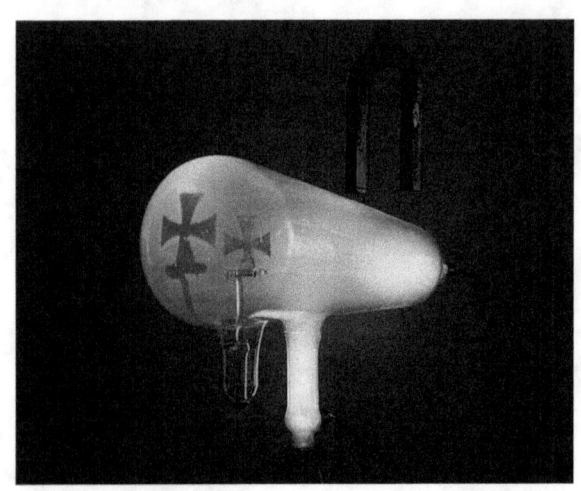

A Crookes tube with a magnetic deflector

The idea that matter consists of smaller particles and that there exists a limited number of sorts of primary, smallest particles in nature has existed in natural philosophy since time immemorial. Such ideas gained physical credibility beginning in the 19th century, but the concept of "elementary particle" underwent some changes in its meaning: notably, modern physics no longer deems elementary particles indestructible. Even elementary particles can decay or collide destructively; they can cease to exist and create (other) particles in result.

Increasingly small particles have been discovered and researched: they include molecules, which are constructed of atoms, that in turn consist of subatomic particles, namely atomic nuclei and electrons. Many more types of subatomic particles have been found. Most such particles (but not electrons) were eventually found to be composed of even smaller particles such as quarks. Particle physics studies these smallest particles and their behaviour under high energies, whereas nuclear physics studies atomic nuclei and their (immediate) constituents: protons and neutrons.

16.1 Early development

Main articles: Atomism and Atomic theory

The idea that all matter is composed of elementary particles dates to at least the 6th century BC.[*][1] The philosophical doctrine of atomism and the nature of elementary particles were studied by ancient Greek philosophers such as Leucippus, Democritus, and Epicurus; ancient Indian philosophers such as Kanada, Dignāga, and Dharmakirti; Muslim scientists such as Ibn al-Haytham, Ibn Sina, and Mohammad al-Ghazali; and in early modern Europe by physicists such as Pierre Gassendi, Robert Boyle, and Isaac Newton. The particle theory of light was also proposed by Ibn al-Haytham, Ibn Sina, Gassendi, and Newton.

Those early ideas were founded through abstract, philosophical reasoning rather than experimentation and empirical observation and represented only one line of thought among many. In contrast, certain ideas of Gottfried Wilhelm Leibniz (see *Monadology*) contradict to almost everything known in modern physics.

In the 19th century, John Dalton, through his work on stoichiometry, concluded that each element of nature was composed of a single, unique type of particle. Dalton and his contemporaries believed those were the fundamental particles of nature and thus named them atoms, after the Greek word *atomos*, meaning "indivisible"[*][2] or "uncut".

16.2 From atoms to nucleons

16.2.1 First subatomic particles

However, near the end of 19th century, physicists discovered that Dalton's atoms are not, in fact, the fundamental particles of nature, but conglomerates of even smaller particles. Electron was discovered between 1879 and 1897 in works of William Crookes, Arthur Schuster, J. J. Thomson, and other physicists; its charge was carefully measured by Robert Andrews Millikan and Harvey Fletcher in their oil drop experiment of 1909. Physicists theorized that negatively charged electrons are constituent part of "atoms", along with some (yet unknown) positively charged substance, and it was later confirmed. Electron became the first elementary, truly fundamental particle discovered.

Studies of the "radioactivity", that soon revealed the phenomenon of radioactive decay, provided another argument against considering chemical elements as fundamental nature's elements. Despite these discoveries, the term *atom* stuck to Dalton's (chemical) atoms and now denotes the smallest particle of a chemical element, not something really indivisible.

16.2.2   Researching particles' interaction

Further information: Rutherford scattering and History of electromagnetic theory

Early 20th-century physicists knew only two fundamental forces: electromagnetism and gravitation, where the latter could not explain the structure of atoms. So, it was obvious to assume that unknown positively charged substance attracts electrons by Coulomb force.

mass concentrated in a tiny atomic nucleus.

16.2.3   Inside the atom

By 1914, experiments by Ernest Rutherford, Henry Moseley, James Franck and Gustav Hertz had largely established the structure of an atom as a dense nucleus of positive charge surrounded by lower-mass electrons.[*][3] These discoveries shed a light to the nature of radioactive decay and other forms of transmutation of elements, as well as of elements themselves. It appeared that atomic number is nothing else than (positive) electric charge of the atomic nucleus of a particular atom. Chemical transformations, governed by electromagnetic interactions, do not change nuclei – that's why elements are chemically indestructible. But when the nucleus change its charge and/or mass (by emitting or capturing a particle), the atom can become the one of another element. Special relativity explained how the *mass defect* is related to the energy produced or consumed in reactions. The branch of physics that studies transformations and the structure of nuclei is now called nuclear physics, contrasted to atomic physics that studies the structure and properties of atoms ignoring most nuclear aspects. The development in the nascent quantum physics, such as Bohr model, led to the understanding of chemistry in terms of the arrangement of electrons in the mostly empty volume of atoms.

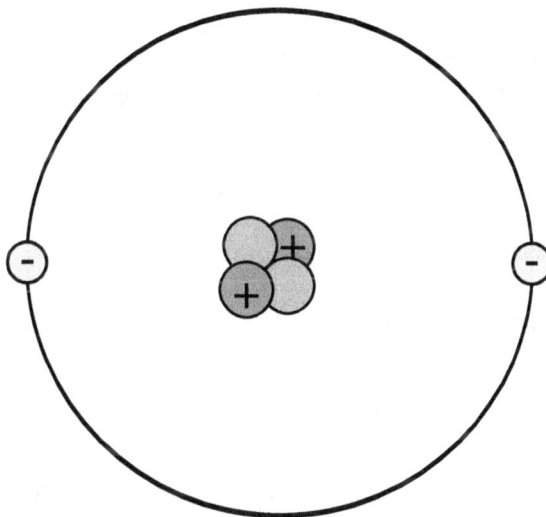

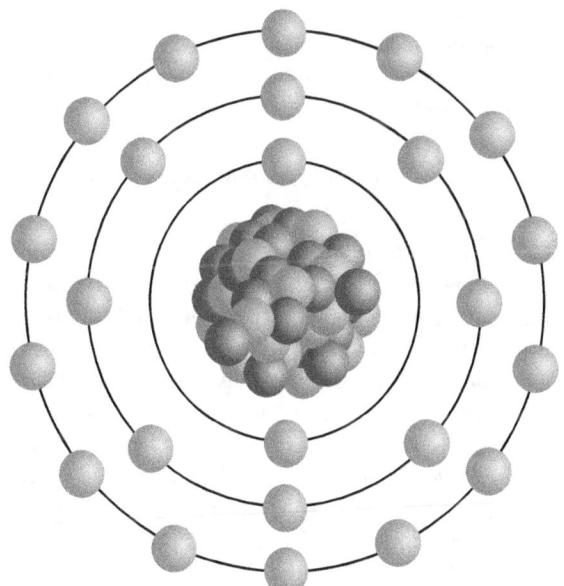

In 1909 Ernest Rutherford and Thomas Royds demonstrated that an alpha particle combines with two electrons and forms a helium atom. In modern terms, alpha particles are doubly ionized helium (more precisely, 4He) atoms. Speculation about the structure of atoms was severely constrained by Rutherford's 1907 gold foil experiment, showing that the atom is mainly empty space, with almost all its

In 1918, Rutherford confirmed that the hydrogen nucleus was a particle with a positive charge, which he named the proton. By then, Frederick Soddy's researches of radioactive elements, and experiments of J. J. Thomson and F.W. Aston conclusively demonstrated existence of isotopes, whose nuclei have different masses in spite of identical atomic numbers. It prompted Rutherford to conjecture that all nuclei other than hydrogen contain charge-less particles, which he named the neutron. Evidences that atomic nuclei consist of some smaller particles (now called *nucleons*) grew; it became obvious that, while protons repulse each other electrostatically, nucleons attract each other by some new force (nuclear force). It culminated in proofs of nuclear fission in 1939 by Lise Meitner (based on experiments by Otto Hahn), and nuclear fusion by Hans Bethe in that same year. Those discoveries gave rise to an active industry of generating one atom from another, even rendering possible (although it will probably never be profitable) the transmutation of lead into gold; and, those same discoveries also led to the development of nuclear weapons.

16.3 Revelations of quantum mechanics

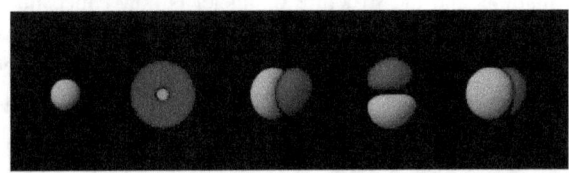

Atomic orbitals of Period 2 elements:
1s 2s 2p (3 items).
All complete subshells (including 2p) are inherently spherically symmetric, but it is convenient to assign to "distinct" p-electrons these two-lobed shapes.

Main article: History of quantum mechanics

Further understanding of atomic and nuclear structures became impossible without improving the knowledge about the essence of particles. Experiments and improved theories (such as Erwin Schrödinger's "electron waves") gradually revealed that there is no fundamental difference between particles and waves. For example, electromagnetic waves were reformulated in terms of particles called *photons*. It also revealed that physical objects do not change their parameters, such as total energy, position and momentum, as continuous functions of time, as it was thought of in classical physics: see atomic electron transition for example.

Another crucial discovery was identical particles or, more generally, quantum particle statistics. It was established that all electrons are identical: although two or more electrons can exist simultaneously that have different parameters, but they do not keep separate, distinguishable histories. This also applies to protons, neutrons, and (with certain differ-

ences) to photons as well. It suggested that there is a limited number of sorts of smallest particles in the universe.

The spin–statistics theorem established that any particle in our spacetime may be either a boson (that means its statistics is Bose–Einstein) or a fermion (that means its statistics is Fermi–Dirac). It was later found that all fundamental bosons transmit forces, like the photon that transmits light. Some of non-fundamental bosons (namely, mesons) also may transmit forces (see below), although non-fundamental ones. Fermions are particles "like electrons and nucleons" and generally comprise the matter. Note that any subatomic or atomic particle composed of even *total* number of fermions (such as protons, neutrons, and electrons) is a boson, so a boson is not necessarily a force transmitter and perfectly can be an ordinary material particle.

The spin is the quantity that distinguishes bosons and fermions. Practically it appears as an intrinsic angular momentum of a particle, that is unrelated to its motion but is linked with some other features like a magnetic dipole. Theoretically it is explained from different types representations of symmetry groups, namely tensor representations (including vectors and scalars) for bosons with their integer (in $\hbar$) spins, and spinor representations for fermions with their half-integer spins.

This culminated in the formulation of ideas of a quantum field theory. The first (and the only mathematically complete) of these theories, quantum electrodynamics, allowed to explain thoroughly the structure of atoms, including the Periodic Table and atomic spectra. Ideas of quantum mechanics and quantum field theory were applied to nuclear physics too. For example, α decay was explained as a quantum tunneling through nuclear potential, nucleons' fermionic statistics explained the nucleon pairing, and Hideki Yukawa proposed certain virtual particles (now knows as π-mesons) as an explanation of the nuclear force.

16.4 Inventory

16.5 Modern nuclear physics

Further information: History of nuclear power, Synthetic element and History of nuclear weapons

Development of nuclear models (such as the liquid-drop model and nuclear shell model) made prediction of properties of nuclides possible. No existing model of nucleon–nucleon interaction can *analytically* compute something more complex than 4He based on principles of quantum mechanics, though (note that complete computation of electron shells in atoms is also impossible yet).

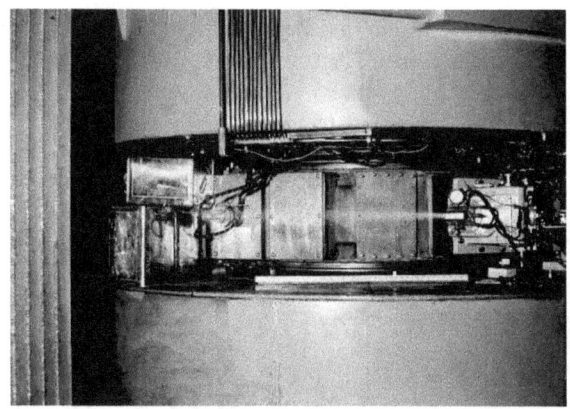

The most developed branch of nuclear physics in 1940s was studies related to nuclear fission due to its military significance. The main focus of fission-related problems is interaction of atomic nuclei with neutrons: a process that occurs in a fission bomb and a nuclear fission reactor. It gradually drifted away from the rest of subatomic physics and virtually became the nuclear engineering. First synthesised transuranium elements were also obtained in this context, through neutron capture and subsequent β^*− decay.

The elements beyond fermium cannot be produced in this way. To make a nuclide with more than 100 protons per nucleus one has to use an inventory and methods of particle physics (see details below), namely to accelerate and collide atomic nuclei. Production of progressively heavier synthetic elements continued into 21st century as a branch of nuclear physics, but only for scientific purposes.

The third important stream in nuclear physics are researches related to nuclear fusion. This is related to thermonuclear weapons (and conceived peaceful thermonuclear energy), as well as to astrophysical researches, such as stellar nucleosynthesis and Big Bang nucleosynthesis.

16.6 Physics goes to high energies

16.6.1 Strange particles and mysteries of the weak interaction

Further information: Resonance (particle physics), Deep inelastic scattering and Strange particle

In the 1950s, with development of particle accelerators and studies of cosmic rays, inelastic scattering experiments on protons (and other atomic nuclei) with energies about hundreds of MeVs became affordable. They created some short-lived resonance "particles", but also *hyperons* and *K-mesons* with unusually long lifetime. The cause of the

latter was found in a new quasi-conserved quantity, named *strangeness*, that is conserved in all circumstances except for the weak interaction. The strangeness of heavy particles and the μ-lepton were first two signs of what is now known as the second generation of fundamental particles.

The weak interaction revealed soon yet another mystery. In 1957 it was found that it does not conserve parity. In other words, the mirror symmetry was disproved as a fundamental symmetry law.

Throughout the 1950s and 1960s, improvements in particle accelerators and particle detectors led to a bewildering variety of particles were found in high-energy experiments and the term *elementary particle* referred to dozens of particles, most of them unstable. It prompted Wolfgang Pauli's remark: "Had I foreseen this, I would have gone into botany". The entire collection was nicknamed the "particle zoo". It became evident that some smaller constituents, yet invisible, form mesons and baryons that counted most of then-known particles.

16.6.2 Deeper constituents of matter

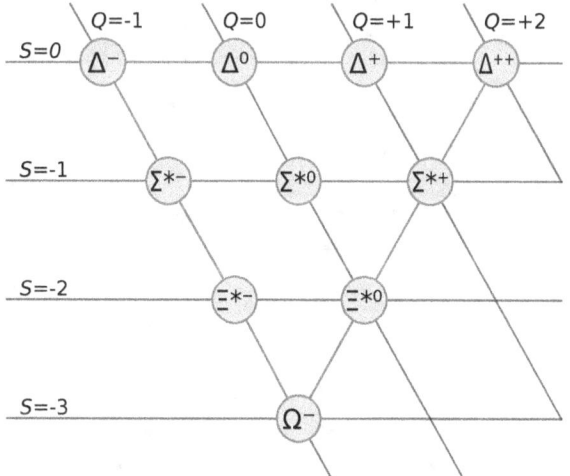

Classification of spin-3/2 baryons known in 1960s

Further information: Hadron

The interaction of these particles by scattering and decay provided a key to new fundamental quantum theories. Murray Gell-Mann and Yuval Ne'eman brought some order to mesons and baryons, the most numerous classes of particles, by classifying them according to certain qualities. It began with what Gell-Mann referred to as the "Eightfold Way", but proceeding into several different "octets" and "decuplets" which could predict new particles, most famously the Ω−, which was detected at Brookhaven National Laboratory in 1964, and which gave rise to the *quark* model of hadron composition. While the quark model at first seemed inadequate to describe strong nuclear forces, allowing the temporary rise of competing theories such as the S-matrix theory, the establishment of quantum chromodynamics in the 1970s finalized a set of fundamental and exchange particles (Kragh 1999). It postulated the fundamental strong interaction, experienced by quarks and mediated by gluons. These particles were proposed as a building material for hadrons (see hadronization). This theory is unusual because individual (free) quarks cannot be observed (see color confinement), unlike the situation with composite atoms where electrons and nuclei can be isolated by transferring ionization energy to the atom.

Then, the old, broad denotation of the term *elementary particle* was deprecated and a replacement term *subatomic particle* covered all the "zoo", with its hyponym "hadron" referring to composite particles directly explained by the quark model. The designation of an "elementary" (or "fundamental") particle was reserved for leptons, quarks, their antiparticles, and quanta of fundamental interactions (see below) only.

16.6.3   Quarks, leptons, and four fundamental forces

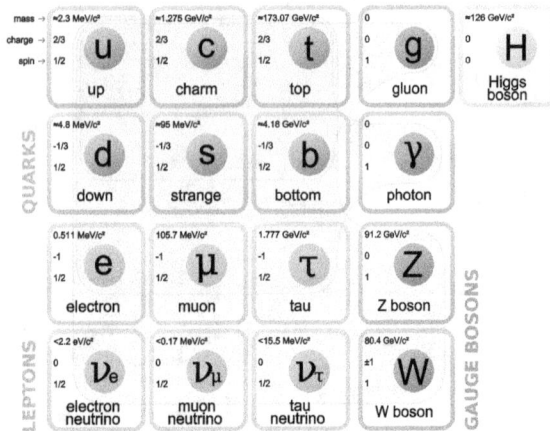

The Standard Model

Main articles: Fundamental interaction and Standard Model

Because the quantum field theory (see above) postulates no difference between particles and interactions, classification of elementary particles allowed also to classify interactions and fields.

Now a large number of particles and (non-fundamental) interactions is explained as combinations of a (relatively) small number of fundamental substances, thought

to be fundamental interactions (incarnated in fundamental bosons), quarks (including antiparticles), and leptons (including antiparticles). As the theory distinguished *several* fundamental interactions, it became possible to see which elementary particles participate in which interaction. Namely:

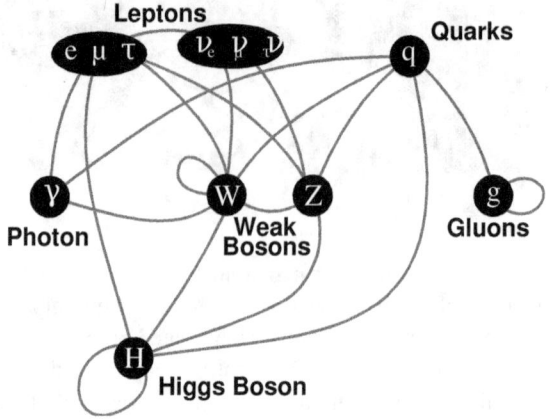

- All particles participate in gravitation.

- All charged elementary particles participate in electromagnetic interaction.

  - As a consequence, neutron participate in it with its magnetic dipole in spite of zero electric charge. This is because it is composed of *charged* quarks whose charges sum to zero.

- All fermions participate in the weak interaction.

- Quarks participate in the strong interaction, along gluons (its own quanta), but not leptons nor any fundamental bosons other than gluons.

The next step was a reduction in number of fundamental interactions, envisaged by early 20th century physicists as the "united field theory". The first successful modern unified theory was the electroweak theory, developed by Abdus Salam, Steven Weinberg and, subsequently, Sheldon Glashow. This development culminated in the completion of the theory called the Standard Model in the 1970s, that included also the strong interaction, thus covering three fundamental forces. After the discovery, made at CERN, of the existence of neutral weak currents,[4][5][6][7] mediated by the Z boson foreseen in the standard model, the physicists Salam, Glashow and Weinberg received the 1979 Nobel Prize in Physics for their electroweak theory.[8] The discovery of the weak gauge bosons (quanta of the weak interaction) through the 1980s, and the verification of their properties through the 1990s is considered to be an age of consolidation in particle physics.

While accelerators have confirmed most aspects of the Standard Model by detecting expected particle interactions at various collision energies, no theory reconciling general relativity with the Standard Model has yet been found, although supersymmetry and string theory were believed by many theorists to be a promising avenue forward. The Large Hadron Collider, however, which began operating in 2008, has failed to find any evidence whatsoever that is supportive of supersymmetry and string theory,[*][9] and appears unlikely to do so, meaning "the current situation in fundamental theory is one of a serious lack of any new ideas at all."[*][10] This state of affairs should not be viewed as a crisis in physics, but rather, as David Gross has said, "the kind of acceptable scientific confusion that discovery eventually transcends."[*][11]

The fourth fundamental force, gravitation, is not yet integrated into particle physics in a consistent way.

16.6.4 Higgs boson

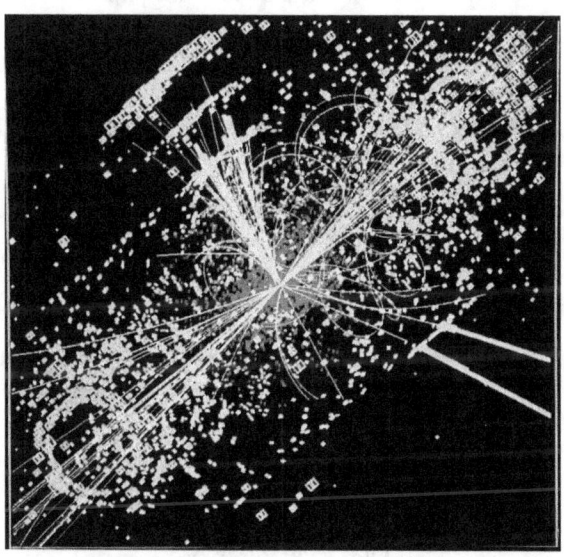

One possible signature of a Higgs boson from a simulated proton–proton collision. It decays almost immediately into two jets of hadrons and two electrons, visible as lines.

Further information: Higgs boson

As of 2011, the Higgs boson, the quantum of a field that is thought to provide particles with rest masses, remained the only particle of the Standard Model to be verified. On July 4, 2012, physicists working at CERN's Large Hadron Collider announced that they had discovered a new subatomic particle greatly resembling the Higgs boson, a potential key to an understanding of why elementary particles have masses and indeed to the existence of diversity and life

in the universe.[*][12] Rolf-Dieter Heuer, the director general of CERN, said that it was too soon to know for sure whether it is an entirely new particle, which weighs in at 125 billion electron volts – one of the heaviest subatomic particles yet – or, indeed, the elusive particle predicted by the Standard Model, the theory that has ruled physics for the last half-century.[*][12] It is unknown if this particle is an impostor, a single particle or even the first of many particles yet to be discovered. The latter possibilities are particularly exciting to physicists since they could point the way to new deeper ideas, beyond the Standard Model, about the nature of reality. For now, some physicists are calling it a "Higgslike" particle.[*][12] Joe Incandela, of the University of California, Santa Barbara, said, "It's something that may, in the end, be one of the biggest observations of any new phenomena in our field in the last 30 or 40 years, going way back to the discovery of quarks, for example."[*][12] The groups operating the large detectors in the collider said that the likelihood that their signal was a result of a chance fluctuation was less than one chance in 3.5 million, so-called "five sigma," which is the gold standard in physics for a discovery. Michael Turner, a cosmologist at the University of Chicago and the chairman of the physics center board, said

> This is a big moment for particle physics and a crossroads —will this be the high water mark or will it be the first of many discoveries that point us toward solving the really big questions that we have posed?
> —Michael Turner, University of Chicago[*][12]

Confirmation of the Higgs boson or something very much like it would constitute a rendezvous with destiny for a generation of physicists who have believed the boson existed for half a century without ever seeing it. Further, it affirms a grand view of a universe ruled by simple and elegant and symmetrical laws, but in which everything interesting in it being a result of flaws or breaks in that symmetry.[*][12] According to the Standard Model, the Higgs boson is the only visible and particular manifestation of an invisible force field that permeates space and imbues elementary particles that would otherwise be massless with mass. Without this Higgs field, or something like it, physicists say all the elementary forms of matter would zoom around at the speed of light; there would be neither atoms nor life. The Higgs boson achieved a notoriety rare for abstract physics.[*][12] To the eternal dismay of his colleagues, Leon Lederman, the former director of Fermilab, called it the "God particle" in his book of the same name, later quipping that he had wanted to call it "the goddamn particle".[*][12] Professor Incandela also stated,

This boson is a very profound thing we have found. We're reaching into the fabric of the universe at a level we've never done before. We've kind of completed one particle's story [...] We're on the frontier now, on the edge of a new exploration. This could be the only part of the story that's left, or we could open a whole new realm of discovery.

—Joe Incandela, University of California*[13]

In quantum theory, which is the language of particle physicists, elementary particles are divided into two rough categories: fermions, which are bits of matter like electrons, and bosons, which are bits of energy and can transmit forces, like the photon that transmits light. Dr. Peter Higgs was one of six physicists, working in three independent groups, who in 1964 invented the notion of the cosmic molasses, or Higgs field. The others were Tom Kibble of Imperial College, London; Carl Hagen of the University of Rochester; Gerald Guralnik of Brown University; and François Englert and Robert Brout, both of Université Libre de Bruxelles.*[12] One implication of their theory was that this Higgs field, normally invisible and, of course, odorless, would produce its own quantum particle if hit hard enough, by the right amount of energy. The particle would be fragile and fall apart within a millionth of a second in a dozen different ways depending upon its own mass. Unfortunately, the theory did not say how much this particle should weigh, which is what made it so difficult to find. The particle eluded researchers at a succession of particle accelerators, including the Large Electron–Positron Collider at CERN, which closed down in 2000, and the Tevatron at the Fermi National Accelerator Laboratory, or Fermilab, in Batavia, Ill., which shut down in 2011.*[12]

Further experiments continued and in March 2013 it was tentatively confirmed that the newly discovered particle was a Higgs Boson.

Although they have never been seen, Higgslike fields play an important role in theories of the universe and in string theory. Under certain conditions, according to the strange accounting of Einsteinian physics, they can become suffused with energy that exerts an antigravitational force. Such fields have been proposed as the source of an enormous burst of expansion, known as inflation, early in the universe and, possibly, as the secret of the dark energy that now seems to be speeding up the expansion of the universe.*[12]

16.6.5   Further theoretical development

Further information: Grand Unified Theory

Modern theoretical development includes refining of the Standard Model, researching in its foundations such as the Yang–Mills theory, and researches in computational methods such as the lattice QCD.

A long-standing problem is quantum gravitation. No solution that is useful for particle physics has been achieved.

16.6.6   Further experimental development

There are researches about quark–gluon plasma, a new (hypothetical) state of matter. There are also some recent experimental evidences that tetraquarks and glueballs exist.

The proton decay is not observed (or, generally, nonconservation of the baryon number), but predicted by the Standard Model, so there are searches for it.

16.7   See also

- Timeline of atomic and subatomic physics

- Golden age of physics

- Subatomic particle#History, authors and dates of important discoveries

- History of string theory

16.8   References

[1] "Fundamentals of Physics and Nuclear Physics" (PDF). Retrieved 2012-07-21.

[2] "Scientific Explorer: Quasiparticles". Sciexplorer.blogspot.com. 2012-05-22. Retrieved 2012-07-21.

[3] Smirnov, B.M. (2003). *Physics of Atoms and Ions*. Springer. pp. 14–21. ISBN 0-387-95550-X.

[4] F. J. Hasert *et al. Phys. Lett.* **46B** 121 (1973).

[5] F. J. Hasert *et al. Phys. Lett.* **46B** 138 (1973).

[6] F. J. Hasert *et al. Nucl. Phys.* **B73** 1(1974).

[7] *The discovery of the weak neutral currents*, CERN courier, 2004-10-04, retrieved 2008-05-08

[8] *The Nobel Prize in Physics 1979*, Nobel Foundation, retrieved 2008-09-10

[9] Woit, Peter (20 October 2013). "Last Links For a While". *Not Even Wrong*. Retrieved 2 November 2013.

[10] Peter Woit (28 May 2013). "A Tale of Two Oxford Talks". Not Even Wrong. Retrieved 19 October 2013.

[11] Peter Byrne (24 May 2013). "Waiting for the Revolution". *Quanta Magazine*. simonsfoundation.org. Retrieved 19 October 2013.

[12] http://www.nytimes.com/2012/07/05/science/cern-physicists-may-have-discovered-higgs-boson-particle.html?pagewanted=3&_r=1&ref=science

[13] Rincon, Paul (2012-07-04). "BBC News - Higgs boson-like particle discovery claimed at LHC". Bbc.co.uk. Retrieved 2013-04-20.

- Kragh, Helge (1999), *Quantum Generations: A History of Physics in the Twentieth Century*, Princeton: Princeton University Press.

Chapter 17

Unified field theory

"Unified theory" redirects here. For the band, see Unified Theory (band).

In physics, a **unified field theory** (**UFT**), occasionally referred to as a **uniform field theory**,*[1] is a type of field theory that allows all that is usually thought of as fundamental forces and elementary particles to be written in terms of a single field. There is no accepted unified field theory, and thus it remains an open line of research. The term was coined by Einstein, who attempted to unify the general theory of relativity with electromagnetism. The "theory of everything" and Grand Unified Theory are closely related to unified field theory, but differ by not requiring the basis of nature to be fields, and often by attempting to explain physical constants of nature.

This article describes unified field theory as it is currently understood in connection with quantum theory. Earlier attempts based on classical physics are described in the article on classical unified field theories.

There may be no *a priori* reason why the correct description of nature has to be a unified field theory. However, this goal has led to a great deal of progress in modern theoretical physics and continues to motivate research.

17.1 Introduction

According to the current understanding of physics, forces are not transmitted directly between interacting objects, but instead are described by intermediary entities called fields. All four of the known fundamental forces are mediated by fields, which in the Standard Model of particle physics result from exchange of gauge bosons. Specifically the four fundamental interactions to be unified are:

- Strong interaction: the interaction responsible for holding quarks together to form hadrons, and holding neutrons and also protons together to form atomic nuclei. The exchange particle that mediates this force is

the gluon.

- Electromagnetic interaction: the familiar interaction that acts on electrically charged particles. The photon is the exchange particle for this force.

- Weak interaction: a short-range interaction responsible for some forms of radioactivity, that acts on electrons, neutrinos, and quarks. It is mediated by the W and Z bosons.

- Gravitational interaction: a long-range attractive interaction that acts on *all* particles. The postulated exchange particle has been named the graviton.

Modern unified field theory attempts to bring these four interactions together into a single framework.

17.2 History

The first successful classical unified field theory was developed by James Clerk Maxwell. In 1820 Hans Christian Ørsted discovered that electric currents exerted forces on magnets, while in 1831, Michael Faraday made the observation that time-varying magnetic fields could induce electric currents. Until then, electricity and magnetism had been thought of as unrelated phenomena. In 1864, Maxwell published his famous paper on a dynamical theory of the electromagnetic field. This was the first example of a theory that was able to encompass previously separate field theories (namely electricity and magnetism) to provide a unifying theory of electromagnetism. By 1905, Albert Einstein had used the constancy of the speed of light in Maxwell's theory to unify our notions of space and time into an entity we now call spacetime and in 1915 he expanded this theory of special relativity to a description of gravity, General Relativity, using a field to describe the curving geometry of four-dimensional spacetime.

In the years following the creation of the general theory, a large number of physicists and mathematicians enthusiastically participated in the attempt to unify the then-known fundamental interactions.[*][2] In view of later developments in this domain, of particular interest are the theories of Hermann Weyl of 1919, who introduced the concept of an (electromagnetic) gauge field in a classical field theory[*][3] and, two years later, that of Theodor Kaluza, who extended General Relativity to five dimensions.[*][4] Continuing in this latter direction, Oscar Klein proposed in 1926 that the fourth spatial dimension be curled up into a small, unobserved circle. In Kaluza–Klein theory, the gravitational curvature of the extra spatial direction behaves as an additional force similar to electromagnetism. These and other models of electromagnetism and gravity were pursued by Albert Einstein in his attempts at a classical unified field theory. By 1930 Einstein had already considered the Einstein–Maxwell–Dirac System [Dongen]. This system is (heuristically) the super-classical [Varadarajan] limit of (the not mathematically well-defined) Quantum Electrodynamics. One can extend this system to include the weak and strong nuclear forces to get the Einstein–Yang–Mills–Dirac System.

17.3 Modern progress

In 1963 American physicist Sheldon Glashow proposed that the weak nuclear force and electricity and magnetism could arise from a partially unified electroweak theory. In 1967, Pakistani Abdus Salam and American Steven Weinberg independently revised Glashow's theory by having the masses for the W particle and Z particle arise through spontaneous symmetry breaking with the Higgs mechanism. This unified theory modeled the electroweak interaction as a force mediated by four particles: the photon for the electromagnetic aspect, and a neutral Z particle and two charged W particles for weak aspect. As a result of the spontaneous symmetry breaking, the weak force becomes short-range and the Z and W bosons acquire masses of 80.4 and 91.2 GeV/c^2, respectively. Their theory was first given experimental support by the discovery of weak neutral currents in 1973. In 1983, the Z and W bosons were first produced at CERN by Carlo Rubbia's team. For their insights, Glashow, Salam, and Weinberg were awarded the Nobel Prize in Physics in 1979. Carlo Rubbia and Simon van der Meer received the Prize in 1984.

After Gerardus 't Hooft showed the Glashow–Weinberg–Salam electroweak interactions to be mathematically consistent, the electroweak theory became a template for further attempts at unifying forces. In 1974, Sheldon Glashow and Howard Georgi proposed unifying the strong and electroweak interactions into Georgi–Glashow model, the first Grand Unified Theory, which would have observable effects for energies much above 100 GeV.

Since then there have been several proposals for Grand Unified Theories, e.g. the Pati–Salam model, although none is currently universally accepted. A major problem for experimental tests of such theories is the energy scale involved, which is well beyond the reach of current accelerators. Grand Unified Theories make predictions for the relative strengths of the strong, weak, and electromagnetic forces, and in 1991 LEP determined that supersymmetric theories have the correct ratio of couplings for a Georgi–Glashow Grand Unified Theory. Many Grand Unified Theories (but not Pati–Salam) predict that the proton can decay, and if this were to be seen, details of the decay products could give hints at more aspects of the Grand Unified Theory. It is at present unknown if the proton can decay, although experiments have determined a lower bound of 10^{35} years for its lifetime.

17.4 Current status

Gravity has yet to be successfully included in a theory of everything. Simply trying to combine the graviton with the strong and electroweak interactions runs into fundamental difficulties since the resulting theory is not renormalizable. Theoretical physicists have not yet formulated a widely accepted, consistent theory that combines general relativity and quantum mechanics. The incompatibility of the two theories remains an outstanding problem in the field of physics. Some theoretical physicists currently believe that a quantum theory of general relativity may require frameworks other than field theory itself, such as string theory or loop quantum gravity. Some models in string theory that are promising by way of realizing our familiar standard model are the perturbative heterotic string models, 11-dimensional M-theory, Singular geometries (e.g. orbifold and orientifold), D-branes and other branes, flux compactification and warped geometry, and non-perturbative type IIB superstring solutions (F-theory).[*][5]

17.5 Notes

[1] *See, e.g., Beyond Art: A Third Culture* page 199. *Compare* Uniform field theory.

[2] See Catherine Goldstein & Jim Ritter (2003) "The varieties of unity: sounding unified theories 1920-1930" in A. Ashtekar, et al. (eds.), *Revisiting the Foundations of Relativistic Physics*, Dordrecht, Kluwer, p. 93-149; Vladimir Vizgin (1994), *Unified Field Theories in the First Third of the 20th Century*, Basel, Birkhäuser; Hubert Goenner On the History of Unified Field Theories.

[3] Erhard Scholtz (ed) (2001), *Hermann Weyl's* Raum - Zeit-Materie *and a General Introduction to His Scientific Work*,

Basel, Birkhäuser.

[4] Daniela Wuensch (2003), "The fifth dimension: Theodor Kaluza's ground-breaking idea", *Annalen der Physik*, vol. 12, p. 519–542.

[5] http://arxiv.org/abs/0812.1372

17.6   References

- Shushi Tomer, *A Possible Connection between Quantum and General Relativity Theories*, SSRN: http://papers.ssrn.com/sol3/papers.cfm?abstract_id= 2538728 (December 15, 2014)

- Jeroen van van Dongen *Einstein's Unification*, Cambridge University Press (July 26, 2010)

- Varadarajan, V.S. *Supersymmetry for Mathematicians: An Introduction (Courant Lecture Notes)*, American Mathematical Society (July 2004)

17.7   External links

- On the History of Unified Field Theories, by Hubert F. M. Goenner

Chapter 18

Physical cosmology

This article is about the branch of astronomy. For other uses, see Cosmology.
"Cosmic Evolution" redirects here. For the book by Eric Chaisson, see Cosmic Evolution (book).

Physical cosmology is the study of the largest-scale structures and dynamics of the Universe and is concerned with fundamental questions about its origin, structure, evolution, and ultimate fate.[1] For most of human history, it was a branch of metaphysics and religion. Cosmology as a science originated with the Copernican principle, which implies that celestial bodies obey identical physical laws to those on Earth, and Newtonian mechanics, which first allowed us to understand those physical laws.

Physical cosmology, as it is now understood, began with the development in 1915 of Albert Einstein's general theory of relativity, followed by major observational discoveries in the 1920s: first, Edwin Hubble discovered that the universe contains a huge number of external galaxies beyond our own Milky Way; then, work by Vesto Slipher and others showed that the universe is expanding. These advances made it possible to speculate about the origin of the universe, and allowed the establishment of the Big Bang Theory, by Georges Lemaitre, as the leading cosmological model. A few researchers still advocate a handful of alternative cosmologies;[2] however, most cosmologists agree that the Big Bang theory explains the observations better.

Dramatic advances in observational cosmology since the 1990s, including the cosmic microwave background, distant supernovae and galaxy redshift surveys, have led to the development of a standard model of cosmology. This model requires the universe to contain large amounts of dark matter and dark energy whose nature is currently not well understood, but the model gives detailed predictions that are in excellent agreement with many diverse observations.[3]

Cosmology draws heavily on the work of many disparate areas of research in theoretical and applied physics. Areas relevant to cosmology include particle physics experiments and theory, theoretical and observational astrophysics, general relativity, quantum mechanics, and plasma physics.

18.1 Subject history

See also: Timeline of cosmology and List of cosmologists

Modern cosmology developed along tandem tracks of theory and observation. In 1916, Albert Einstein published his theory of general relativity, which provided a unified description of gravity as a geometric property of space and time.[4] At the time, Einstein believed in a static universe, but found that his original formulation of the theory did not permit it.[5] This is because masses distributed throughout the universe gravitationally attract, and move toward each other over time.[6] However, he realized that his equations permitted the introduction of a constant term which could counteract the attractive force of gravity on the cosmic scale. Einstein published his first paper on relativistic cosmology in 1917, in which he added this *cosmological constant* to his field equations in order to force them to model a static universe.[7] However, this so-called Einstein model is unstable to small perturbations—it will eventually start to expand or contract.[5] The Einstein model describes a static universe; space is finite and unbounded (analogous to the surface of a sphere, which has a finite area but no edges). It was later realized that Einstein's model was just one of a larger set of possibilities, all of which were consistent with general relativity and the cosmological principle. The cosmological solutions of general relativity were found by Alexander Friedmann in the early 1920s.[8] His equations describe the Friedmann–Lemaître–Robertson–Walker universe, which may expand or contract, and whose geometry may be open, flat, or closed.

In the 1910s, Vesto Slipher (and later Carl Wilhelm Wirtz) interpreted the red shift of spiral nebulae as a Doppler shift that indicated they were receding from Earth.[12][13]

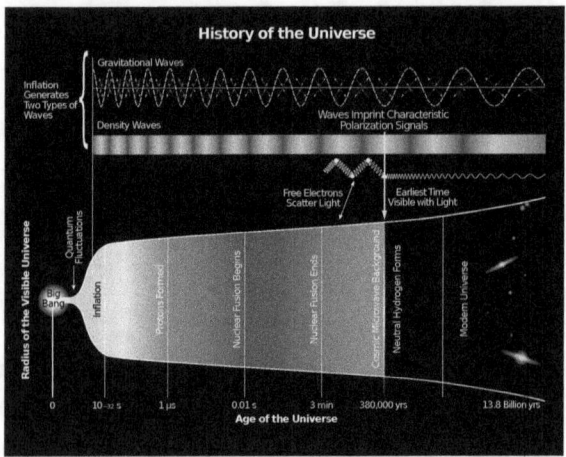

History of the Universe – gravitational waves are hypothesized to arise from cosmic inflation, a faster-than-light expansion just after the Big Bang (17 March 2014).[9]*[10]*[11]*

However, it is difficult to determine the distance to astronomical objects. One way is to compare the physical size of an object to its angular size, but a physical size must be assumed to do this. Another method is to measure the brightness of an object and assume an intrinsic luminosity, from which the distance may be determined using the inverse square law. Due to the difficulty of using these methods, they did not realize that the nebulae were actually galaxies outside our own Milky Way, nor did they speculate about the cosmological implications. In 1927, the Belgian Roman Catholic priest Georges Lemaître independently derived the Friedmann–Lemaître–Robertson–Walker equations and proposed, on the basis of the recession of spiral nebulae, that the universe began with the "explosion" of a "primeval atom"*[14]—which was later called the Big Bang. In 1929, Edwin Hubble provided an observational basis for Lemaître's theory. Hubble showed that the spiral nebulae were galaxies by determining their distances using measurements of the brightness of Cepheid variable stars. He discovered a relationship between the redshift of a galaxy and its distance. He interpreted this as evidence that the galaxies are receding from Earth in every direction at speeds proportional to their distance.*[15] This fact is now known as Hubble's law, though the numerical factor Hubble found relating recessional velocity and distance was off by a factor of ten, due to not knowing about the types of Cepheid variables.

Given the cosmological principle, Hubble's law suggested that the universe was expanding. Two primary explanations were proposed for the expansion. One was Lemaître's Big Bang theory, advocated and developed by George Gamow. The other explanation was Fred Hoyle's steady state model in which new matter is created as the galaxies move away from each other. In this model, the universe is roughly the

same at any point in time.*[16]*[17]

For a number of years, support for these theories was evenly divided. However, the observational evidence began to support the idea that the universe evolved from a hot dense state. The discovery of the cosmic microwave background in 1965 lent strong support to the Big Bang model,*[17] and since the precise measurements of the cosmic microwave background by the Cosmic Background Explorer in the early 1990s, few cosmologists have seriously proposed other theories of the origin and evolution of the cosmos. One consequence of this is that in standard general relativity, the universe began with a singularity, as demonstrated by Roger Penrose and Stephen Hawking in the 1960s.

An alternative view to extend the Big Bang model, suggesting the universe had no beginning or singularity and the age of the universe is infinite, has been presented.*[18]*[19]*[20]

18.2   Energy of the cosmos

Light chemical elements, primarily hydrogen and helium, were created in the Big Bang process *(see Nucleosynthesis)*. The small atomic nuclei combined into larger atomic nuclei to form heavier elements such as iron and nickel, which are more stable *(see Nuclear fusion)*. This caused a *later energy release*. Such reactions of nuclear particles inside stars continue to contribute to *sudden energy releases*, such as in nova stars. Gravitational collapse of matter into black holes is also thought to power the most energetic processes, generally seen at the centers of galaxies *(see Quasar and Active galaxy)*.

Cosmologists cannot explain all cosmic phenomena exactly, such as those related to the accelerating expansion of the universe, using conventional forms of energy. Instead, cosmologists propose a new form of energy called dark energy that permeates all space.*[21] One hypothesis is that dark energy is the energy of virtual particles, which are believed to exist in a vacuum due to the uncertainty principle.

There is no clear way to define the total energy in the universe using the most widely accepted theory of gravity, general relativity. Therefore, it remains controversial whether the total energy is conserved in an expanding universe. For instance, each photon that travels through intergalactic space loses energy due to the redshift effect. This energy is not obviously transferred to any other system, so seems to be permanently lost. On the other hand, some cosmologists insist that energy is conserved in some sense; this follows the law of conservation of energy.*[22]

Thermodynamics of the universe is a field of study that

explores which form of energy dominates the cosmos – relativistic particles which are referred to as radiation, or non-relativistic particles referred to as matter. Relativistic particles are particles whose rest mass is zero or negligible compared to their kinetic energy, and so move at the speed of light or very close to it; non-relativistic particles have much higher rest mass than their energy and so move much slower than the speed of light.

As the universe expands, both matter and radiation in it become diluted. However, the energy densities of radiation and matter dilute at different rates. As a particular volume expands, mass energy density is changed only by the increase in volume, but the energy density of radiation is changed both by the increase in volume and by the increase in the wavelength of the photons that make it up. Thus the energy of radiation becomes a smaller part of the universe's total energy than that of matter as it expands. The very early universe is said to have been 'radiation dominated' and radiation controlled the deceleration of expansion. Later, as the average energy per photon becomes roughly 10 eV and lower, matter dictates the rate of deceleration and the universe is said to be 'matter dominated'. The intermediate case is not treated well analytically. As the expansion of the universe continues, matter dilutes even further and the cosmological constant becomes dominant, leading to an acceleration in the universe's expansion.

18.3 History of the universe

See also: Timeline of the Big Bang

The history of the universe is a central issue in cosmology. The history of the universe is divided into different periods called epochs, according to the dominant forces and processes in each period. The standard cosmological model is known as the Lambda-CDM model.

18.3.1 Equations of motion

Main article: Friedmann–Lemaître–Robertson–Walker metric

The equations of motion governing the universe as a whole are derived from general relativity with a small, positive cosmological constant.[23] The solution is an expanding universe; due to this expansion, the radiation and matter in the universe cool down and become diluted. At first, the expansion is slowed down by gravitation attracting the radiation and matter in the universe. However, as these become diluted, the cosmological constant becomes more

dominant and the expansion of the universe starts to accelerate rather than decelerate. In our universe this happened billions of years ago.

18.3.2 Particle physics in cosmology

Main article: Particle physics in cosmology

Particle physics is important to the behavior of the early universe, because the early universe was so hot that the average energy density was very high. Because of this, scattering processes and decay of unstable particles are important in cosmology.

As a rule of thumb, a scattering or a decay process is cosmologically important in a certain cosmological epoch if the time scale describing that process is smaller than, or comparable to, the time scale of the expansion of the universe. The time scale that describes the expansion of the universe is $1/H$ with H being the Hubble constant, which itself actually varies with time. The expansion timescale $1/H$ is roughly equal to the age of the universe at that time.

18.3.3 Timeline of the Big Bang

Main article: Timeline of the Big Bang

Observations suggest that the universe began around 13.8 billion years ago.[24] Since then, the evolution of the universe has passed through three phases. The very early universe, which is still poorly understood, was the split second in which the universe was so hot that particles had energies higher than those currently accessible in particle accelerators on Earth. Therefore, while the basic features of this epoch have been worked out in the Big Bang theory, the details are largely based on educated guesses. Following this, in the early universe, the evolution of the universe proceeded according to known high energy physics. This is when the first protons, electrons and neutrons formed, then nuclei and finally atoms. With the formation of neutral hydrogen, the cosmic microwave background was emitted. Finally, the epoch of structure formation began, when matter started to aggregate into the first stars and quasars, and ultimately galaxies, clusters of galaxies and superclusters formed. The future of the universe is not yet firmly known, but according to the ΛCDM model it will continue expanding forever.

18.4   Areas of study

Below, some of the most active areas of inquiry in cosmology are described, in roughly chronological order. This does not include all of the Big Bang cosmology, which is presented in *Timeline of the Big Bang*.

18.4.1   Very early universe

The early, hot universe appears to be well explained by the Big Bang from roughly 10^*-33 seconds onwards. But there are several problems. One is that there is no compelling reason, using current particle physics, for the universe to be flat, homogeneous, and isotropic *(see the cosmological principle)*. Moreover, grand unified theories of particle physics suggest that there should be magnetic monopoles in the universe, which have not been found. These problems are resolved by a brief period of cosmic inflation, which drives the universe to flatness, smooths out anisotropies and inhomogeneities to the observed level, and exponentially dilutes the monopoles. The physical model behind cosmic inflation is extremely simple, but it has not yet been confirmed by particle physics, and there are difficult problems reconciling inflation and quantum field theory. Some cosmologists think that string theory and brane cosmology will provide an alternative to inflation.

Another major problem in cosmology is what caused the universe to contain far more matter than antimatter. Cosmologists can observationally deduce that the universe is not split into regions of matter and antimatter. If it were, there would be X-rays and gamma rays produced as a result of annihilation, but this is not observed. Therefore, some process in the early universe must have created a small excess of matter over antimatter, and this (currently not understood) process is called *baryogenesis*. Three required conditions for baryogenesis were derived by Andrei Sakharov in 1967, and requires a violation of the particle physics symmetry, called CP-symmetry, between matter and antimatter. However, particle accelerators measure too small a violation of CP-symmetry to account for the baryon asymmetry. Cosmologists and particle physicists look for additional violations of the CP-symmetry in the early universe that might account for the baryon asymmetry.

Both the problems of baryogenesis and cosmic inflation are very closely related to particle physics, and their resolution might come from high energy theory and experiment, rather than through observations of the universe.

18.4.2   Big Bang Theory

Main article: Big bang nucleosynthesis

Big Bang nucleosynthesis is the theory of the formation of the elements in the early universe. It finished when the universe was about three minutes old and its temperature dropped below that at which nuclear fusion could occur. Big Bang nucleosynthesis had a brief period during which it could operate, so only the very lightest elements were produced. Starting from hydrogen ions (protons), it principally produced deuterium, helium-4, and lithium. Other elements were produced in only trace abundances. The basic theory of nucleosynthesis was developed in 1948 by George Gamow, Ralph Asher Alpher, and Robert Herman. It was used for many years as a probe of physics at the time of the Big Bang, as the theory of Big Bang nucleosynthesis connects the abundances of primordial light elements with the features of the early universe. Specifically, it can be used to test the equivalence principle, to probe dark matter, and test neutrino physics. Some cosmologists have proposed that Big Bang nucleosynthesis suggests there is a fourth "sterile" species of neutrino.

Standard model of Big Bang cosmology

The **ΛCDM** (**Lambda cold dark matter**) or **Lambda-CDM** model is a parametrization of the Big Bang cosmological model in which the universe contains a cosmological constant, denoted by Lambda (Greek Λ), associated with dark energy, and cold dark matter (abbreviated **CDM**). It is frequently referred to as the **standard model** of Big Bang cosmology.

18.4.3   Cosmic microwave background

Main article: Cosmic microwave background

The cosmic microwave background is radiation left over from decoupling after the epoch of recombination when neutral atoms first formed. At this point, radiation produced in the Big Bang stopped Thomson scattering from charged ions. The radiation, first observed in 1965 by Arno Penzias and Robert Woodrow Wilson, has a perfect thermal blackbody spectrum. It has a temperature of 2.7 kelvins today and is isotropic to one part in 10^5. Cosmological perturbation theory, which describes the evolution of slight inhomogeneities in the early universe, has allowed cosmologists to precisely calculate the angular power spectrum of the radiation, and it has been measured by the recent satellite experiments (COBE and WMAP) and many ground and balloon-based experiments (such as Degree Angular Scale Interferometer, Cosmic Background Imager, and Boomerang).

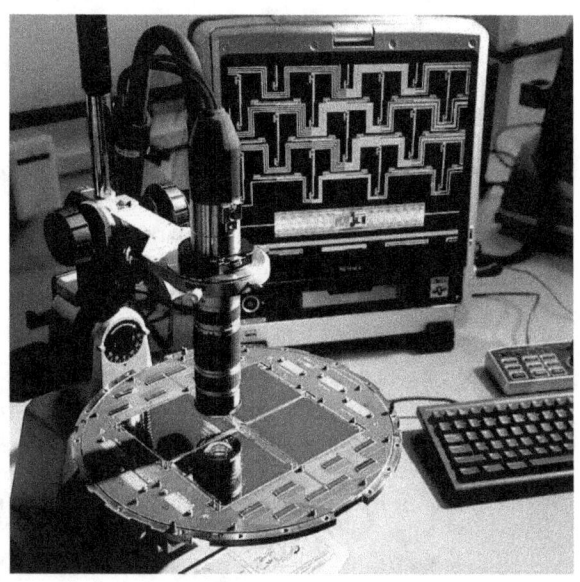

Evidence of gravitational waves in the infant universe may have been uncovered by the microscopic examination of the focal plane of the BICEP2 radio telescope.[9]*[10]*[11]*[25]*

One of the goals of these efforts is to measure the basic parameters of the Lambda-CDM model with increasing accuracy, as well as to test the predictions of the Big Bang model and look for new physics. The recent measurements made by WMAP, for example, have placed limits on the neutrino masses.

Newer experiments, such as QUIET and the Atacama Cosmology Telescope, are trying to measure the polarization of the cosmic microwave background. These measurements are expected to provide further confirmation of the theory as well as information about cosmic inflation, and the so-called secondary anisotropies, such as the Sunyaev-Zel'dovich effect and Sachs-Wolfe effect, which are caused by interaction between galaxies and clusters with the cosmic microwave background.

On 17 March 2014, astronomers at the Harvard–Smithsonian Center for Astrophysics announced the apparent detection of gravitational waves, which, if confirmed, may provide strong evidence for inflation and the Big Bang.*[9]*[10]*[11]*[25] However, on 19 June 2014, lowered confidence in confirming the cosmic inflation findings was reported.*[26]*[27]*[28]

18.4.4 Formation and evolution of large-scale structure

Main articles: Large-scale structure of the cosmos, Structure formation and Galaxy formation and evolution

Understanding the formation and evolution of the largest and earliest structures (i.e., quasars, galaxies, clusters and superclusters) is one of the largest efforts in cosmology. Cosmologists study a model of **hierarchical structure formation** in which structures form from the bottom up, with smaller objects forming first, while the largest objects, such as superclusters, are still assembling. One way to study structure in the universe is to survey the visible galaxies, in order to construct a three-dimensional picture of the galaxies in the universe and measure the matter power spectrum. This is the approach of the *Sloan Digital Sky Survey* and the 2dF Galaxy Redshift Survey.

Another tool for understanding structure formation is simulations, which cosmologists use to study the gravitational aggregation of matter in the universe, as it clusters into filaments, superclusters and voids. Most simulations contain only non-baryonic cold dark matter, which should suffice to understand the universe on the largest scales, as there is much more dark matter in the universe than visible, baryonic matter. More advanced simulations are starting to include baryons and study the formation of individual galaxies. Cosmologists study these simulations to see if they agree with the galaxy surveys, and to understand any discrepancy.

Other, complementary observations to measure the distribution of matter in the distant universe and to probe reionization include:

- The Lyman-alpha forest, which allows cosmologists to measure the distribution of neutral atomic hydrogen gas in the early universe, by measuring the absorption of light from distant quasars by the gas.

- The 21 centimeter absorption line of neutral atomic hydrogen also provides a sensitive test of cosmology

- Weak lensing, the distortion of a distant image by gravitational lensing due to dark matter.

These will help cosmologists settle the question of when and how structure formed in the universe.

18.4.5 Dark matter

Main article: Dark matter

Evidence from Big Bang nucleosynthesis, the cosmic microwave background and structure formation suggests that about 23% of the mass of the universe consists of non-baryonic dark matter, whereas only 4% consists of visible, baryonic matter. The gravitational effects of dark matter are well understood, as it behaves like a cold, non-radiative fluid that forms haloes around galaxies. Dark matter has never been detected in the laboratory, and the particle physics nature of dark matter remains completely unknown. Without observational constraints, there are a number of candidates, such as a stable supersymmetric particle, a weakly interacting massive particle, an axion, and a massive compact halo object. Alternatives to the dark matter hypothesis include a modification of gravity at small accelerations (MOND) or an effect from brane cosmology.

18.4.6 Dark energy

Main article: Dark energy

If the universe is flat, there must be an additional component making up 73% (in addition to the 23% dark matter and 4% baryons) of the energy density of the universe. This is called dark energy. In order not to interfere with Big Bang nucleosynthesis and the cosmic microwave background, it must not cluster in haloes like baryons and dark matter. There is strong observational evidence for dark energy, as the total energy density of the universe is known through constraints on the flatness of the universe, but the amount of clustering matter is tightly measured, and is much less than this. The case for dark energy was strengthened in 1999, when measurements demonstrated that the expansion of the universe has begun to gradually accelerate.

Apart from its density and its clustering properties, nothing is known about dark energy. *Quantum field theory* predicts a cosmological constant (CC) much like dark energy, but 120 orders of magnitude larger than that observed. Steven Weinberg and a number of string theorists (*see string landscape*) have invoked the 'weak anthropic principle': i.e. the reason that physicists observe a universe with such a small cosmological constant is that no physicists (or any life) could exist in a universe with a larger cosmological constant. Many cosmologists find this an unsatisfying explanation: perhaps because while the weak anthropic principle is self-evident (given that living observers exist, there must be at least one universe with a cosmological constant which allows for life to exist) it does not attempt to explain the context of that universe. For example, the weak anthropic principle alone does not distinguish between:

- Only one universe will ever exist and there is some underlying principle that constrains the CC to the value we observe.

- Only one universe will ever exist and although there is no underlying principle fixing the CC, we got lucky.

- Lots of universes exist (simultaneously or serially)

with a range of CC values, and of course ours is one of the life-supporting ones.

Other possible explanations for dark energy include quintessence or a modification of gravity on the largest scales. The effect on cosmology of the dark energy that these models describe is given by the dark energy's equation of state, which varies depending upon the theory. The nature of dark energy is one of the most challenging problems in cosmology.

A better understanding of dark energy is likely to solve the problem of the ultimate fate of the universe. In the current cosmological epoch, the accelerated expansion due to dark energy is preventing structures larger than superclusters from forming. It is not known whether the acceleration will continue indefinitely, perhaps even increasing until a big rip, or whether it will eventually reverse.

18.4.7 Other areas of inquiry

Cosmologists also study:

- Whether primordial black holes were formed in our universe, and what happened to them.

- The GZK cutoff for high-energy cosmic rays, and whether it signals a failure of special relativity at high energies

- The equivalence principle, whether or not Einstein's general theory of relativity is the correct theory of gravitation, and if the fundamental laws of physics are the same everywhere in the universe.

- The increasing complexity of universal structures, an example being the progressively greater energy rate density. *[29]

18.5   See also

- Hubble's law

- Illustris project

- List of cosmologists

- Photon

- Physical ontology

- String cosmology

- Universal Rotation Curve

18.6 References

[1] For an overview, see George FR Ellis (2006). "Issues in the Philosophy of Cosmology". In Jeremy Butterfield & John Earman. *Philosophy of Physics (Handbook of the Philosophy of Science) 3 volume set*. North Holland. pp. 1183*ff*. arXiv:astro-ph/0602280. ISBN 0-444-51560-7.

[2] An Open Letter to the Scientific Community as published in *New Scientist*, May 22, 2004

[3] Beringer, J.; et al. (Particle Data Group) (2012). "2013 Review of Particle Physics" (PDF). *Phys. Rev. D* **86**: 010001. Bibcode:2012PhRvD..86a0001B. doi:10.1103/PhysRevD.86.010001.

[4] "Nobel Prize Biography". *Nobel Prize Biography*. Nobel Prize. Retrieved 25 February 2011.

[5] Liddle, A. *An Introduction to Modern Cosmology*. Wiley. p. 51. ISBN 0-470-84835-9.

[6] Vilenkin, Alex (2007). *Many worlds in one : the search for other universes*. New York: Hill and Wang, A division of Farrar, Straus and Giroux. p. 19. ISBN 978-0-8090-6722-0.

[7] Jones, Mark; Lambourne, Robert (2004). *An introduction to galaxies and cosmology*. Milton Keynes Cambridge, UK New York: Open University Cambridge University Press. p. 228. ISBN 0-521-54623-0.

[8] Jones, Mark; Lambourne, Robert (2004). *An introduction to galaxies and cosmology*. Milton Keynes Cambridge, UK New York: Open University Cambridge University Press. p. 232. ISBN 0-521-54623-0.

[9] Staff (17 March 2014). "BICEP2 2014 Results Release". *National Science Foundation*. Retrieved 18 March 2014.

[10] Clavin, Whitney (17 March 2014). "NASA Technology Views Birth of the Universe". *NASA*. Retrieved 17 March 2014.

[11] Overbye, Dennis (17 March 2014). "Detection of Waves in Space Buttresses Landmark Theory of Big Bang". *New York Times*. Retrieved 17 March 2014.

[12] Slipher, V. M. (1922), Fox, Philip; Stebbins, Joel, eds., "Further Notes on Spectrographic Observations of Nebulae and Clusters", *Publications of the American Astronomical Society* **4**: 284–286, Bibcode:1922PAAS....4..284S

[13] Seitter, Waltraut C.; Duerbeck, Hilmar W. (1999), Egret, Daniel; Heck, Andre, eds., "Carl Wilhelm Wirtz – Pioneer in Cosmic Dimensions", *Harmonizing Cosmic Distance Scales in a Post-Hipparcos Era*, ASP Conference Series **167**: 237–242, Bibcode:1999ASPC..167..237S, ISBN 1-886733-88-0

[14] Lemaître, G. (1927), "Un Univers homogène de masse constante et de rayon croissant rendant compte de la vitesse radiale des nébuleuses extra-galactiques", *Annales de la Société Scientifique de Bruxelles* (in French) **A47**: 49–59, Bibcode:1927ASSB...47...49L

[15] Hubble, Edwin (March 1929), "A Relation between Distance and Radial Velocity among Extra-Galactic Nebulae", *Proceedings of the National Academy of Sciences of the United States of America* **15** (3): 168–173, Bibcode:1929PNAS...15..168H, doi:10.1073/pnas.15.3.168

[16] Hoyle, F. (1948), "A New Model for the Expanding Universe", *Monthly Notices of the Royal Astronomical Society* **108**: 372, Bibcode:1948MNRAS.108..372H

[17] "Big Bang or Steady State?", *Ideas of Cosmology* (American Institute of Physics), retrieved 2015-07-29

[18] Ghose, Tia (26 February 2015). "Big Bang, Deflated? Universe May Have Had No Beginning". *Live Science*. Retrieved 28 February 2015.

[19] Ali, Ahmed Faraq (4 February 2015). "Cosmology from quantum potential". *Physics Letters B* **741**: 276–279. doi:10.1016/j.physletb.2014.12.057. Retrieved 28 February 2015.

[20] Das, Saurya; Bhaduri, Rajat K. (18 November 2014). "Dark matter and dark energy from Bose-Einstein condensate" (PDF). *arXiv*. Retrieved 28 February 2015.

[21] Science 20 June 2003:Vol. 300. no. 5627, pp. 1914 - 1918 Throwing Light on Dark Energy, Robert P. Kirshner. Retrieved December 2006

[22] e.g. Liddle, A. *An Introduction to Modern Cosmology*. Wiley. ISBN 0-470-84835-9. This argues cogently "Energy is always, always, always conserved."

[23] P. Ojeda; H. Rosu (June 2006). "Supersymmetry of FRW barotropic cosmologies". *Internat. J. Theoret. Phys.* (Springer) **45** (6): 1191–1196. arXiv:gr-qc/0510004. Bibcode:2006IJTP...45.1152R. doi:10.1007/s10773-006-9123-2.

[24] "Cosmic Detectives". The European Space Agency (ESA). 2013-04-02. Retrieved 2013-04-25.

[25] Overbye, Dennis (24 March 2014). "Ripples From the Big Bang". *New York Times*. Retrieved 24 March 2014.

[26] Overbye, Dennis (19 June 2014). "Astronomers Hedge on Big Bang Detection Claim". *New York Times*. Retrieved 20 June 2014.

[27] Amos, Jonathan (19 June 2014). "Cosmic inflation: Confidence lowered for Big Bang signal". *BBC News*. Retrieved 20 June 2014.

[28] Ade, P.A.R.; et al. (BICEP2 Collaboration) (19 June 2014). "Detection of B-Mode Polarization at Degree Angular Scales by BICEP2" (PDF). *Physical Review Letters* **112**: 241101. arXiv:1403.3985. Bibcode:2014PhRvL.112x1101A. doi:10.1103/PhysRevLett.112.241101. PMID 24996078. Retrieved 20 June 2014.

[29] Chaisson, Eric (1987-01-01). "The life ERA: cosmic selection and conscious evolution". *Faculty Publications*.

18.7 Further reading

18.7.1 Popular

- Brian Greene (2005). *The Fabric of the Cosmos*. Penguin Books Ltd. ISBN 0-14-101111-4.

- Alan Guth (1997). *The Inflationary Universe: The Quest for a New Theory of Cosmic Origins*. Random House. ISBN 0-224-04448-6.

- Hawking, Stephen W. (1988). *A Brief History of Time: From the Big Bang to Black Holes*. Bantam Books, Inc. ISBN 0-553-38016-8.

- Hawking, Stephen W. (2001). *The Universe in a Nutshell*. Bantam Books, Inc. ISBN 0-553-80202-X.

- Ostriker, Jeremiah P.; Mitton, Simon (2013). *Heart of Darkness: Unraveling the mysteries of the invisible Universe*. Princeton, NJ: Princeton University Press. ISBN 978-0-691-13430-7.

- Simon Singh (2005). *Big Bang: The Origin of the Universe*. Fourth Estate. ISBN 0-00-716221-9.

- Steven Weinberg (1993) [First published 1978]. *The First Three Minutes*. Basic Books. ISBN 0-465-02437-8.

18.7.2 Textbooks

- Cheng, Ta-Pei (2005). *Relativity, Gravitation and Cosmology: a Basic Introduction*. Oxford and New York: Oxford University Press. ISBN 0-19-852957-0. Introductory cosmology and general relativity without the full tensor apparatus, deferred until the last part of the book.

- Dodelson, Scott (2003). *Modern Cosmology*. Academic Press. ISBN 0-12-219141-2. An introductory text, released slightly before the WMAP results.

- Grøn, Øyvind; Hervik, Sigbjørn (2007). *Einstein's General Theory of Relativity with Modern Applications in Cosmology*. New York: Springer. ISBN 978-0-387-69199-2.

- Harrison, Edward (2000). *Cosmology: the science of the universe*. Cambridge University Press. ISBN 0-521-66148-X. For undergraduates; mathematically gentle with a strong historical focus.

- Kutner, Marc (2003). *Astronomy: A Physical Perspective*. Cambridge University Press. ISBN 0-521-52927-1. An introductory astronomy text.

- Kolb, Edward; Michael Turner (1988). *The Early Universe*. Addison-Wesley. ISBN 0-201-11604-9. The classic reference for researchers.

- Liddle, Andrew (2003). *An Introduction to Modern Cosmology*. John Wiley. ISBN 0-470-84835-9. Cosmology without general relativity.

- Liddle, Andrew; David Lyth (2000). *Cosmological Inflation and Large-Scale Structure*. Cambridge. ISBN 0-521-57598-2. An introduction to cosmology with a thorough discussion of inflation.

- Mukhanov, Viatcheslav (2005). *Physical Foundations of Cosmology*. Cambridge University Press. ISBN 0-521-56398-4.

- Padmanabhan, T. (1993). *Structure formation in the universe*. Cambridge University Press. ISBN 0-521-42486-0. Discusses the formation of large-scale structures in detail.

- Peacock, John (1998). *Cosmological Physics*. Cambridge University Press. ISBN 0-521-42270-1. An introduction including more on general relativity and quantum field theory than most.

- Peebles, P. J. E. (1993). *Principles of Physical Cosmology*. Princeton University Press. ISBN 0-691-01933-9. Strong historical focus.

- Peebles, P. J. E. (1980). *The Large-Scale Structure of the Universe*. Princeton University Press. ISBN 0-691-08240-5. The classic work on large-scale structure and correlation functions.

- Rees, Martin (2002). *New Perspectives in Astrophysical Cosmology*. Cambridge University Press. ISBN 0-521-64544-1.

- Weinberg, Steven (1971). *Gravitation and Cosmology*. John Wiley. ISBN 0-471-92567-5. A standard reference for the mathematical formalism.

- Weinberg, Steven (2008). *Cosmology*. Oxford University Press. ISBN 0-19-852682-2.

- Benjamin Gal-Or, "Cosmology, Physics and Philosophy", Springer Verlag, 1981, 1983, 1987, ISBN 0-387-90581-2, ISBN 0-387-96526-2.

18.8 External links

18.8.1 From groups

- Cambridge Cosmology- from Cambridge University (public home page)

- Cosmology 101 - from the NASA WMAP group

- Center for Cosmological Physics. University of Chicago, Chicago.

- Origins, Nova Online - Provided by *PBS*.

18.8.2 From individuals

- Gale, George, "Cosmology: Methodological Debates in the 1930s and 1940s", *The Stanford Encyclopedia of Philosophy*, Edward N. Zalta (ed.)

- Madore, Barry F., "*Level 5 : A Knowledgebase for Extragalactic Astronomy and Cosmology*". Caltech and Carnegie. Pasadena, California, USA.

- Tyler, Pat, and Phil Newman "*Beyond Einstein*". Laboratory for High Energy Astrophysics (LHEA) NASA Goddard Space Flight Center.

- Wright, Ned. "*Cosmology tutorial and FAQ*". Division of Astronomy & Astrophysics, UCLA.

- George Musser (February 2004). "Four Keys to Cosmology". *Scientific American* (Scientific American). Retrieved 22 March 2015.

- Cliff Burgess; Fernando Quevedo (November 2007). "The Great Cosmic Roller-Coaster Ride". *Scientific American* (print). pp. 52–59. (subtitle) Could cosmic inflation be a sign that our universe is embedded in a far vaster realm?

Chapter 19

Timeline of fundamental physics discoveries

19.1 See also

- Timeline of developments in theoretical physics

19.2 References

- American Physical Society

19.3 Text and image sources, contributors, and licenses

19.3.1 Text

- **History of physics** *Source:* https://en.wikipedia.org/wiki/History_of_physics?oldid=687132862 *Contributors:* AxelBoldt, CYD, Bryan Derksen, The Anome, AstroNomer~enwiki, RobLa, Jeronimo, Ap, Css, Andre Engels, XJaM, Rgamble, Fredbauder, DavidLevinson, Heron, Camembert, Kickaha~enwiki, Michael Hardy, Isomorphic, Ixfd64, Mcarling, Looxix~enwiki, Ahoerstemeier, Smack, Hike395, Charles Matthews, Timwi, Reddi, Dandrake, Zoicon5, Tpbradbury, Djungelurban, Fibonacci, Jni, Twang, Gentgeen, Gandalf61, TimR, Anthony, Alan Liefting, Ancheta Wis, Netoholic, Art Carlson, Fastfission, Risk one, Everyking, SteffenB~enwiki, Matthead, Bobblewik, Wmahan, Beland, Karol Langner, Wikimol, Kesac, Sam Hocevar, Lumidek, Mramirez, Grstain, Rich Farmbrough, Dbachmann, Brian0918, Rgdboer, Laurascudder, Art LaPella, West London Dweller, Wareh, CDN99, Bobo192, Wood Thrush, Beige Tangerine, I9Q79oL78KiL0QTFHgyc, SpeedyGonsales, Passw0rd, Jumbuck, Keenan Pepper, Ricky81682, Isfisk, Linas, Fbcunha, Carcharoth, Ruud Koot, Wikiklrsc, Doric Loon, Allen3, Mandarax, Magister Mathematicae, BD2412, Qwertyus, Jshadias, Rjwilmsi, DonSiano, FlaBot, Ewlyahoocom, Srleffler, Chobot, DVdm, Digitalme, Whosasking, Daduzi, YurikBot, Wavelength, Borgx, RobotE, RussBot, Gaius Cornelius, Rjensen, Howcheng, JFD, EEMIV, DeadEyeArrow, Ejl, Pawyilee, Claygate, MaNeMeBasat, Chaiken, Katieh5584, Philip Stevens, Qero, KnightRider~enwiki, SmackBot, FocalPoint, Incnis Mrsi, TestPilot, Vald, KocjoBot~enwiki, Jagged 85, Eskimbot, Pedrose, Harald88, Edgar181, Saros136, Jayanta Sen, MalafayaBot, Nbarth, Colonies Chris, Adrian Baker, Leinad-Z, Berland, Nixeagle, Cybercobra, Dreadstar, Bejnar, FlyHigh, John, Korval, Maziar fayaz, Zarniwoot, ManiF, NongBot~enwiki, Gavin6942, Waggers, AdultSwim, Will Thomas, Peter M Dodge, Newone, Delta x, Buckyboy314, Ouishoebean, CmdrObot, Mhklein, Lemmio, ShelfSkewed, Ken Gallager, Myasuda, Logicus, Gtxfrance, Gogo Dodo, Rracecarr, Tawkerbot4, Quibik, Benford R, Ssilvers, Editor at Large, SteveMcCluskey, Thijs!bot, Qwyrxian, Oerjan, Headbomb, D.H, Nick Number, Icep, AntiVandalBot, Gnixon, Rickybuchanan, Naveen Sankar, Darklilac, Kent Witham, Dan D. Ric, MER-C, Matthew Fennell, Grant Gussie, Thenub314, Magioladitis, Hroðulf, TARBOT, Bfiene, SwiftBot, KConWiki, Cardamon, Allstarecho, Robin S, NikNaks, Anaxial, R'n'B, CommonsDelinker, Pharaoh of the Wizards, Hans Dunkelberg, Lantonov, Michael Daly, Peppergrower, SmallPaul, Warut, Fountains of Bryn Mawr, KylieTastic, Squids and Chips, TXiKiBoT, Voorlandt, Littlealien182, Wikiisawesome, Kwakkie~enwiki, Synthebot, AlleborgoBot, Logan, EmxBot, Thony C., Tiddly Tom, Karaboom, Til Eulenspiegel, Flyer22 Reborn, Arjen Dijksman, Miguel.mateo, Sunrise, Akarkera, Mojoworker, Capitalismojo, Chi-Eloka, Athenean, Tomasz Prochownik, Sfan00 IMG, ClueBot, DFRussia, J8079s, Niceguyedc, ChandlerMapBot, Estirabot, SchreiberBike, Bleeben, GlasGhost, Mlaffs, Thingg, Roflmywaffles, XLinkBot, Saeed.Veradi, Little Mountain 5, Daniel andersson, Eleven even, MystBot, Addbot, Xp54321, Vero.Verite, DOI bot, Fyrael, Fgnievinski, Vchorozopoulos, Redheylin, Favonian, LemmeyBOT, LinkFA-Bot, Legobot, Luckas-bot, Yobot, EdwardLane, Bunnyhop11, AnomieBOT, Ulric1313, Materialscientist, Citation bot, LilHelpa, Obersachsebot, Xqbot, Parkyere, Wghanem, Crzer07, BookWormHR, Omnipaedista, Alexscara, RibotBOT, Mathonius, Shadowjams, FrescoBot, Machine Elf 1735, Citation bot 1, Tkuvho, Pinethicket, Istcol, Trappist the monk, Zhernovoi, Oswaldo Zapata, DARTH SIDIOUS 2, RjwilmsiBot, Generalboss3, Koothrappali, EmausBot, John of Reading, WikitanvirBot, WittyMan1986, Hhhippo, JSquish, ZéroBot, Darkraid1, Captain Screebo, Chewings72, Adib Khaled, Aadivinus, ClueBot NG, Widr, Helpful Pixie Bot, Calabe1992, PhnomPencil, Harizotoh9, The Traditionalist, Physmuseum, HankW512, StarryGrandma, Pratyya Ghosh, Ninmacer20, ChrisGualtieri, Gdrg22, Dexbot, Numbermaniac, Blackredstart, محمد شعیب, Thepalerider2012, Reatlas, Srjl, LudicrousTripe, NycLightsBH, Dinosaurz54321, Sol1, Noyster, JaconaFrere, Qwerty123uiop, Chewhawka, Filedelinkerbot, Sariphys, The Erudite Philosopher, C1776M, Neverbuckets22, Knife-in-the-drawer and Anonymous: 225

- **Classical physics** *Source:* https://en.wikipedia.org/wiki/Classical_physics?oldid=683274661 *Contributors:* Zundark, Patrick, Michael Hardy, Albertplanck, Kevin Baas, Jeff Relf, David Shay, Sverdrup, Wile E. Heresiarch, Decumanus, Matt Gies, Bensaccount, Wmahan, Antandrus, Onco p53, Karol Langner, Lumidek, Juan Ponderas, Hidaspal, Vsmith, Neko-chan, Joanjoc~enwiki, Army1987, Keenan Pepper, OwenX, Localh77, SeventyThree, Magister Mathematicae, EcoMan, Mkuehn10, Desdinova, Ttwaring, FlaBot, Srleffler, Chobot, Roboto de Ajvol, YurikBot, Jpbowen, Lankhorst, SmackBot, Jsheyl~enwiki, Mets501, Samael775~enwiki, IvanLanin, RekishiEJ, CarlosPS, Mjohnrussell, JRSpriggs, Gregbard, Michael C Price, Christian75, Headbomb, Slaweks, Jomoal99, Sherbrooke, Austin Maxwell, North Shoreman, Gökhan, JAnDbot, MartinBot, HEL, AlnoktaBOT, ConfusedGremlin, Thurth, Philip Trueman, SieBot, Hxhbot, Lisatwo, Curtdbz, Loren.wilton, ClueBot, Sesameball, DragonBot, Djr32, Estirabot, Jotterbot, Crowsnest, Nick84, The Rationalist, Addbot, Glane23, Digifreak194, Numbo3-bot, AnomieBOT, Materialscientist, ArthurBot, Xqbot, St.nerol, NOrbeck, Crzer07, Mumbai terrorist bomber, Erik9bot, FrescoBot, Sandgem Addict, Rajputize, TobeBot, TjBot, EmausBot, Hhhippo, JSquish, Maschen, ClueBot NG, BG19bot, PhnomPencil, Cadiomals, Zedshort, Popeye mendoza, Dexbot, Mogism, Jemclear, JaconaFrere, Qpdatabase, Dtaylor369, KasparBot and Anonymous: 49

- **Modern physics** *Source:* https://en.wikipedia.org/wiki/Modern_physics?oldid=681485798 *Contributors:* Dcljr, Mxn, Rich Farmbrough, Vsmith, ESkog, TigerShark, Wafry, Nihiltres, Srleffler, Chobot, YurikBot, Stephenb, Pred, Skizzik, Cybercobra, IronGargoyle, Ben Moore, RekishiEJ, CarlosPS, Rowellcf, Meno25, Goldencako, Christian75, Thijs!bot, Headbomb, Bongwarrior, Lilac Soul, Uncle Dick, SoCalSuperEagle, VolkovBot, TJollans, Falcon8765, Spinningspark, AlleborgoBot, Puuropyssy, OKBot, ClueBot, The Thing That Should Not Be, Djr32, Cristi215, Kamleshkohli, Saeed.Veradi, Addbot, Leszek Jańczuk, SpBot, Numbo3-bot, Citation bot, Obersachsebot, MauritsBot, Xqbot, Sketchmoose, Crzer07, Omnipaedista, W Nowicki, LiamSP, Diannaa, Ripchip Bot, EmausBot, Amratia519, Oliverlyc, Hhhippo, JSquish, Ejmarley, Traxs7, Llightex, ClueBot NG, SusikMkr, Helpful Pixie Bot, Vritt, Klilidiplomus, Mdann52, Gdrg22, M0532062613, Garuda0001, PaulOcuana, Big Henry Scientist, Vnorov, SpookyGhostMan, BethNaught, Mmorris701, Isambard Kingdom, Delmaq, KasparBot, Crosleybendix and Anonymous: 49

- **Aristotelian physics** *Source:* https://en.wikipedia.org/wiki/Aristotelian_physics?oldid=685776832 *Contributors:* Ed Poor, Ewen, Edward, Fred Bauder, SebastianHelm, EALacey, Alan Liefting, Beland, Doops, DragonflySixtyseven, Discospinster, Rgdboer, Art LaPella, Wareh, Giraffedata, Jeltz, Ahruman, Stemonitis, Woohookitty, BD2412, Rjwilmsi, Koavf, Bgwhite, Dialectric, Moe Epsilon, Syrthiss, Sardanaphalus, SmackBot, TestPilot, J-beda, C.Fred, Neptunius, Jagged 85, EVula, Zvis~enwiki, Rigadoun, Fatworm, JHunterJ, Rinnenadtrosc, Iridescent, RekishiEJ, Chris55, הסרפד, Headbomb, Colin MacLaurin, Yurei-eggtart, Giggy, JaGa, R'n'B, J.delanoy, Belovedfreak, DadaNeem, Layzner, Treisijs, Deor, Anonymous Dissident, Jungegift, The Mad Genius, SieBot, Likebox, Dabomb87, Mr. Granger, Ideal gas equation, J8079s, SuperHamster, Auntof6, MelonBot, Johnuniq, DumZiBoT, Jack Bauer00, JKeck, Addbot, LightSpectra, Juan During, Lightbot, Ccaarft, Rubinbot, Almabot, GrouchoBot, Patrizio18, Ibinthinkin, Machine Elf 1735, Tetraedycal, Citation bot 1, Haida19, Jean-François Clet, Angstorm, Thinking of England, DARTH SIDIOUS 2, Magmalex, RjwilmsiBot, Thomas Peardew, DASHBot, Syncategoremata, Thywob, Wayne Slam, Jacobisq, Donner60, JFB80, Doctorambient, ClueBot NG, Helpful Pixie Bot, Bibcode Bot, أبو حمزة, BG19bot, WithSelet, BattyBot, Gebars, ChrisGualtieri, Khazar2, Mmxiicybernaut, IndigoDeberry, Bieber74, Ioannes Piscator and Anonymous: 75

- **History of science and technology in China** *Source:* https://en.wikipedia.org/wiki/History_of_science_and_technology_in_China?oldid= 682162179 *Contributors:* William Avery, Michael Hardy, GTBacchus, Timc, Huangdi, Dzhuo, Alan Liefting, Ianeiloart, Ancheta Wis, Per Honor et Gloria, TJSwoboda, Rich Farmbrough, Bender235, Coolcaesar, Smalljim, Alansohn, Gary, Mrholybrain, Wtmitchell, Danaman5, Jheald, RyanGerbil10, Duncan.france, BD2412, Chenxlee, Josh Parris, Rjwilmsi, Yug, RexNL, Gurch, Eric.dane~enwiki, Benjwong, Kaku-rady, Wester, Deeptrivia, RussBot, Stephenb, Ksyrie, Dialectric, Anetode, Wknight94, BazookaJoe, Yvwv, Sardanaphalus, SmackBot, Far-nak, Lawrencekhoo, Jagged 85, Bwithh, Lds, Gilliam, Hmains, Allen Riddell, TheDarkArchon, Cattus, Grimhelm, Neo-Jay, Xiner, Tesseran, Ergative rlt, Jinnai, Tazmaniacs, SilkTork, J Milburn, Lazulilasher, Jac16888, Grahamec, Mato, Hanfresco, Doug Weller, Alaibot, Kozuch, Epbr123, Marek69, Bobblehead, Ideogram, Robert Ullmann, AntiVandalBot, Darklilac, Hermant patel, Mcorazao, Hut 8.5, VoABot II, Misheu, KConWiki, Robotman1974, Fang 23, DerHexer, Gun Powder Ma, S3000, Utc-100, Paul Gard, MartinBot, Jonathan Hall, R'n'B, Em Mitchell, Fconaway, J.delanoy, Trusilver, Ginsengbomb, Polenth, BrokenSphere, DadaNeem, KylieTastic, STBotD, Dorftrottel, Squids and Chips, The-Pointblank, CardinalDan, Funandtrvl, Malik Shabazz, VolkovBot, ABF, Lop.dong, Armetrek, Philip Trueman, Andres rojas22, GcSwRhIc, Qxz, PDFbot, Master of the Oríchalcos, Dark Dragon Sword, PericlesofAthens, Eragonlvr117, Wraithdart, Nyikto, Enkyo2, StAnselm, Dusti, Keilana, Exert, HkCaGu, Oxymoron83, MagnusF, ClueBot, Srudes2, Laurarosethomson, Niceguyedc, Blanchardb, Cirt, Waltigs, Aua, Ktr101, Naerii, KevinX94, Techfast50, Hadoooookin, Zappa711, Snake66, 40fifw0, Black Knight takes White Queen, Slashem, Addbot, Proofreader77, DOI bot, Jncraton, Njaelkies Lea, Ka Faraq Gatri, Krano, Ptbotgourou, AnomieBOT, Punnybunny, Mountwolseley, Jim1138, Galoubet, Ul-ric1313, Materialscientist, Citation bot, Madalibi, Xqbot, Sionus, DSisyphBot, Crzer07, Frosted14, Resident Mario, Amaury, Shadowjams, Prezbo, Samwb123, Honza.havlicek, GT5162, FrescoBot, LucienBOT, Doremo, Steve Quinn, Yanajin33, Weetoddid, Chen19711, Citation bot 1, Elockid, RedBot, MCeighternity, Rpwikiman, Hai398, Zhonghuo~enwiki, Zbayz, Callanecc, ZhBot, Chipmunkdavis, Generalboss3, Wild-Bot, EmausBot, Immunize, GoingBatty, Tommy2010, Mckinnona, Xanchester, ClueBot NG, Andrei S, Rurik the Varangian, Helpful Pixie Bot, Curb Chain, Lowercase sigmabot, BG19bot, Frze, Cold Season, Mark Arsten, Utesfan99, Insidiae, EricEnfermero, Acadēmica Orien-tālis, Khazar2, Dexbot, Zeeyanwiki, Hmainsbot1, Numbermaniac, Rajmaan, Rui24114, Spankerer, Eerreerr, ElHef, Scottie111, Fyddlestix, Engorgium, ExperiencedArticleFixer and Anonymous: 216

- **History of science and technology in the Indian subcontinent** *Source:* https://en.wikipedia.org/wiki/History_of_science_and_technology_ in_the_Indian_subcontinent?oldid=686655285 *Contributors:* The Anome, SimonP, Michael Hardy, Shyamal, Charles Matthews, Justin Ba-con, Timrollpickering, Sundar, Utcursch, Pgan002, Gazpacho, Mike Rosoft, CALR, Discospinster, Dbachmann, Brian0918, CanisRufus, Ced-ders, Pearle, ClementSeveillac, Alansohn, Richard Harvey, Gene Nygaard, Stemonitis, Woohookitty, Agthorr, Toussaint, Prashanthns, BD2412, Pranathi, Melesse, Dpv, Jshadias, Josh Parris, Rjwilmsi, Smithfarm, Bhadani, Ian Pitchford, CalJW, Mark83, Alphachimp, OpenToppedBus, DVdm, Siddhant, Deeptrivia, Jlittlet, RussBot, Gaius Cornelius, Wimt, Wiki alf, Dialectric, Steelhead, Ragesoss, Sdsouza, Wujastyk, Deville, Closedmouth, Dspradau, Qero, Yvwv, BonsaiViking, SmackBot, Jagged 85, Kedar damle, Edgar181, Hmains, Holy Ganga, Anwar saadat, Chris the speller, Bazonka, J. Spencer, Colonies Chris, Zachorious, Rama's Arrow, Can't sleep, clown will eat me, OrphanBot, RedHillian, Sljaxon, GourangaUK, N Shar, Zedall, DO11.10, John, Shyamsunder, Bless sins, Skinsmoke, DabMachine, Rayfield, Bsskchaitanya, Trialsanderrors, FairuseBot, Eastlaw, Joostvandeputte~enwiki, Searles2sels, Jamoche, Abczyx, Bakanov, Kar403, JFreeman, DIGIwarez, Gv365, Tawkerbot4, Doug Weller, DumbBOT, Savitr, SteveMcCluskey, Barticus88, Sagaciousuk, Pjvpjv, Nick Number, AbcXyz, Khened, Trakesht, Kedarg6500, Kaobear, Dsp13, Chanakyathegreat, IanOsgood, Kerotan, VoABot II, Sodabottle, KConWiki, MCG, Heliac, Ekotkie, Mdsats, Gun Powder Ma, MartinBot, I am searching, R'n'B, CommonsDelinker, Rajathk, Abecedare, Fowler&fowler, Polenth, Salih, Michael Daly, Katalaveno, Bil-bobee, RB972, Redtigerxyz, Signalhead, Sured07, Philip Trueman, Pahari Sahib, Kww, Tameeria, Ann Stouter, IPSOS, WereSpielChequers, WRK, Flyer22 Reborn, Radon210, 3rdAlcove, Athenean, Okstasy, ClueBot, The Thing That Should Not Be, Kafka Liz, Lawrence Cohen, Ep-silon60198, Bschhikara, EnigmaMcmxc, Walrasiad, Blossoms789, Jayantanth, SchreiberBike, BOTarate, Aitias, EdChem, XLinkBot, Sac1891, WikHead, Addbot, Ronhjones, Favonian, Issyl0, Tide rolls, Lightbot, JSR, Luckas-bot, Yobot, Bunnyhop11, Ptbotgourou, Guy1890, TestE-ditBot, AnomieBOT, DarklyCute, Materialscientist, Citation bot, Dewan357, Crzer07, J04n, WebCiteBOT, FrescoBot, OgreBot, Elockid, Avinashsharma.iitm, Onel5969, RjwilmsiBot, Salvio giuliano, Dewritech, Thecheesykid, White Trillium, Rhadamanthus222, ClueBot NG, Heathfield89, Peter James, Rajaram Sarangapani, Pposluszny51, Ashishtgb, Rurik the Varangian, Helpful Pixie Bot, Proudtobeindian007, Drspaz, Jyoti Woodhouse, PhnomPencil, Dzlinker, Solomon7968, Areapeaslol, Yearsclosehaveto, Abilngeorge, Gyrodoor33, Cpt.a.haddock, Hmainsbot1, Delljvc, Tyler95MI, Ibrahim Husain Meraj, Bladesmulti, VeryCrocker, Pingili anilkumar, Vreswiki, Bongan, Jombiecutter, Ran-jith Raja Simha, Shaash317 and Anonymous: 166

- **Physics in the medieval Islamic world** *Source:* https://en.wikipedia.org/wiki/Physics_in_the_medieval_Islamic_world?oldid=667975868 *Contributors:* Ed Poor, William M. Connolley, Bearcat, Rich Farmbrough, Dbachmann, Kwamikagami, Woohookitty, Rjwilmsi, Koavf, Jmcc150, Srleffler, Dialectric, Grafen, Ospalh, SmackBot, Jagged 85, Chris the speller, Jerome Charles Potts, Cybercobra, AdultSwim, TheTito, Hemlock Martinis, Headbomb, Marek69, Acroterion, Magioladitis, JaGa, Rasel70, Schmloof, R'n'B, CommonsDelinker, Uncle Dick, Oxguy3, Katharineamy, Fountains of Bryn Mawr, Hugo999, Willshaman, Neophaze, Sunrise, Mohummy, Tomasz Prochownik, Nsk92, J8079s, SamuelTheGhost, Chrono1084, Al-Andalusi, Johnuniq, Xiquet, Tkech, ZooFari, Addbot, DOI bot, Luckas-bot, Legobot II, AnomieBOT, Ci-tation bot, LilHelpa, Xqbot, Cavila, Fragma08, Machine Elf 1735, Citation bot 1, Alph Bot, Dr Philip Toop, Syncategoremata, Cobaltcigs, Someone65, ClueBot NG, Bibcode Bot, BG19bot, Contact '97, Mughal Lohar, Jamesmcmahon0, Monochrome Monitor, Monkbot, Duncan mccollum, Denn4657, C1776M, 468SM, Tugay354, Hershdchabra and Anonymous: 18

- **Science in the medieval Islamic world** *Source:* https://en.wikipedia.org/wiki/Science_in_the_medieval_Islamic_world?oldid=685644307 *Contributors:* Ed Poor, Michael Hardy, Dcljr, Sannse, William M. Connolley, Ugen64, Bogdangiusca, Cimon Avaro, WhisperToMe, Jerzy, Kizor, Goethean, Arkuat, Dina, Graeme Bartlett, DocWatson42, Lethe, Zigger, Everyking, Zora, Per Honor et Gloria, Quadell, Antandrus, CJCurrie, Sam Hocevar, Zeeshanhasan, Rich Farmbrough, Dpm64, Bender235, Eric Forste, Kwamikagami, Iqu, Ruyn, Thuresson, Bobo192, Flammifer, Famousdog, Mdd, Passw0rd, OneGuy, Orzetto, Alansohn, Jheald, WhatWouldEmperorNortonDo, Patito, Zereshk, Recury, Tariqab-jotu, Stemonitis, Woohookitty, Mindmatrix, Guy M, Aaron McDaid, Unixer, Ruud Koot, Jeff3000, Striver, Wayward, Farhansher, Mandarax, BD2412, Amir85, Dpv, Rjwilmsi, Koavf, Jweiss11, Jmcc150, Yuber, FayssalF, Gurch, AnthonyA7, Gwernol, Spacepotato, X42bn6, Russ-Bot, Anonymous editor, Gaius Cornelius, CambridgeBayWeather, Zeno of Elea, Aftermath, Dialectric, Ragesoss, Bestofmed, Ridiculous fish, Bucketsofg, Gadget850, Cheese Sandwich, Igiffin, Ali K, Closedmouth, Fram, Curpsbot-unicodify, RG2, DVD R W, Sardanaphalus, Jbalint, SmackBot, Prodego, NZUlysses, C.Fred, Jim62sch, Jagged 85, Delldot, Salmaakbar, Wzhao553, Commander Keane bot, Gilliam, Ohnoit-sjamie, Hmains, Ferix, Cplakidas, OrphanBot, Bolivian Unicyclist, Ddon, Pepsidrinka, Savidan, RandomP, Adnanmuf, Bejnar, Chaldean, Vanished user 9i39j3, John, Tktktk, Mitso Bel, Deviathan~enwiki, Ckatz, Bless sins, Aeluwas, Godfrey Daniel, Infantrymarine25, Sheep81, Hu12, DabMachine, Kernow, Donmac, Imad marie, RekishiEJ, JLCA, ALM scientist, Kaischwartz, Jibran1, Merzbow, Kurtan~enwiki, JForget, CmdrObot, CBM, Drinibot, Mattreo, Itaqallah, Hemlock Martinis, Gregbard, Cydebot, Yolocavo, Katherine Tredwell, Sa.vakilian, B, Christian75, Paddles, SteveMcCluskey, Nishidani, Thijs!bot, Epbr123, Nezzadar, Puntori, Madbehemoth, Myanw, Dsp13, GurchBot, Demophon, Yahel Guhan, Mardavich, Magioladitis, VoABot II, Rich257, Indon, Joe hill, Aziz1005, Frotz, DerHexer, JaGa, LinkLink~enwiki, Zeeshan Arshad, Szczepan1990, Gun Powder Ma, David J Wilson, TheEgyptian, R'n'B, CommonsDelinker, Proabivouac, 7day, AlphaEta, Arrow740, J intela, Ian.thomson, Speed8ump, Shawn in Montreal, Jeepday, Fountains of Bryn Mawr, AA, Kansas Bear, Intothefire, Signalhead, Hugo999, Hiromiando, BoogaLouie, Philip Trueman, Aymatth2, Jackfork, Wikiisawesome, Graymornings, Cnilep, Wavehunter, LOTRrules, Abunadine, OKBot, Mccujo, Altzinn, M2Ys4U, JL-Bot, Jobas, ImageRemovalBot, Tomasz Prochownik, YSSYguy, ClueBot, Richtig27, Iza9, Mild Bill Hiccup, J8079s, BIG BOOTY5, Singinglemon~enwiki, BlueAmethyst, SamuelTheGhost, DragonBot, Excirial, Eeekster, Tweetlebeetle367, WillMcD999, Aurora2698, Jaguar14, HssanKachal, Poodledog, Al-Andalusi, Mczack26, Johnuniq, Beroal, DumZiBoT, Polysynaptic, XLinkBot, Xiquet, Dany4175, Mavigogun, Cradel, ZooFari, MystBot, Lemmey, Nabuchadnessar, Addbot, DOI bot, Dawynn, Ronhjones, Aratak80, Lightbot, ماني, Samuel Pepys, Middayexpress, Luckas-bot, Ptbotgourou, Fraggle81, AnomieBOT, Daniele Pugliesi, Jim1138, Citation bot, Racconish, GB fan, Frankenpuppy, Xqbot, Coutasji, Ebu Katada, JimVC3, Cavila, Loveless, J04n, Xashaiar, Zincox, Sharveet, Thehelpfulbot, FrescoBot, Grinevitski, Ghost det, Machine Elf 1735, Citation bot 1, SpacemanSpiff, Bluey1361, BRUTE, Meamwye, علی ویکی, Tbhotch, Sideways713, RjwilmsiBot, Ripchip Bot, Msin10, EmausBot, John of Reading, WikitanvirBot, Berlew, Syncategoremata, Aquib American Muslim, Dreameq, ZxxZxxZ, Tommy2010, Wikipelli, Azeemrags, Josve05a, Muhib3000, Someone65, Illbased, Factfinderz, RockMagnetist, Karlchat, Khestwol, Helpsome, ClueBot NG, Marechal Ney, Helpful Pixie Bot, The Mark of the Beast, Dzlinker, Metricopolus, Silvrous, Arosby, Dohezarsersdah, Kuzuryu66, Greenknight dv, Andyrule0, Darylgolden, GAYousefSaanei, YFdyh-bot, Khazar2, EuroCarGT, Filursiax, Dexbot, Hishampgm, Sriharsh1234, Afaz warsi, IAMAM2012Expert, François Robere, Atef54321, Monkbot, Mwaseem1, Assyriandude, Thssn1234, Hihfhfhfhfhf, Hurvashtahumvata888 and Anonymous: 242

- **History of classical mechanics** *Source:* https://en.wikipedia.org/wiki/History_of_classical_mechanics?oldid=662305943 *Contributors:* Charles Matthews, Greenrd, BenRG, Giftlite, Lethe, Andycjp, DanielDemaret, Laurascudder, I9Q79oL78KiL0QTFHgyc, Pjohanneson, Dodiad, Tlroche, Rjwilmsi, DonSiano, Deeptrivia, RussBot, NawlinWiki, SmackBot, Melchoir, Jagged 85, TenPoundHammer, Loodog, BeenAroundAWhile, Myasuda, AndrewDressel, Headbomb, AbcXyz, Paste, Lfstevens, RickyCayley, JaGa, David J Wilson, Lradrama, Neparis, Oxymoron83, Tomasz Prochownik, The Thing That Should Not Be, J8079s, Jusdafax, Gnowor, Yobot, Citation bot, Sionus, Machine Elf 1735, RjwilmsiBot, JustMyThoughts, DolphinL, GoingBatty, RockMagnetist, Cntras, CitationCleanerBot, Doctor Universalis, Harizotoh9, Hmainsbot1, Belief action, Shadout mapes, संजीव कुमार and Anonymous: 28

- **Philosophiæ Naturalis Principia Mathematica** *Source:* https://en.wikipedia.org/wiki/Philosophi%C3%A6_Naturalis_Principia_ Mathematica?oldid=685592689 *Contributors:* AxelBoldt, Mav, Bryan Derksen, Zundark, Diatarn_iv, SimonP, Montrealais, Bdesham, Michael Hardy, Mic, TakuyaMurata, Looxix~enwiki, Mortene, Suisui, Julesd, Disdero, Charles Matthews, Zoicon5, Steinsky, Tpbradbury, Saltine, Bevo, JorgeGG, Robbot, Chrism, Merovingian, SimonMeacham, Blainster, AsianAstronaut, Alan Liefting, Ancheta Wis, Giftlite, ShaunMacPherson, Curps, Solipsist, Matt Crypto, Rparle, Jackol, LucasVB, Antandrus, Beland, Piotrus, APH, DragonflySixtyseven, Johnflux, Icairns, Almit39, Neutrality, MakeRocketGoNow, Garrison, Dbachmann, Altmany, Bender235, Jpittman, Kwamikagami, Causa sui, Bobo192, John Vandenberg, JW1805, DaveGorman, Oop, Alansohn, Gary, Borisblue, JoaoRicardo, Kocio, UnHoly, Mysdaao, Saga City, Dirac1933, Voxadam, Tafinucane, Pcpcpc, Cleonis, Sneakums, Graham87, Qwertyus, Kbdank71, Ciroa, Casey Abell, Tim!, R.e.b., MapsMan, Johnrpenner, FlaBot, Wikiliki, RobertG, Mathbot, Gurch, Wars, Sonitus, Srleffler, Joonasl, Idaltu, Chobot, Jaraalbe, YurikBot, Wavelength, Borgx, Bambaiah, StuffOfInterest, Jensboot, Jpbowen, Killerfox, Beanyk, Ospalh, CLW, Zzuuzz, Nightryder84, JoanneB, Mais oui!, Flowersofnight, Curpsbot-unicodify, Appleseed, Theroachman, SmackBot, Lestrade, Alex1011, J7, KocjoBot~enwiki, Jfgrcar, Harald88, Edgar181, Alsandro, Skizzik, Quadratic, Rmosler2100, Woofboy, Miquonranger03, Rhtcmu, Colonies Chris, Modest Genius, Can't sleep, clown will eat me, Hermzz, Snowmanradio, SundarBot, Savidan, Andrew c, Wikiklaas, Ohconfucius, Lambiam, Xdamr, Thedoj, Rigadoun, VirtualDave, RandomCritic, Loadmaster, MaximvsDecimvs, BillFlis, LaMenta3, Iridescent, Kencf0618, Newone, Thepossumdance, Drinibot, Thomasmeeks, Gregbard, Cydebot, Languagehat, M a s, Gfbs, Thijs!bot, Epbr123, Kubanczyk, TonyTheTiger, Nonagonal Spider, Headbomb, EdJohnston, Dgies, D.H, Nick Number, RobotG, AstroLynx, NBeale, Pipedreamergrey, 100110100, Bencherlite, Bongwarrior, VoABot II, Catslash, Nyq, Gboweswhitton, KConWiki, Crunchy Numbers, David Eppstein, Khalid Mahmood, Gwern, Keith D, J.delanoy, Hu Totya, Ajjusoni, Maurice Carbonaro, Knight of BAAWA, Siebengang, Nwbeeson, Alnokta, Fylwind, Janderie, Hugo999, King Lopez, CWii, Mrh30, Roarshocker, Scoop100, Entropy1963, Finnrind, Redragon104, SieBot, Zorkmid24, JohnWarnock, Spobmur, Arpose, Mrw7, Hxhbot, Isaacnewton333, Markcymru, Ubermission, Randy Kryn, Marmite disaster, Martarius, ClueBot, The Thing That Should Not Be, Uxorion, ChandlerMapBot, Contado83ll, Bwwm, Danscottgraham, Brews ohare, Revgraves, Johnuniq, Palnot, JKeck, XLinkBot, Terry0051, Stickee, Felix Folio Secundus, Bunich, Addbot, Grayfell, AVand, Ronhjones, FiriBot, Favonian, LinkFA-Bot, Numbo3-bot, Zorrobot, Flopy, Legobot, Yobot, Pigetrational, Nocturnalsleeper, AnomieBOT, Kingpin13, Materialscientist, Jmundo, Anonymous from the 21st century, Omnipaedista, Kyng, Auréola, Edgars2007, CES1596, FrescoBot, DoostdarWKP, Pedaub, Kwiki, TalphaLyra, Kusluj, Tom.Reding, Skyerise, BigDwiki, RedBot, SpringSummerAutumn, Vhsatheeshkumar, Scholium, Puzl bustr, BaldBoris, Andemora, Kggy, عبقری2009, 7kingis, Mevami, Slon02, EmausBot, NotAnonymous0, Wikipelli, Anirudh Emani, Solomonfromfinland, JSquish, Calstan7, Josve05a, A2soup, Landemor, Danlevy100, DMichael6, Zdorovo, ClueBot NG, Ian gordon bruce, Jorgecarleitao, MerllwBot, Pefue, Helpful Pixie Bot, BG19bot, NaturalPhilos, Knowledge Examiner, Transportabelle, JohnChrysostom, MusikAnimal, WithSelet, Harizotoh9, Manoguru, Cengime, Jcf139er, Justincheng12345-bot, Acc60, Locust21, EdwardWilsonLee, Mykhaylo Balush, Dexbot, Mogism, Chrisalex0207, LunarPhyla, Ephemeratta, WrenLibrary, USNorseman, Isaac Newton The Person, CV9933, Maxisaurus, KasparBot, Thedarklady154 and Anonymous: 279

- **Classical mechanics** *Source:* https://en.wikipedia.org/wiki/Classical_mechanics?oldid=686928544 *Contributors:* AxelBoldt, CYD, Mav, Tarquin, AstroNomer~enwiki, Ap, Josh Grosse, XJaM, William Avery, Roadrunner, Peterlin~enwiki, Maury Markowitz, FlorianMarquardt, Camembert, Isis~enwiki, Lir, Patrick, Michael Hardy, Tim Starling, Grahamp, Bcrowell, TakuyaMurata, Looxix~enwiki, Stevenj, Lupinoid, Glenn, Bogdangiusca, Rossami, Denny, Pizza Puzzle, Charles Matthews, Aravindet, Reddi, Dandrake, The Anomebot, Jeepien, Furrykef, Phys, Raul654, BenRG, RadicalBender, Phil Boswell, Robbot, F3meyer, Mayooranathan, Moink, Hadal, Papadopc, Fuelbottle, Anthony, Tobias Bergemann, Giftlite, Wolfkeeper, Tom harrison, Wwoods, Wgmccallum, Jorge Stolfi, Dan Gardner, PlatinumX, Mobius, Quadell, Antandrus, Beland, Karol Langner, APH, Gauss, Icairns, Zfr, Muijz, Guanabot, FT2, Dave souza, Paul August, SpookyMulder, Bender235, JoeSmack, Brian0918, MBisanz, Surachit, Bobo192, Nigelj, John Vandenberg, BrokenSegue, Haham hanuka, LucaB~enwiki, Mlessard, Sun King, Batmanand, Orionix, Velella, Evil Monkey, Dirac1933, Woodstone, Gene Nygaard, RandomWalk, Oleg Alexandrov, Nuno Tavares, Linas, StradivariusTV, Drostie, Ruud Koot, Dodiad, Jeff3000, Ulcph, Mayz, XaosBits, Phlebas, Leapfrog314, Graham87, Magister Mathematicae, Qwertyus, FreplySpang, Yurik, Seidenstud, Kinu, MarSch, Thechamelon, RE, Bhadani, Cethegus, DirkvdM, FlaBot, Mathbot, RexNL, Srleffler, Chobot, Krishnavedala,

Sharkface217, Sanpaz, Gwernol, Wavelength, Hairy Dude, Deeptrivia, Retodon8, RussBot, Carl T, JabberWok, David R. Ingham, Johann Wolfgang, Ragesoss, Chichui, Enormousdude, Covington, Thou shalt not have any gods before Willy on Wheels, RG2, Timothyarnold85, Sbyrnes321, SmackBot, Tom Lougheed, Hydrogen Iodide, Jagged 85, Ptpare, Harald88, Squiddy, Frédérick Lacasse, Saros136, Bluebot, TimBentley, SMP, Pieter Kuiper, Silly rabbit, Complexica, DHN-bot~enwiki, Salmar, Foxjwill, Berland, Rsm99833, Cybercobra, Chrylis, Dr. Sunglasses, Sure kr06~enwiki, Vgy7ujm, Loodog, Farid2053, Phancy Physicist, Xunex, SirFozzie, Mets501, Ssiruuk25, Anjor, Tawkerbot2, RSido, Sketch051, BeenAroundAWhile, Matthew Auger, Gregbard, Logicus, Cydebot, Rushbie, Rracecarr, Thijs!bot, Barticus88, TonyTheTiger, AndrewDressel, Kahriman~enwiki, MrXow, Imusade, Headbomb, James086, Memayer, Austin Maxwell, Seaphoto, Storkk, JAnDbot, CosineKitty, Db099221, Yill577, Magioladitis, VoABot II, Ling.Nut, Dfalcantara, Ryeterrell, David Eppstein, User A1, MarcusMaximus, JaGa, Ekotkie, Euneirophrenia, Rohan Ghatak, Nigholith, AtholM, Bcartolo, C quest000, CompuChip, Juliancolton, Treisijs, Useight, Idioma-bot, Pafcu, VolkovBot, JohnBlackburne, TXiKiBoT, The Original Wildbear, BertSen, GroveGuy, Hqb, Sankalpdravid, Anna Lincoln, Costela, Windrixx, BotKung, Amd628, Gnf1, Tom Atwood, Synthebot, AlleborgoBot, Neparis, SieBot, ToePeu.bot, JerrySteal, Paolo.dL, Lisatwo, Duae Quartunciae, Tomasz Prochownik, ClueBot, DeepBlueDiamond, Luke490, CyrilThePig4, Razimantv, Mild Bill Hiccup, Niceguyedc, Djr32, Excirial, Jomsborg, Gulmammad, Brews ohare, Arjayay, PhySusie, BOTarate, Crowsnest, XLinkBot, Rror, Saeed.Veradi, Andeasling, Truthnlove, Cholewa, Addbot, Willking1979, Atethnekos, Dgroseth, Njaelkies Lea, Fluffernutter, SpillingBot, Cst17, EconoPhysicist, Bassbonerocks, CUSENZA Mario, LinkFA-Bot, Tassedethe, Tide rolls, Lightbot, Lrrasd, Luckas-bot, Bunnyhop11, Tannkrem, AnomieBOT, Rubinbot, Keithbob, Jpc4031, Citation bot, Xqbot, Tripodian, Amareto2, Charvest, Aaron Kauppi, Thehelpfulbot, Dan6hell66, LucienBOT, Tobby72, Steve Quinn, Machine Elf 1735, Pinethicket, Codwiki, SpaceFlight89, Corinne68, TobeBot, Trappist the monk, Wdanbae, Lotje, Dinamik-bot, JLincoln, Diannaa, Onel5969, RjwilmsiBot, EmausBot, Syncategoremata, Elementaro, Amrator, Wikipelli, JSquish, Cogiati, Knight1993, Stanford96, Empty Buffer, Vramasub, Maschen, ChuispastonBot, RockMagnetist, Wakebrdkid, ClueBot NG, Satellizer, SusikMkr, Enopet, Frietjes, Braincricket, Widr, ساجد ساجد امجد, Lincoln Josh, Helpful Pixie Bot, ಬಿಱ ಕಾಂಡಲೆ಼, IzackN, Prof McCarthy, Brian Tomasik, Blue Mist 1, Sparkie82, Snow Blizzard, Miszatomic, StopTheCrackpots, YFdyh-bot, Khazar2, Dexbot, Fragapanagos, Thatguy1234352, Rahulsehwag, Reatlas, Devinray1991, Howicus, Fidasty, Jburnett63, Arachmen, Kiranbg12, ElectronicKing888, Peterzipfel37, Mars wanderer, Jarjarbinks123455555, FeatheredOrcian, KasparBot, Kafishabbir, Qazplmqwertybchefbiusdb wieuhdcbiqhedcbwhuid and Anonymous: 231

- **History of thermodynamics** *Source:* https://en.wikipedia.org/wiki/History_of_thermodynamics?oldid=680385706 *Contributors:* Collabi, Lumos3, Arkuat, Gandalf61, Cutler, Karol Langner, Eric Forste, PAR, Marianika~enwiki, Carcharoth, Benbest, Rjwilmsi, Ligulem, Srleffler, Chobot, Gaius Cornelius, CambridgeBayWeather, Ragesoss, Dhollm, Moe Epsilon, Rayc, Tropylium, SmackBot, Jagged 85, TimBentley, Colonies Chris, A.R., DMacks, Ligulembot, Mion, Sadi Carnot, Pilotguy, JzG, JorisvS, Peterlewis, Special-T, AdultSwim, Lottamiata, Myasuda, FilipeS, Gtxfrance, Doug Weller, M karzarj, Barticus88, D.H, Greg L, EdJogg, VoABot II, Cardamon, Jtir, Inwind, ElinorD, Riick, Enviroboy, Radagast3, Natox, SieBot, I Love Pi, Anchor Link Bot, Tomasz Prochownik, MCCRogers, Taroaldo, J8079s, Djr32, CohesionBot, Eeekster, XLinkBot, Saeed.Veradi, Ariconte, Kwjbot, Addbot, Lightbot, Wikkidd, Luckas-bot, Yobot, Ptbotgourou, Ajh16, AnomieBOT, Citation bot, ArthurBot, Xqbot, J04n, GrouchoBot, SassoBot, Geraldo61, Fortdj33, Machine Elf 1735, Citation bot 1, TobeBot, Marie Poise, Syncategoremata, ClueBot NG, Helpful Pixie Bot, Bibcode Bot, Ludi Romani, Bfong2828, Belief action, Nerlost, CleanEnergyPundit and Anonymous: 32

- **History of special relativity** *Source:* https://en.wikipedia.org/wiki/History_of_special_relativity?oldid=678914256 *Contributors:* Edward, William M. Connolley, Charles Matthews, Robbot, Lowellian, Giftlite, Wolfkeeper, Marcika, Dratman, Muzzle, Alvestrand, Dvavasour, Jo-Jan, Rich Farmbrough, Pjacobi, Vapour, Bender235, Rgdboer, Shadow demon, Teorth, Haham hanuka, Tony Sidaway, Allen McC.~enwiki, Drbreznjev, Daniel Case, Carcharoth, Robert K S, Ketiltrout, Rjwilmsi, Nightscream, Arabani, Ems57fcva, Alphachimp, Tedder, DVdm, Bgwhite, Hillman, Pip2andahalf, Gaius Cornelius, ErkDemon, Schlafly, Tony1, 2over0, Meegs, Infinity0, SmackBot, Harald88, MediaMangler, Chris the speller, JCSantos, Colonies Chris, Bob Denny, Squigish, CWesling, Leonard Dickens, Pgf, Reade, Dr Greg, Iridescent, Eluchil404, DickBrook, CmdrObot, Myasuda, Cydebot, Michael C Price, Martin Hogbin, Headbomb, James086, D.H, Dawnseeker2000, Tim Shuba, Igodard, Harmanjitsingh, JamesBWatson, Email4mobile, CommonsDelinker, Vanwhistler, Adavidb, Maurice Carbonaro, Lantonov, Gemini1980, Alan U. Kennington, JohnBlackburne, Nxavar, Joshb2, Grant.Alpaugh, Paradoctor, Likebox, Punkvijay, Renata500, Blue bear sd, Roibeird, RAmesbury, Paulcmnt, Spirals31, Yizhenwilliam, ChrisHodgesUK, El bot de la dieta, MystBot, Addbot, Leszek Jańczuk, Delaszk, Mwse87, CosmiCarl, OlEnglish, CountryBot, Luckas-bot, Drdonzi, Yobot, Kilom691, AnomieBOT, Citation bot, Xqbot, NOrbeck, Cantons-de-l'Est, Charvest, Ignoranteconomist, Shadowjams, CES1596, Fortdj33, Agnon5, Citation bot 1, Wandering-teacher, Pinethicket, RedBot, Foobarnix, Trappist the monk, Mauricewa, Flegelpuss, Cardinality, ZoneW, Iphegenia, JLincoln, Earthandmoon, RjwilmsiBot, Ripchip Bot, Blighcapn, John of Reading, Bethnim, ZéroBot, Relativa, JFB80, Lom Konkreta, ClueBot NG, BobHK, Frietjes, Euty, Lincoln Josh, Helpful Pixie Bot, Bibcode Bot, BG19bot, Benzband, BattyBot, Justincheng12345-bot, Niltone, ChrisGualtieri, Dexbot, Mogism, Flau98bert, Joeinwiki, FeliksK, Soll, Ginsuloft, Anrnusna, Monkbot, Xerberos, Jwinder47 and Anonymous: 87

- **History of general relativity** *Source:* https://en.wikipedia.org/wiki/History_of_general_relativity?oldid=686338729 *Contributors:* Malcolm Farmer, Michael Hardy, Bcrowell, Charles Matthews, Robbot, Fropuff, Alvestrand, Karol Langner, Rich Farmbrough, ThomasK, Kipton, Rgdboer, Plumbago, RJFJR, Capecodeph, Billhpike, Linas, PatGallacher, Carcharoth, Mpatel, Joke137, Christopher Thomas, Rjwilmsi, Mike Peel, Ligulem, Ems57fcva, Tedder, DVdm, Hillman, KSmrq, Sardanaphalus, SmackBot, Hbackman, Edgar181, Chris the speller, Colonies Chris, Georg-Johann, GeorgeMoney, Addshore, William Ackerman, E4mmacro, Fuhghettaboutit, Ligulembot, Ohconfucius, Lambiam, JR-Spriggs, Pgr94, Tonyle, D.H, Rico402, Easchiff, DAGwyn, Mollwollfumble, CommonsDelinker, Vanwhistler, TheSeven, Lantonov, Tcisco, JohnBlackburne, Hqb, Wmpearl, SockPuppetForTomruen, Addbot, AnomieBOT, Materialscientist, Citation bot, Xqbot, Spartan S58, Troglo, Steve Quinn, SkinnyPrude, Full-date unlinking bot, Lotje, Earthandmoon, EmausBot, Stibu, RockMagnetist, Will Beback Auto, Helpful Pixie Bot, Bibcode Bot, BG19bot, Mark Arsten, Harizotoh9, Mcn999, ChrisGualtieri, Filedelinkerbot, CV9933, Anooo Pandey and Anonymous: 35

- **History of quantum mechanics** *Source:* https://en.wikipedia.org/wiki/History_of_quantum_mechanics?oldid=659448203 *Contributors:* J-Wiki, Topbanana, BenRG, Alan Liefting, Graeme Bartlett, Dratman, Kusunose, Chris Howard, Discospinster, Demaag, Laurascudder, John Vandenberg, Voxadam, StradivariusTV, Wikiklrsc, BD2412, Kbdank71, Nanite, Rjwilmsi, Strait, Lmatt, DVdm, Okedem, Cryptic, Grafen, Welsh, Ospalh, Modify, Itub, Yvwv, SmackBot, Jagged 85, Gilliam, Betacommand, Bduke, Colonies Chris, LuchoX, Jbergquist, Sadi Carnot, Khazar, Lottamiata, Rhetth, Geremia, BeenAroundAWhile, Myasuda, Jac16888, Boardhead, Michael C Price, PKT, Headbomb, D.H, DuncanHill, Matthew Fennell, Yill577, Easchiff, Grimlock, JaGa, Coppertwig, Yym1997, VolkovBot, Moose-32, Truthanado, Biscuittin, YohanN7, Fadesga, Mild Bill Hiccup, Niceguyedc, Agge1000, Iohannes Animosus, SchreiberBike, BOTarate, Riversider2008, PSimeon, Dthomsen8, Stephen Poppitt, Addbot, Lightbot, Luckas-bot, Yobot, Citation bot, Bci2, GrouchoBot, Topherwhelan, Ignoranteconomist, Shadowjams, Nagualdesign, Citation bot 1, Dinamik-bot, Bookalign, GoingBatty, JSquish, Dagko, RockMagnetist, Mpkannan, ClueBot NG, CocuBot, Dharmaraj.S, Oddbodz,

Helpful Pixie Bot, Wbm1058, Bibcode Bot, Dstuck, BattyBot, InformationvsInjustice, Vanamonde93, Monkbot, Garfield Garfield, Archiloc and Anonymous: 37

- **History of subatomic physics** *Source:* https://en.wikipedia.org/wiki/History_of_subatomic_physics?oldid=666567740 *Contributors:* BD2412, Bubba73, Incnis Mrsi, Chris the speller, Magioladitis, Hedwig in Washington, Favonian, MeDrewNotYou, Dwalsh3, Maschen, Hmainsbot1 and Anonymous: 2

- **Unified field theory** *Source:* https://en.wikipedia.org/wiki/Unified_field_theory?oldid=685825252 *Contributors:* M~enwiki, Michael Hardy, Ahoerstemeier, William M. Connolley, Charles Matthews, Reddi, Dysprosia, Pakaran, SJRubenstein, Josh Cherry, Academic Challenger, Rursus, Dbenbenn, JamesMLane, Xerxes314, StargateX1, Gzornenplatz, Karol Langner, Gscshoyru, Vitaleyes, Svdb, Caroline Thompson, Brianhe, H0riz0n, Cfailde, DPFJr, Pjacobi, Paul August, Dmr2, ESkog, Lentando~enwiki, BenjBot, Oldsoul, Etimbo, Noren, Bobo192, Ablathanalba, Mordemur, John Vandenberg, I9Q79oL78KiL0QTFHgyc, Nsaa, Hdeasy, Count Iblis, Falcorian, Blaze Labs Research, Simetrical, Linas, Mindmatrix, Decrease789, Savantnavas, Mpatel, Knuckles, Marudubshinki, BD2412, Nightscream, Macumba, R.e.b., Protez, Srleffler, Chobot, Moocha, GangofOne, YurikBot, RadioFan2 (usurped), Rsrikanth05, NawlinWiki, Trovatore, Syrthiss, Steve G~enwiki, KasugaHuang, SmackBot, McGeddon, Gilliam, Chris the speller, Bluebot, Jjalexand, Crazy8s, Jeysaba, Silly rabbit, Timneu22, Redattore, Colonies Chris, QFT, InnocentMind, Xyzzyplugh, Jgwacker, Corby, RolandR, Marcus Brute, Sadi Carnot, Vina-iwbot~enwiki, Lambiam, Nishkid64, ArglebargleIV, Titus III, John, Gobonobo, Kevlarmry, Ckatz, Slakr, Rainwarrior, Trounce, Twas Now, Bridg, Courcelles, Shedsan, Prof.Maque, JForget, Will314159, Friendly Neighbour, NickW557, Gregbard, Peripitus, Gogo Dodo, Michael C Price, Qwyrxian, Roger Anderton, Headbomb, Marek69, Twcjr, KrakatoaKatie, AntiVandalBot, Emeraldcityserendipity, Tim Shuba, Verticordia~enwiki, Alphachimpbot, AndreasWittenstein, Blaine Steinert, Yill577, Wasell, VoABot II, Tobogganoggin, Arrowcatcher, DAGwyn, Catgut, Web-Crawling Stickler, Ours18, Skylights76, Jke310, Stephenchou0722, R'n'B, J.delanoy, Captain panda, Trusilver, Eliz81, Dogstar11, CardinalDan, VolkovBot, Butwhatdoiknow, Fennmeister, Alphanon, Cheffoxx, Betanon, ARUNKUMAR P.R, Deanlsinclair, PaddyLeahy, Gaelen S., Travisbmoore, Gammanon, Zharradan.angelfire, Anakin101, De728631, ClueBot, NossB, EhJJ, Bhushan foryou, Versus22, Qwfp, Pandanator75, Arthur chos, Addbot, Cxz111, Bobtron5000, Bte99, MrOllie, Deepthought137, Favonian, AtheWeatherman, Whitematter, Scientryst, MuZemike, Yobot, TaBOT-zerem, AnomieBOT, 9258fahsflkh917fas, Kanat Abildinov, Materialscientist, JohnnyB256, Addihockey10, Shirik, Mathonius, Pereant antiburchius, Natural Cut, Shadowjams, Lovelylilian, A. di M., Paine Ellsworth, Ottokar~enwiki, Knowandgive, RandomStringOfCharacters, Grandunifier, Tennant uk, Miracle Pen, Afteread, RA0808, Tommy2010, Wikipelli, Hhhippo, Susfele, AvicAWB, S.Lenane, DanielBurnstein, Barendjacobus, Davidaedwards, Mkh025, Dudge1983, ClueBot NG, StevenPower, Theopolisme, MerlIwBot, Mophedd, Orphadeus, Peter Donald Rodgers, Joe0x7F, Dilaton, Astralbound, LynnetteA11, GabeIglesia, Jamesmcmahon0, NfrHtp, KEVIN123456789, Froglich, BuilderE, Michelle1881, Balbinder1706, Stowcalj, KasparBot and Anonymous: 195

- **Physical cosmology** *Source:* https://en.wikipedia.org/wiki/Physical_cosmology?oldid=682803174 *Contributors:* AxelBoldt, CYD, Wesley, Bryan Derksen, The Anome, Koyaanis Qatsi, Ed Poor, Andre Engels, XJaM, DavidLevinson, Youandme, Modemac, Edward, Boud, Michael Hardy, GABaker, IZAK, Looxix~enwiki, Andrewa, Glenn, Bogdangiusca, Andres, Evercat, Rob Hooft, Reddi, Dragons flight, Phys, Wetman, BenRG, Jni, Phil Boswell, Robbot, Craig Stuntz, Merovingian, Lsy098~enwiki, Sverdrup, Rursus, Meelar, Fuelbottle, Giftlite, Ich, Curps, Gracefool, Node ue, Bobblewik, Pgan002, Andycjp, Spatch, BozMo, Beland, OverlordQ, APH, RetiredUser2, Icairns, Urhixidur, Centroyd, Mike Rosoft, Discospinster, Rich Farmbrough, Oliver Lineham, Pjacobi, Vsmith, Jpk, Mani1, SpookyMulder, RJHall, El C, Shanes, Megaton~enwiki, Dralwik, Mtruch, Fritz freiheit, I9Q79oL78KiL0QTFHgyc, Nk, Geschichte, Jumbuck, Erichwanh, Truth seeker, Gilgameshfuel, Wtmitchell, Orionix, Tycho, Gene Nygaard, Mattbrundage, Ceyockey, Nuno Tavares, FeanorStar7, Rocastelo, Kzollman, Mpatel, GregorB, Joke137, Graham87, Nanite, Drbogdan, Rjwilmsi, Mayumashu, Mike Peel, Maxim Razin, TeaDrinker, DVdm, Gwernol, Wavelength, JabberWok, Trious, Salsb, SEWilcoBot, Welsh, Dna-webmaster, Jayamohan, Pawyilee, Leptictidium, JonathanD, Syd Midnight, Smoggyrob, Josh3580, Wsiegmund, Kungfuadam, Infinity0, Sardanaphalus, SmackBot, Reedy, KnowledgeOfSelf, Vald, Fractions, Hbackman, Edgar181, Alsandro, Hmains, Chris the speller, Bluebot, Silly rabbit, Droll, Colonies Chris, Hallenrm, Hve, Vanished User 0001, Stangbat, Nakon, John D. Croft, Zadignose, Lambiam, Vgy7ujm, Philosophus, Mgiganteus1, Cyberstrike2000x, Ckatz, Dan Gluck, Aeternus, Pathosbot, Raystorm, Kurtan~enwiki, RCS, Gregbard, DumbBOT, Legis, Abtract, Shocktherapy, Letranova, Coelacan, Oliver202, Headbomb, Peter Gulutzan, Seaphoto, Movses, Zhengdabei, Tim Shuba, CosineKitty, Jenattiyeh, Wikidudeman, NotACow, KConWiki, Catgut, Elentirmo, Kevinwiatrowski, NatureA16, MartinBot, R'n'B, HEL, J.delanoy, Neonguru, Maurice Carbonaro, Lantonov, Skier Dude, Plasticup, Utad3~enwiki, CWii, Milenita~enwiki, Zrallo, WWEUNDERTAKER, Clarince63, Vendrov, BotKung, Maxim, James McBride, PHiZiX, Undead warrior, Coffee, MaynardClark, Lightmouse, Ccwth, Anchor Link Bot, ClueBot, VQuakr, Agge1000, Djr32, Excirial, CohesionBot, Alan268, Brews ohare, Scog, Askahrc, Hercule, Panos84, TimothyRias, Avoided, Truthnlove, Dazza79, Addbot, Atethnekos, Tassedethe, Astro-norte, AstroBjorn, Greyhood, VP-bot, Ttoolow, Yinweichen, Luckas-bot, Yobot, Boyer the destroyer, Legobot II, Aldebaran66, Mmonne, Ccraccnam, Amble, KamikazeBot, Keeratura, AnomieBOT, Rubinbot, 1exec1, Captain Quirk, Flewis, Materialscientist, Citation bot, Icosmology, Rightly, Hexadecima, Xqbot, Pra1998, Nasa-verve, Зvípi, FrescoBot, Zionist agent, Tavernsenses, SF88, Geomet9, Tom.Reding, Pmokeefe, Naturehead, Geogene, Hoo man, Aknochel, Caribibble, Begoon, GGT, Chronulator, Earthandmoon, Marie Poise, EmausBot, Italia2006, SporkBot, David J Johnson, Eparksbuckeye, Donner60, HCPotter, Mentibot, ClueBot NG, Lakithunderboom, Braincricket, X-men2011, Helpful Pixie Bot, Bibcode Bot, Bm gub2, Juro2351, Marsambe, Marsambe1, Harizotoh9, Seshavatharam.bhc, MathewTownsend, Cyberbot II, ChrisGualtieri, Louey37, Soulbust, Rhlozier, Von Numinous, Wjs64, Alexander1257, Rfassbind, Praemonitus, Imaloverboy12345, Wrrdsck, CuirassierX, Prokaryotes, PirtleShell, Jwratner1, Vinny Lam, Mfb, Vsilv, Monkbot, Neeraj Bhakta, Stefania.deluca, Tetra quark, KasparBot, Huritisho and Anonymous: 186

- **Timeline of fundamental physics discoveries** *Source:* https://en.wikipedia.org/wiki/Timeline_of_fundamental_physics_discoveries?oldid=678982366 *Contributors:* Rmhermen, Selket, Fastfission, Karol Langner, Urvabara, Slicky, Dominic, Christopher Thomas, Srleffler, Vald, Jagged 85, Thumperward, Colonies Chris, Radagast83, Mets501, JeffW, Cloudguitar, Ken Gallager, PKT, Headbomb, Alphachimpbot, Cardamon, 272727, Lantonov, Fountains of Bryn Mawr, Wing gundam, StewartMH, ClueBot, Skyler.cohen, Alexius08, Addbot, Favonian, Yobot, Aaron Kauppi, OriumX, Deanmullen09, DexDor, Mathewsyriac, Bookalign, Hhhippo, Mougin.nicolas, ClueBot NG, David C Bailey, CaroleHenson, Taaruvot, Andyhowlett, I am One of Many, Ybidzian, TheWereToaster and Anonymous: 19

19.3.2 Images

- **File:10_Quantum_Mechanics_Masters.jpg** *Source:* https://upload.wikimedia.org/wikipedia/commons/7/79/10_Quantum_Mechanics_Masters.jpg *License:* CC BY-SA 3.0 *Contributors:*

- **File:Galileo.arp.300pix.jpg** *Source:* https://upload.wikimedia.org/wikipedia/commons/c/cc/Galileo.arp.300pix.jpg *License:* Public domain *Contributors:*

  http://www.nmm.ac.uk/mag/pages/mnuExplore/ViewLargeImage.cfm?ID=BHC2700

  Original artist: Justus Sustermans

- **File:Ganesha_ink.jpg** *Source:* https://upload.wikimedia.org/wikipedia/commons/e/e4/Ganesha_ink.jpg *License:* Public domain *Contributors:* India, Karnataka, Mysore. Bhagavata Purana (Ancient Stories of the Lord) Manuscript. Museum Number M.91.349.4, LACMA [1] *Original artist:* Unknown

- **File:GodfreyKneller-IsaacNewton-1689.jpg**   *Source:*   https://upload.wikimedia.org/wikipedia/commons/3/39/ GodfreyKneller-IsaacNewton-1689.jpg *License:* Public domain *Contributors:* http://www.newton.cam.ac.uk/art/portrait.html *Original artist:* Sir Godfrey Kneller

- **File:Gottfried_Wilhelm_von_Leibniz.jpg** *Source:*   https://upload.wikimedia.org/wikipedia/commons/6/6a/Gottfried_Wilhelm_von_ Leibniz.jpg *License:* Public domain *Contributors:* /gbrown/philosophers/leibniz/BritannicaPages/Leibniz/LeibnizGif.html *Original artist:* Christoph Bernhard Francke

- **File:Hand-propelled_wheel_cart_from_Indus_Valley_Civilization.GIF** *Source:* https://upload.wikimedia.org/wikipedia/commons/9/9e/ Hand-propelled_wheel_cart_from_Indus_Valley_Civilization.GIF *License:* CC BY-SA 3.0 *Contributors:* http://www.nationalmuseumindia. gov.in/phis_ill.html *Original artist:* Photo of a wheeled cart in Indian National Museum.

- **File:Heisenberg_10.jpg** *Source:* https://upload.wikimedia.org/wikipedia/commons/b/b0/Heisenberg_10.jpg *License:* Public domain *Contributors:* MacTutor *Original artist:* Unknown

- **File:Hendrik_Antoon_Lorentz,_in_1916_geschilderd_door_Menso_Kamelingh_Onnes.jpg** *Source:*   https://upload.wikimedia.org/ wikipedia/commons/2/26/Hendrik_Antoon_Lorentz%2C_in_1916_geschilderd_door_Menso_Kamelingh_Onnes.jpg *License:* Public domain *Contributors:* ? *Original artist:* ?

- **File:Hindu-arabic1.jpg** *Source:* https://upload.wikimedia.org/wikipedia/commons/3/34/Hindu-arabic1.jpg *License:* Public domain *Contributors:*

- Transferred from en.wikipedia by SreeBot *Original artist:* User:JSR

- **File:History_of_the_Universe.svg** *Source:* https://upload.wikimedia.org/wikipedia/commons/d/db/History_of_the_Universe.svg *License:* CC BY-SA 3.0 *Contributors:* Own work *Original artist:* Yinweichen

- **File:Ice-calorimeter.jpg** *Source:* https://upload.wikimedia.org/wikipedia/commons/3/35/Ice-calorimeter.jpg *License:* Public domain *Contributors:* originally uploaded http://en.wikipedia.org/wiki/Image:Ice-calorimeter.jpg *Original artist:* Originally en:User:Sadi Carnot

- **File:Ilc_9yr_moll4096.png** *Source:* https://upload.wikimedia.org/wikipedia/commons/3/3c/Ilc_9yr_moll4096.png *License:* Public domain *Contributors:* http://map.gsfc.nasa.gov/media/121238/ilc_9yr_moll4096.png *Original artist:* NASA / WMAP Science Team

- **File:Image-Al-Kitāb_al-muḫtaṣar_fī_ḥisāb_al-ğabr_wa-l-muqābala.jpg** *Source:* https://upload.wikimedia.org/wikipedia/commons/2/ 23/Image-Al-Kit%C4%81b_al-mu%E1%B8%ABta%E1%B9%A3ar_f%C4%AB_%E1%B8%A5is%C4%81b_al-%C4%9Fabr_wa-l-muq% C4%81bala.jpg *License:* Public domain *Contributors:* John L. Esposito. *The Oxford History of Islam.* Oxford University Press. ISBN 0195107993. *Original artist:* Muhammad ibn Musa al-Khwarizmi

- **File:J.J_Thomson.jpg** *Source:* https://upload.wikimedia.org/wikipedia/commons/c/c1/J.J_Thomson.jpg *License:* Public domain *Contributors:* First World War.com *Original artist:* Not Mentioned

- **File:James_Clerk_Maxwell.png** *Source:* https://upload.wikimedia.org/wikipedia/commons/5/57/James_Clerk_Maxwell.png *License:* Public domain *Contributors:* Frontpiece in James Maxwell, *The Scientific Papers of James Clerk Maxwell.* Ed: W. D. Niven. New York: Dover, 1890. *Original artist:* George J. Stodart

- **File:Jantar_Mantar,_Delhi,_1826.jpg** *Source:*   https://upload.wikimedia.org/wikipedia/commons/7/71/Jantar_Mantar%2C_Delhi%2C_ 1826.jpg *License:* Public domain *Contributors:* British Library *Original artist:* Anonymous

- **File:Jesuites_en_chine.jpg** *Source:* https://upload.wikimedia.org/wikipedia/commons/e/e8/Jesuites_en_chine.jpg *License:* Public domain *Contributors:* Originally from fr.wikipedia; description page is (was) here *Original artist:* Jean-Baptiste Du Halde

- **File:Jingangjing.jpg** *Source:* https://upload.wikimedia.org/wikipedia/commons/d/d2/Jingangjing.jpg *License:* Public domain *Contributors:* Zoomable image from the British Library's Online Gallery. Originally uploaded to en:Wikipedia (log) in January 2008 by Fconaway (talk) and in November 2009 by Earthsound (talk). *Original artist:* The colophon, at the inner end, reads: Reverently [caused to be] made for universal free distribution by Wang Jie on behalf of his two parents on the 13th of the 4th moon of the 9th year of Xiantong [i.e. 11th May, CE 868].

- **File:Jupiter_and_the_Galilean_Satellites.jpg** *Source:* https://upload.wikimedia.org/wikipedia/commons/f/fe/Jupiter_and_the_Galilean_ Satellites.jpg *License:* Public domain *Contributors:* NASA planetary photojournal, borders removed by Daniel Arnold *Original artist:* NASA/JPL/DLR

- **File:Justus_Sustermans_-_Portrait_of_Galileo_Galilei,_1636.jpg** *Source:*   https://upload.wikimedia.org/wikipedia/commons/d/d4/ Justus_Sustermans_-_Portrait_of_Galileo_Galilei%2C_1636.jpg *License:* Public domain *Contributors:* http://www.nmm.ac.uk/mag/pages/ mnuExplore/PaintingDetail.cfm?ID=BHC2700 *Original artist:* Justus Sustermans

- **File:Katódsugarak_mágneses_mezőben(3).jpg** *Source:* https://upload.wikimedia.org/wikipedia/commons/a/a9/Kat%C3%B3dsugarak_m% C3%A1gneses_mez%C5%91ben%283%29.jpg *License:* CC BY-SA 3.0 *Contributors:* Own work *Original artist:* Zátonyi Sándor, (ifj.)

- **File:Kepler-solar-system-2.gif** *Source:* https://upload.wikimedia.org/wikipedia/commons/1/1d/Kepler-solar-system-2.gif *License:* Public domain *Contributors:* ? *Original artist:* ?

- **File:Libr0310.jpg** *Source:* https://upload.wikimedia.org/wikipedia/commons/2/24/Libr0310.jpg *License:* Public domain *Contributors:* ? *Original artist:* ?

- **File:Lord_Kelvin_photograph.jpg** *Source:* https://upload.wikimedia.org/wikipedia/commons/a/a0/Lord_Kelvin_photograph.jpg *License:* Public domain *Contributors:* http://www.sil.si.edu/digitalcollections/hst/scientific-identity/CF/by_scientist_display_results.cfm?scientist= kelvin *Original artist:* ?

- **File:Manchester_John_Rylands_Library_Isaac_Newton_16-10-2009_13-54-26.JPG** *Source:* https://upload.wikimedia.org/wikipedia/ commons/e/e9/Manchester_John_Rylands_Library_Isaac_Newton_16-10-2009_13-54-26.JPG *License:* CC BY-SA 3.0 *Contributors:* Own work *Original artist:* Paul Hermans

- **File:Maquina_vapor_Watt_ETSIIM.jpg** *Source:* https://upload.wikimedia.org/wikipedia/commons/9/9e/Maquina_vapor_Watt_ETSIIM. jpg *License:* CC-BY-SA-3.0 *Contributors:* Enciclopedia Libre *Original artist:* Nicolás Pérez

- **File:Mariecurie.jpg** *Source:* https://upload.wikimedia.org/wikipedia/commons/d/d9/Mariecurie.jpg *License:* Public domain *Contributors:* http://www.mlahanas.de/Physics/Bios/MarieCurie.html *Original artist:* Unknown

- **File:Max_Planck_1878.GIF** *Source:* https://upload.wikimedia.org/wikipedia/commons/6/62/Max_Planck_1878.GIF *License:* Public domain *Contributors:* http://alephwww.physik.uni-siegen.de/~{}brandt/jubil/index.html *Original artist:* Unknown

- **File:Max_planck.jpg** *Source:* https://upload.wikimedia.org/wikipedia/commons/d/d7/Max_planck.jpg *License:* Public domain *Contributors:* http://clendening.kumc.edu/dc/pc/planck.jpg (Clendening History of Medicine Library, University of Kansas Medical Center. http://clendening. kumc.edu/dc/) *Original artist:* Unknown

- **File:Max_von_Laue.jpg** *Source:* https://upload.wikimedia.org/wikipedia/commons/0/0e/Max_von_Laue.jpg *License:* Public domain *Contributors:* Les Prix Nobel 1914 *Original artist:* Nobel foundation

- **File:Mir_Sayyid_Ali_-_Portrait_of_a_Young_Indian_Scholar.jpg** *Source:* https://upload.wikimedia.org/wikipedia/commons/b/b9/Mir_ Sayyid_Ali_-_Portrait_of_a_Young_Indian_Scholar.jpg *License:* Public domain *Contributors:* ? *Original artist:* ?

- **File:Modernphysicsfields.svg** *Source:* https://upload.wikimedia.org/wikipedia/commons/5/56/Modernphysicsfields.svg *License:* CC BY-SA 3.0 *Contributors:* Self-made based on Physicsdomains.jpg. *Original artist:* This vector image was created with Inkscape.

- **File:Neon_orbitals.JPG** *Source:* https://upload.wikimedia.org/wikipedia/commons/2/2d/Neon_orbitals.JPG *License:* Public domain *Contributors:* ? *Original artist:* ?

- **File:NewtonsPrincipia.jpg** *Source:* https://upload.wikimedia.org/wikipedia/commons/4/41/NewtonsPrincipia.jpg *License:* CC BY-SA 2.0 *Contributors:* ? *Original artist:* ?

- **File:Newtons_cradle_animation_book_2.gif** *Source:* https://upload.wikimedia.org/wikipedia/commons/d/d3/Newtons_cradle_animation_ book_2.gif *License:* CC-BY-SA-3.0 *Contributors:* Image:Newtons cradle animation book.gif *Original artist:* DemonDeLuxe (Dominique Toussaint)

- **File:Newtons_proof_of_Keplers_second_law.gif** *Source:* https://upload.wikimedia.org/wikipedia/commons/6/60/Newtons_proof_of_ Keplers_second_law.gif *License:* Public domain *Contributors:* Own work *Original artist:* Lucas V. Barbosa

- **File:Nikolaus_Kopernikus.jpg** *Source:* https://upload.wikimedia.org/wikipedia/commons/f/f2/Nikolaus_Kopernikus.jpg *License:* Public domain *Contributors:* http://www.frombork.art.pl/Ang10.htm *Original artist:* Unknown

- **File:NuclearReaction.png** *Source:* https://upload.wikimedia.org/wikipedia/commons/7/7d/NuclearReaction.png *License:* CC BY-SA 3.0 *Contributors:* Own work *Original artist:* Michalsmid

- **File:Nuvola_apps_kalzium.svg** *Source:* https://upload.wikimedia.org/wikipedia/commons/8/8b/Nuvola_apps_kalzium.svg *License:* LGPL *Contributors:* Own work *Original artist:* David Vignoni, SVG version by Bobarino

- **File:Office-book.svg** *Source:* https://upload.wikimedia.org/wikipedia/commons/a/a8/Office-book.svg *License:* Public domain *Contributors:* This and myself. *Original artist:* Chris Down/Tango project

- **File:Orbital_motion.gif** *Source:* https://upload.wikimedia.org/wikipedia/commons/4/4e/Orbital_motion.gif *License:* GFDL *Contributors:*

- Earth derived from this image (public domain) *Original artist:* Own work

- **File:PIA17993-DetectorsForInfantUniverseStudies-20140317.jpg** *Source:* https://upload.wikimedia.org/wikipedia/commons/1/1a/ PIA17993-DetectorsForInfantUniverseStudies-20140317.jpg *License:* Public domain *Contributors:* http://photojournal.jpl.nasa.gov/jpeg/ PIA17993.jpg *Original artist:* NASA/JPL-Caltech

- **File:People_icon.svg** *Source:* https://upload.wikimedia.org/wikipedia/commons/3/37/People_icon.svg *License:* CC0 *Contributors:* OpenClipart *Original artist:* OpenClipart

- **File:Photoelectric_effect.svg** *Source:* https://upload.wikimedia.org/wikipedia/commons/f/f5/Photoelectric_effect.svg *License:* CC-BY-SA- 3.0 *Contributors:* en:Inkscape *Original artist:* Wolfmankurd

- **File:Physicsdomains.svg** *Source:* https://upload.wikimedia.org/wikipedia/commons/f/f0/Physicsdomains.svg *License:* CC BY-SA 3.0 *Contributors:* Own work *Original artist:* Loodog (talk), SVG conversion by User:Surachit

- **File:Poincare.jpg** *Source:* https://upload.wikimedia.org/wikipedia/commons/e/ed/Poincare.jpg *License:* Public domain *Contributors:* fr:Image: Poincare.jpg *Original artist:* Eugène Pirou (1841–1909) [1]

- **File:Portal-puzzle.svg** *Source:* https://upload.wikimedia.org/wikipedia/en/f/fd/Portal-puzzle.svg *License:* Public domain *Contributors:* ? *Original artist:* ?

- **File:Principia_Page_1726.jpg** *Source:* https://upload.wikimedia.org/wikipedia/commons/d/d0/Principia_Page_1726.jpg *License:* Public domain *Contributors:* ? *Original artist:* Isaac Newton

19.3.3 Content license

www.ingramcontent.com/pod-product-compliance
Lightning Source LLC
Chambersburg PA
CBHW080807180526
45168CB00006B/2358